209
Topics in Current Chemistry

Springer

Berlin
Heidelberg
New York
Barcelona
Hong Kong
London
Milan
Paris
Singapore
Tokyo

Biosynthesis

Aromatic Polyketides, Isoprenoids, Alkaloids

Volume Editors: Finian J. Leeper,
John C. Vederas

With contributions by
R. Croteau, E. M. Davis, T. Hartmann,
T. Hemscheidt, D. Ober, J. F. Sanz-Cervera,
B. Shen, E. M. Stocking, R. M. Williams

Springer

The series *Topics in Current Chemistry* presents critical reviews of the present and future trends in modern chemical research. The scope of coverage includes all areas of chemical science including the interfaces with related disciplines such as biology, medicine and materials science. The goal of each thematic volume is to give the non-specialist reader, whether at the university or in industry, a comprehensive overview of an area where new insights are emerging that are of interest to a larger scientific audience.

As a rule, contributions are specially commissioned. The editors and publishers will, however, always be pleased to receive suggestions and supplementary information. Papers are accepted for *Topics in Current Chemistry* in English.

In references *Topics in Current Chemistry* is abbreviated Top. Curr. Chem. and is cited as a journal.

Springer WWW home page: http://www.springer.de
Visit the TCC home page at http://link.springer.de/series/tcc/
or http://link.springer-ny.com/series/tcc/

ISSN 0340-1022
ISBN 3-540-66573-0
Springer-Verlag Berlin Heidelberg New York

Library of Congress Catalog Card Number 74-644622

Cover design: Friedhelm Steinen-Broo, Barcelona; MEDIO, Berlin
Typesetting: Fotosatz-Service Köhler GmbH, 97084 Würzburg

SPIN: 10681141 02/3020 ra – 5 4 3 2 1 0 – Printed on acid-free paper

Volume Editor

Dr. Finian J. Leeper
Department of Chemistry
University of Cambridge
Lensfield Road
Cambridge CB2 1EW, Great Britain
E-mail: fjl1@cam.ac.uk

Prof. Dr. John C. Vederas
Department of Chemistry
University of Alberta
Edmonton, Alberta
Canada T6G 2G2
E-mail: john.vederas@ualberta.ca

Editorial Board

Topics in Current Chemistry Now Also Available Electronically

For all customers with a standing order for Topics in Current Chemistry we offer the electronic form via LINK free of charge. Please contact your librarian who can receive a password for free access to the full articles by registration at:

http://link.springer.de/series/tcc/reg_form.htm

If you do not have a standing order you can nevertheless browse through the table of contents of the volumes and the abstracts of each article at:

http://link.springer.de/series/tcc

There you will also find information about the

- Editorial Board
- Aims and Scope
- Instructions for Authors

Preface

The present book is the second of two volumes that provide state of the art expert reviews of central topics in modern natural products chemistry and secondary metabolism. Using specific examples, the previous volume emphasized two revolutions in experimental techniques that completely transformed the field of natural products chemistry from what it was in the 1950s. These were the use of stable isotopes in conjunction with modern NMR and mass spectrometry, and more recently, the development of molecular biological techniques to identify, purify and manipulate the enzymes responsible for the intricate series of steps to complex natural compounds. The previous volume specifically covered the use of isotopes in biosynthetic research and the formation of enzyme cofactors, vitamin B_{12} and reduced polyketides.

This second volume describes the application of the same approaches (isotope methodology and molecular biology) to the biosynthesis of aromatic (unreduced) polyketides, enzymes responsible for cyclization of terpenoids (isoprenoids), and biochemical generation of selected classes of alkaloids (prenylated tryptophan, tropane, pyrrolizidine). The knowledge of the metabolic pathways and the techniques to elucidate them opens the door to combinatorial biosynthesis as well as to the production of targeted pharmaceutical agents utilizing a combination of chemistry, molecular biology and protein biochemistry. Recent advances suggest that it may soon be possible to rationally manipulate biochemical pathways to produce any target molecule, including non-natural variants, in substantial quantity. Genetically modified organisms containing mix-and-match combinations of biosynthetic enzymes (natural and/or mutated) will allow formation of large numbers of new compounds for biological evaluation. In addition, the availability of vast arrays of specialized enzymes in pure form may provide new reagents for combinatorial chemistry and parallel synthesis in drug discovery programs.

In the current volume, *Ben Shen* begins with a review of the assembly of unreduced polyketides leading to aromatic compounds. The current understanding of the functions and interactions of the enzymes involved is gradually providing the rules for designing new compounds of this class as well as affording a basis for biosynthesis of flavonoids via chalcone synthases. In chapter 2, *Edward Davis* and *Rodney Croteau* describe the terpenoid synthases responsible for formation of the huge array of mono-, sesqui- and diterpenes (over 30,000 terpene derivatives are known). In particular, detailed structural and functional evaluation of four representative terpene synthases is provided. In the third chapter, *Robert Williams, Emily Stocking and Juan Sanz-Cervera* review the biosynthesis of prenylated indole alkaloids and related substances derived from tryptophan. Many of these compounds are potent mycotoxins that contaminate food, but some, such as the ergot alkaloids (e.g. ergotamine, ergonovine) see extensive application in medicine. In chapter 4 *Thomas Hemscheidt* describes the current state of knowledge on the biosynthesis of tropane and related alkaloids, including cocaine. This well illustrates the difficulties that can be faced in elucidating the sequence of reactions involved in a biosynthetic pathway, especially when the intermediates are unstable and produced in very low amounts. The last chapter, by Thomas Hartmann covers the pyrrolizidine alkaloids. It not only describes the chemistry of the biosynthetic pathway but also gives an account of the physiology and ecology involved in the distribution and elaboration of the alkaloids within the producing plant, in the insects which eat the plants, and even in the animals which eat the insects. This shows us that for many natural products an understanding of how they are made is only a part of the whole story.

Cambridge, January 2000

Finian J. Leeper
John C. Vederas

Contents

Contents of Volume 195

Biosynthesis – Polyketides and Vitamins

Volume Editors: Finian J. Leeper, John C. Vederas

ISBN 3-540-63418-5

Biosynthesis of Aromatic Polyketides

Ben Shen

Department of Chemistry, University of California, One Shields Avenue, Davis, CA 95616, USA. *E-mail: shen@chem.ucdavis.edu*

Aromatic polyketides differ from other polyketides by their characteristic polycyclic aromatic structures. These polyketides are widely distributed in bacteria, fungi, and plants, and many of them are clinically valuable agents or exhibit other fascinating biological activities. Analogous to fatty acids and reduced polyketide biosynthesis, aromatic polyketide biosynthesis is accomplished by the polyketide synthases that catalyze sequential decarboxylative condensation between the starter and extender units to yield a linear poly-β-ketone intermediate. The latter undergoes regiospecific reduction, aromatization, or cyclization to furnish the polycyclic aromatic structures, which are further modified by tailoring enzymes to imbue them with various biological activities. This review begins with a brief discussion on the architectural organizations among various polyketide synthase genes and genetic contributions to understanding polyketide synthases. It then presents a comprehensive account of the most recent advances in the biochemistry and enzymology of bacterial, fungal, and plant polyketide synthases, with emphasis on in vitro studies. It concludes with a cautious summary of the so-called design-rules to guide rational engineering of polyketide synthases for the synthesis of novel aromatic polyketides.

Keywords. Aromatic polyketides, Bacterial polyketide synthase, Engineered biosynthesis, Fungal polyketide synthase, Plant polyketide synthase

Topics in Current Chemistry, Vol. 209

List of Abbreviations

ACP	acyl carrier protein
ACPS	holo-acyl carrier protein synthase
Act	actinorhodin
AF	aflatoxin
ARO	aromatase
AT	acyl transferase
CHS	chalcone synthase
CLF	chain length factor
CoA	coenzyme A
CYC	cyclase
DMAC	3,8-dihydroxy-1-methyl anthraquinone-2-carboxylic acid
FAS	fatty acid synthase
KR	ketoreductase
KS	β-ketoacyl: ACP synthase
MAT	malonyl CoA: ACP acyltransferase
6MSAS	6-methylsalicyclic acid synthase
NAC	*N*-acetylcysteamine
Orf	open reading frame
PCP	peptidyl carrier protein
PCR	polymerase chain reaction
PKS	polyketide synthase
PMSF	phenylmethylsulfonyl fluoride
PPTase	phosphopantetheinyl transferase

2PS	2-pyrone synthase
SDS-PAGE	sodium dodecyl sulfate-polyacrylamide gel electrophoresis
ST	sterigmatocystin
STS	stilbene synthase
Tcm	tetracenomycin
TE	thioesterase

1 Introduction

Polyketide metabolites are found in bacteria, fungi, and plants and represent one of the largest groups of natural products [1–4]. They are structurally classified into four major groups: aromatics (e.g., doxorubicin and tetracycline), macrolides (e.g., erythromycin and rapamycin), polyethers (e.g., monensin and salinomycin), and polyenes (e.g., amphotericin and candicidin), many of which are clinically valuable antibiotics or chemotherapeutic agents, or exhibit other pharmacological activities [5–8]. Despite their apparent structural diversity, polyketides share a common mechanism of biosynthesis. The carbon backbone of a polyketide results from sequential condensation of short fatty acids like acetate, propionate, or butyrate, in a manner resembling fatty acid biosynthesis, and this process is catalyzed by polyketide synthases (PKSs). Much of the current research on polyketide biosynthesis is driven by the unprecedented biochemistry and enzymology of the PKSs that provide an excellent model for elucidating the structure-function relationship of complex multienzyme systems and by the great potential of generating novel polyketide libraries via combinatorial biosynthesis with engineered PKSs [7, 9–21].

Following the convention of fatty acid synthases (FASs) [22–25], PKSs have been classified into two types according to their enzyme architecture and gene organization. Type I PKSs are multifunctional proteins consisting of domains for individual enzyme activities and have been found in bacteria as well as in fungi and plants. Type II PKSs are multienzyme complexes consisting of discrete proteins that are largely monofunctional and have so far only been found in bacteria. Although early isotope labeling experiments clearly demonstrated that FASs and PKSs use similar substrates, it is the recent cloning of PKS genes and the biochemical characterization of PKS enzymes that have provided a mechanistic explanation of how PKSs achieve the vast structural diversity during polyketide biosynthesis by varying the similar biosynthetic reactions of FASs. Thus, unlike fatty acid biosynthesis, in which the β-ketone group of the growing fatty acid intermediate **1** undergoes full reduction to a methylene group **2**, **3** during each cycle of elongation (pathway A in Fig. 1), the β-ketone group of the growing polyketide intermediate **4** could either be left untouched (**5**, **6**), leading to aromatic polyketides (pathway B in Fig. 1), or be subjected to no, partial, or full reduction (**7**–**10**), depending on a given cycle of elongation, leading to macrolides, polyethers, or polyenes (pathway C in Fig. 1). The latter forms the mechanistic basis for grouping macrolides, polyethers, and polyenes together as complex or reduced polyketides. It has now been well established that the biosynthesis of

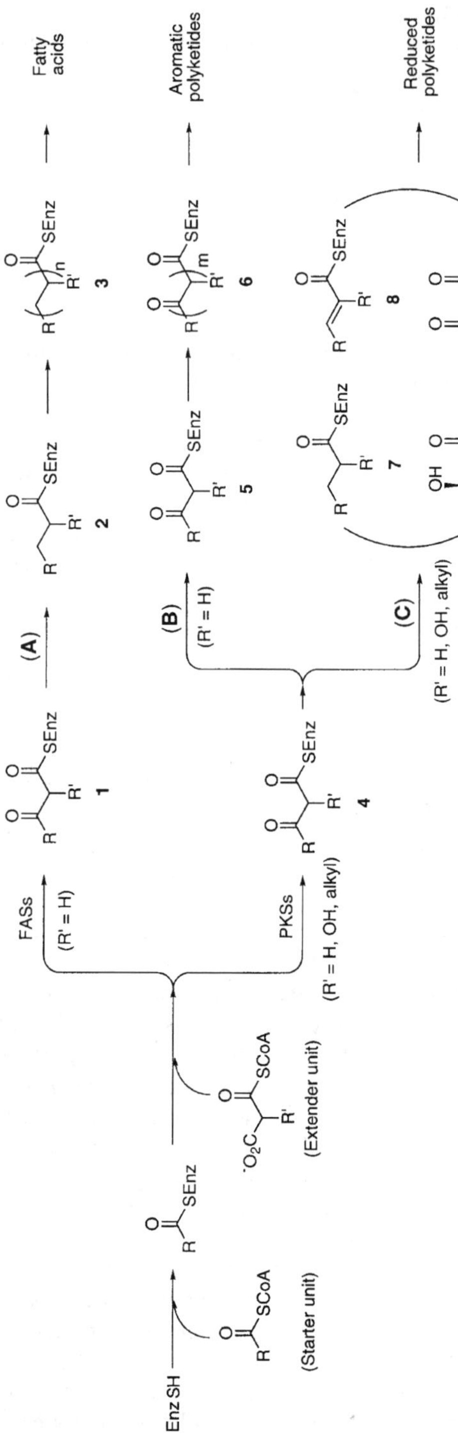

Fig. 1. Biosynthetic pathways showing the requirement of starter unit and extender unit and the processing of the elongating intermediates: (A) for fatty acids, (B) for aromatic polyketides, and (C) for reduced polyketides

reduced polyketides is catalyzed by the noniterative type I PKSs, which control product structural variation by evolving a set of noniteratively used domains that are arranged in a linear order mirroring the biosynthetic sequence of the metabolite. For comprehensive coverage on reduced polyketide biosynthesis, readers are referred to several excellent reviews appearing in the recent literature [10, 15, 25–34], including the one by Staunton and Wilkinson in Volume I of this two-volume series [34].

A

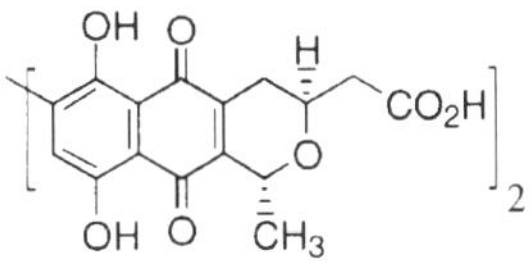

Actinorhodin, **11**

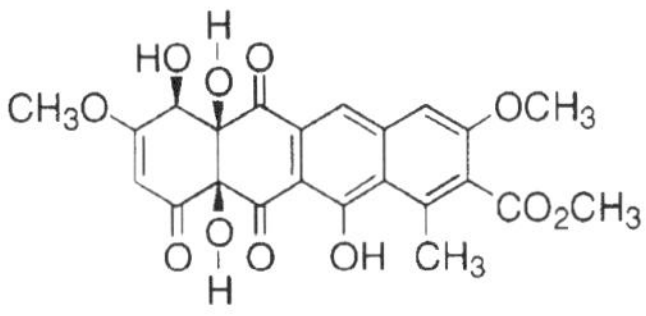

Tetracenomycin C, **12**

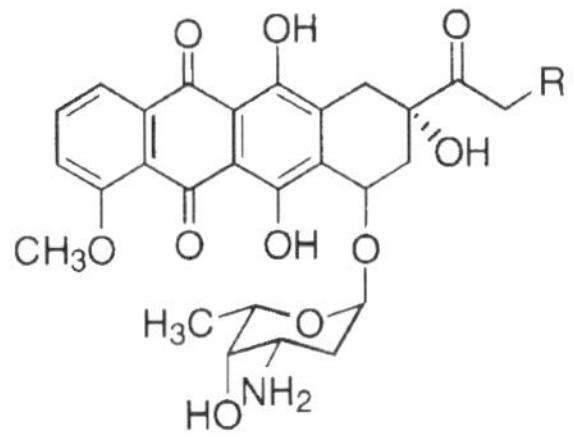

Daunorubicin, **13**, R = H
Doxorubicin, **14**, R = OH

B

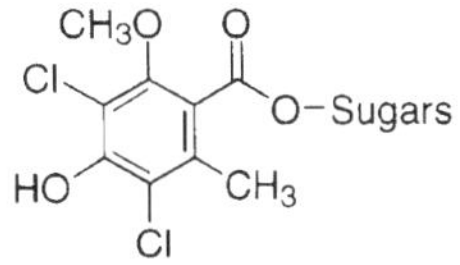

The aromatic moiety of avilamycin A, **15**

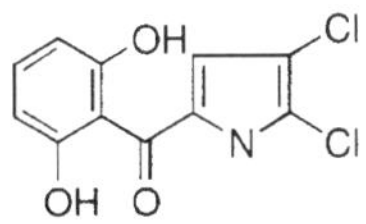

Pyoluteorin, **16**

C

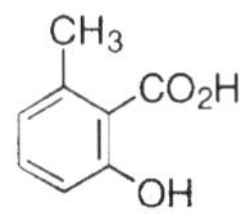

6-Methylsalicyclic acid, **17**

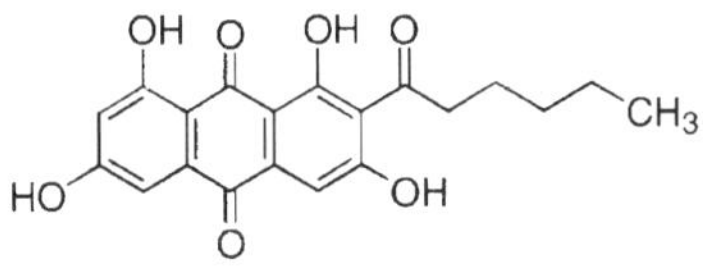

Norsolorinic acid, **18**

D

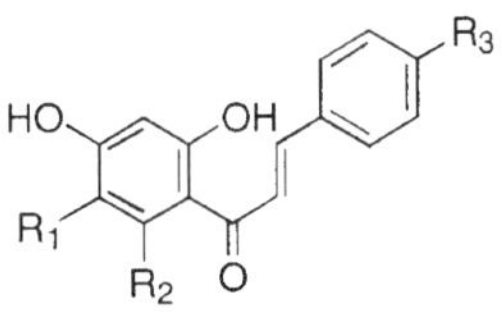

Chalcone, **19**, R_1 = H, R_2 = R_3 = OH
6'-Deoxychalcone, **20**, R_1 = R_2 = H, R_3 = OH
Methylchalcone, **21**, R_1 = CH_3, R_2 = OH, R_3 = H

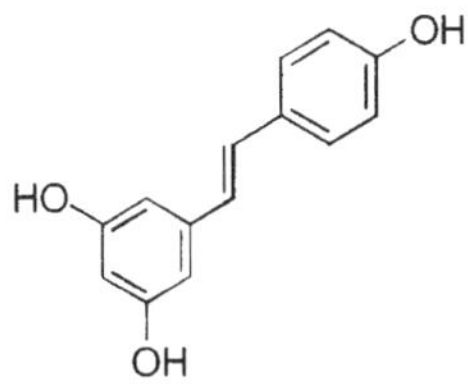

Resveratrol, **22**

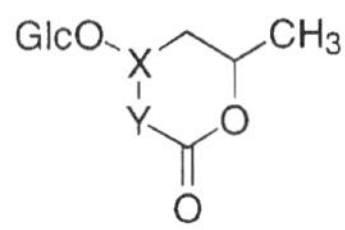

Gerberin, **23**, X-Y = C=CH
Parasorboside, **24**, X-Y = CH-CH_2

Fig. 2A–D. Representatives of aromatic polyketide metabolites produced by aromatic polyketide synthases from: **A, B** bacteria; **C** fungi; **D** plants

The mechanisms by which aromatic PKSs control structural diversity are distinct from those for noniterative type I PKSs for reduced polyketide biosynthesis. Aromatic PKSs catalyze the biosynthesis of various polycyclic, mostly aromatic, polyketides (Fig. 2), involving a linear poly-β-ketone intermediate such as **5** or **6**, and using malonyl coenzyme A (CoA) as an extender unit exclusively (pathway B in Fig. 1). The principal challenges faced by aromatic PKSs are to choose a starter unit, to determine the number of extensions, to control the folding of the linear poly-β-ketone intermediates, and to carry out regiospecific reduction, aromatization, and cyclizations of the correctly folded polyketide intermediates into polycyclic metabolites.

Genetic advances in *Streptomyces* species and related actinomycetes in the last decade have made it possible to apply genetic techniques in addressing these challenges. Since the cloning of the first two sets of aromatic PKSs, *act* PKS for **11** and *tcm* PKS for **12**, from *Streptomyces coelicolor* [35, 36] and *Streptomyces glaucescens* [37, 38], respectively, and the discovery of extensive cross-hybridizations among aromatic PKS genes in different actinomycetes [39] a decade ago, numerous aromatic PKS genes have now been cloned [40–62]. These studies revealed that aromatic polyketides in bacteria are assembled by type II PKSs that consist of several discrete proteins carrying a set of iteratively used enzyme activities (Fig. 3 A), reminiscent of bacterial FASs. However, exceptions have been noticed, in which either a iterative type I PKS [63] or noniterative type I PKS [64] is involved for the biosynthesis of an aromatic structure (Fig. 3B). In fungi, only type I PKSs have been found for aromatic polyketide biosynthesis, which is compose of iteratively used domains (Fig. 3 C) [65–69], reminiscent of vertebrate and fungal FASs. Plant aromatic PKS genes are classified as iterative type I enzymes in general but differ substantially from bacterial and fungal PKSs and lack the acyl carrier protein (ACP) domain (Fig. 3 D) [70–76]. They are often discussed together with bacterial and fungal PKSs because they share the characteristics of PKS chemistry of linking acyl CoA by repetitive decarboxylative condensation.

It became apparent upon sequencing that aromatic PKSs consist of proteins or domains that bear similar activities as FASs or noniterative type I PKSs for reduced polyketide biosynthesis, such as ACP, acyl transferase (AT), β-ketoacyl:ACP synthase or ketosynthase (KS), and ketoreductase (KR), reinforcing the mechanistic relationship between fatty acid and polyketide biosynthesis. However, the mere cloning and sequencing of gene clusters for aromatic polyketide biosynthesis fall short of explaining how aromatic PKSs control product structures, because only one set of very similar active sites has been identified for all aromatic PKSs. Consequently aromatic PKSs must act iteratively and recognize different intermediates in every cycle of elongations to build the linear poly-β-ketone chains and in the subsequent folding and cyclization processes. Insights into the mechanisms of aromatic PKSs came primarily from genetic manipulation of the PKS genes in vivo and, more recently, from biochemical characterization of the PKS enzymes in vitro.

The goal of this review is therefore to present our current knowledge of aromatic polyketide biosynthesis with emphasis on the most recent advances in the biochemistry and enzymology of PKSs. The so-called "tailoring enzymes", which modify the initial products synthesized by PKSs to imbue them with

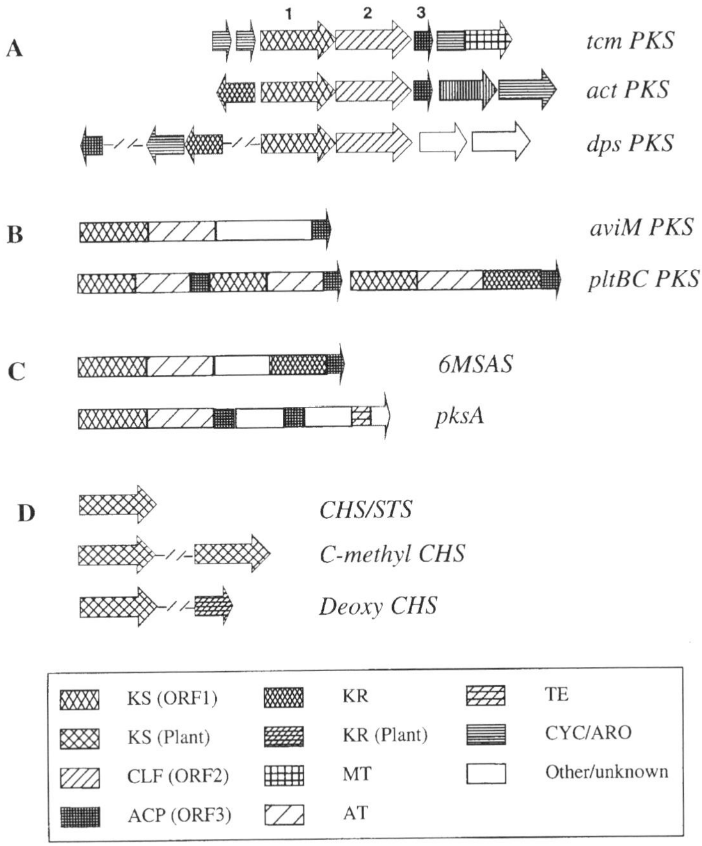

Fig. 3 A–D. Pictorial representation of structural organizations of selected aromatic polyketide synthase and associated genes from: **A, B** bacteria; **C** fungi; **D** plants. References to these polyketide synthases are given in the text

various biological activities, are not discussed here because they are not considered to be unique to aromatic polyketide biosynthesis. While most of the work on genetic engineering of aromatic PKSs has been aimed at elucidating the structure and function of PKS subunits or domains, engineered PKSs will be reviewed in such a way as to provide general design-rules for the biosynthesis of novel aromatic polyketides. Readers are encouraged to consult several other excellent reviews on various aspects of aromatic polyketide biosynthesis that have already appeared in the recent literature [15, 25–27, 29, 77–79].

2
Aromatic Polyketide Synthase Genes

Hopwood has recently reviewed exhaustively the discovery, the cloning, and the architecture of genes encoding aromatic polyketide biosynthesis in bacteria, fungi, and plants [29]. Selected in this section are only a few examples, illustrating organizational variations and structural features of aromatic PKSs.

2.1
Bacterial Polyketide Synthase

The type II nature of aromatic PKSs was established immediately upon sequencing of several gene clusters for aromatic polyketide biosynthesis [36, 38, 58]. The most striking features of various aromatic PKS genes are their high sequence homology and remarkably conserved gene organization [15, 29]. As illustrated in Fig. 3A, most of the aromatic PKS genes studied so far consist of three core open reading frames (orf), encoding the minimal PKSs [80], along with a few additional orfs encoding KR [49, 50, 81–84], aromatase (ARO) [53, 85–88], and cyclase (CYC) [36, 49, 53, 85–100]. While sequence comparison with FAS genes readily revealed that orf1 and orf3 of the minimal PKS genes encode KS and ACP, respectively, orf 2 is unique to type II aromatic PKSs whose function as chain length factor (CLF) was established by genetic manipulation of the minimal PKS genes [101]. The minimal PKS genes of orf1,2,3 have invariably been found in the same organization with orf1,2 being translationally coupled and orf3 lying immediately downstream of the orf1,2 pair. The *dps/dau* cluster is the only exception – orf3 is located 6.8 kb upstream of orf1,2 [49, 84]. The number and position of other orfs associated with the minimal PKSs were less predictable, although they always flank the minimal PKS genes. These orfs are unique to aromatic PKSs, whose functions as KR [83, 101, 102, 161], ARO [85–88], and CYC [84–88, 92, 94–96, 100] were established only after in vivo and in vitro experiments.

The striking resemblance in overall architecture among the various clusters has led to the paradigm that iterative type II PKSs are responsible for aromatic polyketide biosynthesis in bacteria, dramatically facilitating the cloning of new type II PKS genes and the characterization of type II PKS enzymes. However, since most of the type II PKSs studied so far were cloned with the *act* PKS as probe and limited to *Streptomyces* species and related actinomycetes [14, 28], caution has to be taken in applying this paradigm for aromatic polyketide biosynthesis to bacteria in general. In fact, in the study of the biosynthesis of avilamycin in *Streptomyces viridochromogens* Tü57 [63], Bechthold and coworkers cloned the *avi* gene cluster using a deoxysugar gene as probe that was amplified by the polymerase chain reaction (PCR) method [103]. Intriguingly, sequence analysis of the *avi* cluster revealed the *aviM* gene consisting of the characteristic domains of KS, AT, and ACP (Fig. 3B). They named *aviM* as the orsellinic acid synthase because expression of *aviM* in either *Streptomyces lividans* TK24 or *S. coelicolor* CH999 resulted in the production of orsellinic acid. AviM is therefore responsible for the biosynthesis of the aromatic moiety of avilamycin

A (**15**), representing the first iterative type I PKS for aromatic polyketide biosynthesis in bacteria. Equally unexpected was the PKS genes encoding pyoluteorin (**16**) biosynthesis in *Pseudomonas fluorescens* PF-5 [64]. Kraus and Loper identified the *plt* gene cluster by Tn5 transposon mutagenesis and established *pltB* and *pltC* as the PKS genes responsible for the biosynthesis of the aromatic polyketide moiety of **16** [104]. Sequence analysis revealed [64] surprisingly that PltB and PltC consist of characteristic domains of KS, AT, ACP, or KR in the same order as bacterial noniterative type I PKS for reduced polyketides, representing the first noniterative type I PKS for aromatic polyketide biosynthesis in bacteria (Fig. 3B). Although it is yet to be established how broadly these exceptions exist in *Streptomyces* species and related actinomycetes as well as in other bacteria, it certainly causes some doubt in predicting uncharacterized PKSs as type I or type II structure based solely on the chemical structure of either aromatic or reduced polyketides.

2.2 Fungal Polyketide Synthase

Fungal aromatic PKSs are iterative type I enzymes. 6-Methylsalicyclic acid synthase (6MSAS), the first microbial PKS purified [105], is a classical example of fungal PKSs, although the 6MSAS gene has only been cloned recently from *Penicillium patulum* [65, 106]. A homologous 6MSAS gene, *atX*, has also been cloned from *Aspergillus terreus* [66], with the KS domain of the 6MSAS of *P. patulum* as a probe. Sequence determination of the 6MSAS gene from both *P. patulum* [65] and *A. terreus* [66] revealed a single orf that consists of characteristic KS, AT, KR, and ACP domains, confirming the type I nature of 6MSAS (Fig. 3C). Since the synthesis of 6-methylsalicyclic acid (**17**) requires three steps of condensation, 6MSAS must have utilized the KS, AT, and ACP activities iteratively.

The PKS genes, *pksA* and *stcA*, for norsolorinic acid (**18**), an octaketide intermediate of aflatoxin (AF) and sterigmatocystin (ST) biosynthesis [68, 107–109], were also determined recently from *Aspergillus parasiticus* [110–112] and *Aspergillus nidulans* [113, 114], respectively. The *pksA*, also known as *pksL1* [111], and *stcA* genes are highly homologous and consist of only one set of characteristic KS, AT, and ACP domains, along with an additional thioesterase (TE) domain, reinforcing the notion that fungal PKSs are iterative type I enzymes (Fig. 3C). Interestingly, two functionally distinct FAS genes have been identified that are involved in primary and secondary metabolism, respectively [115]. In *A. nidulans*, mutants of FAS genes for secondary metabolism, *stcJ* and *stcK*, grew normally but cannot synthesize ST, yet supplementation of the mutants with hexanoic acid restored their ability to produce ST. Similarly, in *A. parasiticus*, the two specialized FASs were identified as *fas-1A* and *fas-2A* that showed high sequence homology to the yeast FASβ and FASα subunits, respectively [108, 116, 117], and insertional inactivation of *fas-1A* gave mutants that were unable to incorporate acetate into AF pathway [117]. These results strongly suggested the functional requirement for FAS in the biosynthesis of the C-6 starter unit that is further elongated by PksA and StcA during the biosyntheses of AF and ST, respectively. Other noteworthy features of fungal PKS genes include the apparent lack of ARO or CYC in all known fungal PKSs, the tandem structure of the two ACP domains in *stcA*, and the distinct TE domain in *pksA* and *stcA* genes. The TE domain is known to off-load the fully processed polyketide product from the noniterative type I PKS in reduced polyketide biosynthesis [118–120] but has not been found in all other known aromatic PKSs. Intriguingly, while sequence analysis of the *wA* gene from *Aspergillus nidulans*, which encodes the naphthopyrone YWA1 PKS, clearly shows that it is highly homologous to *pksA* and *stcA* with a TE domain at its C-terminus, overexpression of *wA* in *Aspergillus orzyae* suggested that the TE domain acts as a CYC, in addition to off-loading the polyketide product from the PKS enzyme [265, 266].

2.3
Plant Polyketide Synthase

Chalcone synthases (CHS) and stilbene synthase (STS) are closely related plant PKSs that catalyze the stepwise condensation between acyl CoA esters in the biosynthesis of flavonoids, stilbenes, and other related plant aromatic polyketides [70, 72, 75]. Sequences for numerous CHS as well as a few STS have been determined from various plants [72–76, 122–131]. They all possess a highly conserved cysteine (Cys) residue that is essential for PKS activity, although the sequences in this Cys motif have no apparent similarity to that of KS of bacterial and fungal PKSs [132]. The plant PKSs are essentially condensing enzymes, lack the ACP domain, and use the acyl CoA esters directly as substrate for the condensing reactions [72, 75].

Plant PKSs are iterative type I PKSs, as exemplified by CHS and STS (Fig. 3D). CHS and STS utilize the Cys active site iteratively by selecting a phenylpropanoid CoA as the starter unit and catalyzing three sequential condensations with malonyl CoA to synthesize a tetraketide intermediate that subsequently folds regiospecifically into aromatic structures like chalcone (**19**) [73, 124–126, 128, 132, 133] or resveratrol (**22**) [74, 127, 132–134]. An exception to this paradigm is the *gchs2* gene from the ornamental plant *Gerbera hybrida* [71, 123, 135]. While the GCHS2 protein is 73% identical to the CHS enzymes, it uses acetyl CoA, not phenylpropanoid CoA, as the starter unit and catalyzes two condensations with malonyl CoA to form the pyrone backbone of the gerberin (**23**) or parasorboside (**24**) aglycone [135]. Accordingly, the *gchs2* gene was renamed as *g2ps1* on the basis that the GCHS2 protein is in fact a 2-pyrone synthase (2PS). This finding revealed a new pathway to polyketide biosynthesis in plants and suggested that CHS-related enzymes are involved in the biosynthesis of a much larger range of plant products than was previously realized for **19**, **22**, and their analogs [135].

Discrete plant KRs have been identified that interact with CHS for the biosynthesis of 6′-deoxychalcone (**20**) (Fig. 3D) [133, 136–139]. Interestingly, the plant KR has no similarity with those that catalyze the reduction of the poly-β-ketone intermediates in polyketide or fatty acid biosynthesis [49, 50, 81–84]. Instead it is similar to various aldo/keto-reductases, mostly from carbohydrate metabolism [139–141], and contains a leucine zipper motif known to be involved in protein-protein interaction [142]. Plant *O*-methyltransferases are well known [143], but enzymes for C-methylation have not been described. The methyl group of *C*-methylated chalcone (**21**) is in fact derived from methylmalonyl CoA, suggesting that an aromatic PKS utilizes a methylmalonyl CoA as an extender unit. Cloning of CHSs from *Pinus strobus* indeed resulted in the identification of two highly homologous proteins, PStrCHS1 and PStrCHS2 (87.6% identity). These two enzymes interact with each other in the biosynthesis of **21**, with PstrCHS2 catalyzing specifically the condensation step with methylmalonyl CoA (Fig. 3D) [76]. The latter feature is distinct from all known aromatic PKSs that utilize malonyl CoA as an extender units exclusively. Methyl side chains in reduced polyketides are introduced by the use of methylmalony CoA as an extender unit, determined by the AT domain of the noniterative type I PKS [144–147]. PstCHS2 represents a novel mechanism by which a condensing enzyme can select an extender unit to control structural diversity in polyketide biosynthesis.

3 Polyketide Synthase Biochemistry and Enzymology

3.1 Bacterial Type II Polyketide Synthase

It has been a long sought after goal to study the biochemistry and enzymology of the bacterial type II PKSs in vitro, because the mechanisms by which the PKSs assemble the carbon skeleton of aromatic polyketides eventually have to be studied using purified enzymes. The two major challenges in this endeavor are to obtain the entire functional PKS complex in sufficient quantity and to monitor the synthesis of the inherently labile poly-β-ketone intermediates [148], which are thought to remain enzyme-bound throughout the elongation processes until the carbon chain reaches its full length. Retrospectively, early attempts at purifying and assaying bacterial PKS could not be successful because of the lack of the fundamental information regarding the PKS complex such as (1) the nature of individual subunits, (2) the minimal number of subunits that constitute a functional complex, and (3) the involvement of ARO or CYC that make it possible to measure the PKS activity by converting the poly-β-ketone intermediate into stable products. The cloning of the PKS genes provided many of these missing links, dramatically accelerating the pace for mechanistic investigations of bacterial PKSs by both in vivo and in vitro experiments. Functions of individual components of a PKS complex have been examined either in vivo by mutagenesis or expression of the native, mutant, or recombinant PKS genes in various polyketide producers or nonproducers [64, 81, 83–93, 95, 99–102, 149–166], or in vitro by overexpression and biochemical characterization of the recombinant proteins [94, 96, 167–180]. Furthermore, cell-free systems for in vitro synthesis of aromatic polyketides have been developed, making it possible to purify individual components of a functional PKS complex [181, 182]. The latter successfully led to the in vitro reconstitution of both the actinorhodin (Act) [173, 183] and tetracenomycin (Tcm) PKSs [184] from individually purified components, setting the stage to address the molecular basis of the structure and function relationship of type II aromatic PKSs.

3.1.1 *Phosphopantetheinyl Transferase*

Both the extender unit of malonyl CoA and the growing poly-β-ketone intermediates are covalently tethered to the ACP subunit of a bacterial PKS in an acyl thioester linkage to the phosphopantetheinyl moiety during polyketide biosynthesis [26, 29, 177, 185, 186]. The latter prosthetic group is introduced posttranslationally by the phosphopantetheinyl transferase (PPTase), which transfers the 4′-phosphopantetheine moiety of CoA (**25**) to the highly conserved Ser residue of apo-ACP (**26**), converting **26** into holo-ACP (**27**) (Fig. 4). Phosphopantetheinylation is absolutely necessary for polyketide biosynthesis, without which **26** is nonfunctional. Consequently, each FAS or PKS must have associated with it a PPTase competent to carry out its modification. The fact that both the PKS ACP [17] and FAS ACP [178] purified from *S. glaucescens* were phosphopantetheinylated supported the presence of such enzymes in *Streptomyces* species. Indeed, the FAS apo-ACP in *E. coli* is phosphopantetheinylated on Ser-36 to give a holo-ACP, and this modification is catalyzed by the holo-ACP synthase (ACPS) [24, 187], a member of the recently characterized superfamily of PPTases [188–191]. Intriguingly, sequence analysis of the gene clusters known for aromatic polyketide biosynthesis failed to reveal any orf, equivalent to ACPS, which is apparently needed for the modification of PKS ACPs, despite the fact that genes for secondary metabolite biosynthesis are clustered in one region of the microbial chromosome [26, 29]. It is, therefore, very tempting to propose that a PPTase, equivalent to ACPS in *E. coli*, associated primarily with fatty acid biosynthesis in a polyketide-producing organism, could modify both the FAS and PKS apo-ACPs, providing a functional connection between fatty acid and polyketide biosynthesis. While such PPTase is yet to be characterized from these polyketide-producing *Streptomyces* species, ACPS homologous genes have been identified during genome sequencing projects from *Mycobacterium tuberculosis* [262] and *S. coelicolor* [263].

The *E. coli* ACPS indeed displayed very broad substrate specificity for both the ACP and CoA substrates. Co-expression of *Streptomyces* genes encoding various PKS ACPs in *E. coli* strains overproducing ACPS led to high levels of holo-ACP production, indicative of in vivo phosphopantetheinylation by the ACPS protein [167]. Various ACPs have also been phosphopantetheinylated in vitro with the purified *E. coli* ACPS [173, 184, 188, 189, 192]. The decrease in catalytic efficiency of a *Streptomyces* PKS ACP as a substrate for the *E coli* ACPS has been correlated with a decrease in the overall negative charge in the highly anionic ACP family [188]. ACPS also catalyzes the transfer of acyl CoAs as well as several CoA analogs to apo-ACP, and a steady state kinetic study showed that acetyl CoA (**28**) is as efficient an ACPS substrate as CoA [188, 192]. Enzymatically synthesized acetyl-ACPs (**29**) were shown to be efficient substrates for the Act PKS, indicating that acetyl-ACP is a chemically competent intermediate of aromatic polyketide biosynthesis [192]. Together, these methods provide a way to generate preparative quantities of holo-ACPs, acyl ACP, as well as holo-ACP analogs with phosphopantetheine groups of altered compositions, length, and reactivity. The availability of these ACP derivatives opened the way to studying

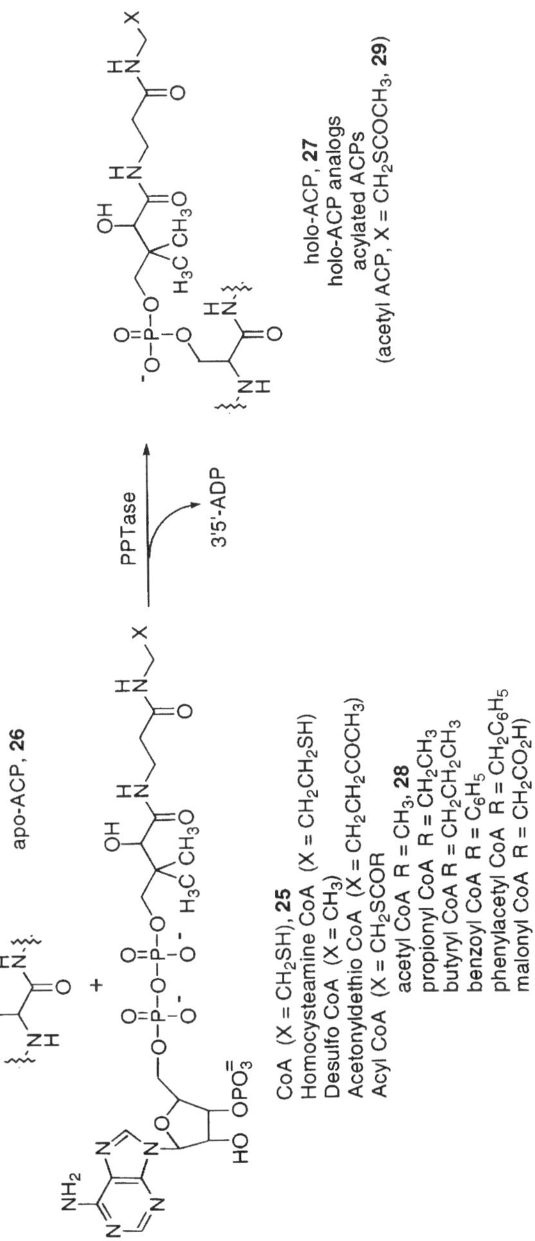

Fig. 4. Phosphopantetheinyl transferase catalyzed posttranslational modification of apo-ACP (**26**) into holo-ACP (**27**) or holo-ACP analogs

the role of the phosphopantetheine group in subsequent reactions of acyltransfer, condensation, reduction, dehydration, and cyclization in fatty acid and polyketide biosynthesis.

It is noteworthy that nonribosomal peptide synthetase is similarly posttranslationally modified by covalent attachment of the 4′-phosphopantetheine group to the peptidyl carrier protein (PCP) [193–198]. While the ACPS can modify various apo-ACPs [167, 173, 187–189, 191, 192], it failed to modify PCPs from a variety of peptide synthetases [189]. This led to the discovery of the second family of PPTases [189], such as EntD from *E. coli* [189, 199, 200], Sfp from *Bacillus subtilis* [189, 200–204], PptT from *M. tuberculosis* [264], and Gsp from *B. brevis* [189, 205, 206], required for the biosynthesis of enterobactin, surfactin, mycobactin, and gramicidin S, respectively. In contrast to ACPS, proteins in the latter family, such as Sfp, showed broader substrate specificity, modifying apo-PCPs, apo-ACPs, as well as apo-aryl carrier proteins and utilizing both CoA, acyl CoAs, and CoA analogs [204].

3.1.2
Acyl Carrier Protein

ACP is a central component of type II PKS and is involved in possibly all reactions of bacterial aromatic polyketide biosynthesis. The starter unit (before its transfer to the KS subunit), the extender units, the growing poly-β-ketone intermediates, as well as the full length linear poly-β-ketone product, are covalently bound to ACP in a thioester linkage to the terminal sulfhydryl of the 4′-phosphopantetheine prosthetic group (see Sect. 3.1.1). These ACP thioesters must have been recognized as substrates by type II PKSs and other associated enzymes such as KR, CYC, or ARO to build and process the poly-β-ketone intermediates into aromatic polyketides.

The first sets of data demonstrating the necessity of ACP for the function of a bacterial type II PKS came from targeted gene replacements in the *act* cluster [155, 156]. Either deletion or introduction of a frame shift mutation into the ACP gene of the *act* cluster in *S. coelicolor* resulted in mutants whose polyketide production was completely abolished [155]. Site-directed mutagenesis further confirmed specifically that Ser-42 of Act ACP is essential for the PKS activity, which is the attachment site for the 4′-phosphopantetheine group [155]. Similar experiments have already been carried out for FAS ACPs [207]. Alternatively, overexpression of the *tcmM* gene in *S. glaucescens* dramatically stimulated the production of various biosynthetic precursors of **12**, presumably by increasing the malonyl-ACP concentration, and hence the overall activity of the Tcm PKS complex [151]. These conclusions were further supported by in vitro biochemical investigations. Indeed, removal of the TcmM ACP by immunoprecipitation from a Tcm PKS cell-free preparation reduced 50% of the PKS activity; addition of purified TcmM ACP [171, 177] to this ACP-deficient PKS fully restored the PKS activity [182]. Interestingly, the TcmM ACP cannot be substituted in vitro by either *S. glaucescens* FAS ACP or the *E. coli* ACP [182], despite the fact that the three ACPs have a high amino acid sequence similarity and are equally active in the in vitro malonyl CoA:ACP transacylase assay [182]. The latter results underscore the ambiguity of assaying the individual biochemical activity of the PKS components without considering the interaction among them during their in vitro reconstitution. It was unfortunate in these early studies that few ACPs in holo-form were available in sufficient quantity. Therefore, despite the fact the Act ACP can be replaced in vivo in *S. coelicolor* by various type II PKS ACPs, including Tcm ACP [155, 156], the effect of supplementing other PKS ACPs to the Tcm ACP-deficient PKS was not determined. It has now been well established that the ACPs of bacterial type II PKSs can be interchanged in vivo without compromising or altering the PKS activity and specificity [80, 85, 88, 92, 95, 96, 101, 161, 162]. Very recently, the Act ACP was even replaced in vivo by an ACP domain of the rat FAS [179], suggesting the structural tolerance or conformational flexibility of the type II PKS enzymes. It will be very interesting to repeat these experiments by interchanging the ACP subunit as well as other components of the PKS complex to explore the structure and function relationship of type II PKSs. The easy access to various holo-ACPs or acyl ACPs by either in vivo or in vitro phosphopantetheinylations [167, 184, 187–192] and the successful reconstitution of both the Act PKS [173, 184] and Tcm PKS [183] from individually purified components will surely facilitate such investigations.

Functional assay for PKS ACPs was developed [168, 177] long before the development of a cell-free system to assay the entire PKS complex [181, 182]. ACPs can be assayed as a substrate to accept a malonate from malonyl CoA, catalyzed by a crude malonyl CoA:ACP acyltransferase (MAT) preparation [168, 174, 175, 177, 178]. Only holo-ACPs were biochemically active in the malonyl CoA:ACP transacylase assay, confirming that 4′-phosphopantetheinylation is a prerequisite for loading the malonyl CoA extender unit to ACP [168, 177, 186].

The role ACP plays in loading the starter unit is less clear. It has been proposed that the starter unit be transiently loaded to the ACP as a thioester before it is transferred to the KS subunit to initiate polyketide biosynthesis. Acetyl CoA is assumed to be the starter unit for aromatic polyketides with an acetate starter. However, no such acetyl CoA:ACP acyltransferase has been identified. Early sequence analysis did reveal a -G-H-S-motif in the KS subunit of the bacterial type II PKS [36, 38], which was believed to constitute the AT active site with Ser for substrate attachment [22, 23]. The recognition of this motif led to the hypothesis that KS is bifunctional, with this AT domain involved in catalyzing the transfer of acetyl CoA, via an acetyl-ACP intermediate, to load the acetate starter to KS. However, mutations of Ser351Ala in *tcm* KS [164] and Ser347Leu in *act* KS [158] failed to abolish the production of aromatic polyketides, disproving the AT function of the Ser residue within this -G-H-S- motif of KS. In contrast, malonyl CoA alone was sufficient to support polyketide biosynthesis in vitro by cell-free preparations of both Act PKS [181] and Tcm PKS [182]. More recent studies with both the Act PKS [173, 184] and Tcm PKS [183] reconstituted from individually purified subunits demonstrated that acetyl CoA is not required for the production of SEK 4 (**30**) and SEK 4b (**31**), precursor of **11**, and of Tcm F2 (**32**), precursor of **12**, respectively. These results suggested that the acetate starter in fact arises from decarboxylation of malonyl-ACP, a reaction that has been noted in FAS [208], bacterial type I PKS [209], and plant PKS [124, 134]. It has been indeed suggested that type II PKSs for aromatic polyketides with starter units other than acetate may require dedicated enzymes to initiate polyketide synthesis, since these starter units cannot be produced directly from malonyl CoA by a simple decarboxylation [152, 153, 163].

The solution structure of the Act PKS apo-ACP has been determined using ^{1}H NMR spectroscopy, representing the first PKS component for which detailed structural information has been obtained [169, 170]. The holo-ACP assumed essentially the same structure as the apo-ACP, on which the structural studies were solely centered [170]. The Act PKS ACP showed strong secondary structure homology with the *E. coli* FAS ACP, whose structure was also determined by ^{1}H NMR [210–212]. Like *E. coli* FAS ACP that exists as a four helix bundle (49% α-helical content) with a hydrophobic core that forms the presumptive binding site for a growing fatty acid chain, the Act PKS ACP is composed of four principally similar helices (51% α-helical content) [170]. However, subtle structural differences have been noticed that may confer binding specificity to these proteins. Thus, although the Act PKS ACP contains a hydrophobic core, there are also a number of buried hydrophilic groups, principally Arg-72 and Asn-79, both of which are 100% conserved in PKS ACPs but not in FAS ACPs [170]. Such a polar environment within the core may provide a mechanism for PKS ACPs to stabilize the extremely reactive poly-β-ketone intermediates via hydrogen bondings, and Arg-72 and Asn-79 may be suitably positioned for such interaction [170].

The structure of the Act PKS ACP also supported the early prediction that the -D-S-L- motif of ACPs forms an important molecular recognition site for ACPS [168, 187–189, 191]. As discussed in Sect. 3.1.1, ACPS can phosphopantetheinylate various FAS ACPs and PKS ACPs at the Ser residue within this motif. Important areas of homology indeed exist around this motif in both the *E. coli* FAS ACP and the Act PKS ACP. None of the residues within this motif is significantly buried in the final ACP structure, suggesting that they may all be available to form important interactions with ACPS. Indeed, a related PKS ACP with an -E-S-L- motif was not modified by ACPS, suggesting that even a conservative mutation like D/E within this motif may prevent molecular recognition despite the similar secondary structure of these ACPs [170].

3.1.3
Malonyl CoA:Acyl Carrier Protein Transacylase

MAT activates the extender unit of malonyl CoA into malonyl-ACP for type II FASs in fatty acid biosynthesis. *E. coli* FAS MAT has been purified [213, 214], various aspects of the enzymatic mechanism have been studied [215, 216], the *fabD* gene encoding the MAT enzyme has been cloned [217, 218], and a mutant defective in the enzyme has been isolated [216, 219]. Together, these studies conclusively established the essential role of MAT in fatty acid biosynthesis and demonstrated the remarkable conservation of the active site motif of -G-H-S- between MAT and other acyltransferase proteins or domains [26, 29]. However, genetic analysis has shown that all bacterial type II PKS gene clusters studied so far apparently lack genes encoding such MAT proteins [29]. This has led to an intensive search for the MAT enzyme involved in polyketide biosynthesis [174,

177, 178, 183, 184] and to the current controversy on how PKS ACPs are acylated by malonyl CoA [172, 173, 183, 184].

On the assumption that a MAT must have existed in *Streptomyces* species for either fatty acid or polyketide biosynthesis, Hutchinson and co-workers detected a MAT activity in a crude cell-free preparation from *S. glaucescens* that produces **12**, using malonyl CoA and the *E. coli* FAS ACP as substrates [177]. They then used this MAT preparation to assay the presence of an ACP and subsequently purified two ACPs from *S. glaucescens* [177, 178]. Amino acid sequence analysis revealed that one was the Tcm PKS ACP [177], encoded by the *tcmM* gene that had already been cloned early along with the rest of the *tcm* gene cluster [38]. The other ACP led to the cloning of a gene cluster consisting of four orfs, including *fabC* that encodes the purified FAS ACP and *fabD* that is homologous to the *E. coli fabD* gene [178]. Based on the fact that FabC was the principal ACP produced constitutively in *S. glaucescens* [178] and that the *S. glaucescens fabD* gene was able to complement an *E. coli fabD* mutant [216, 219], this cluster was assigned to encode a type II FAS in *S. glaucescens*. Remarkably, the FabD MAT of *S. glaucescens* was fully competent to charge the Tcm PKS ACP with malonate, suggesting PKS and FAS might share the MAT protein for polyketide and fatty acid biosynthesis in *S. glaucescens* [178].

To address the role FabD MAT played in polyketide biosynthesis directly, disruption of the *fabD* gene in *S. glaucescens* was attempted. However, under all conditions tried, the organism strongly resisted loss of the *fabD* function, suggesting that *fabD* is essential for growth, as might be expected for an FAS enzyme [178]. Alternatively, Bao et al. overexpressed and purified the FabD MAT and examined its effect on the Tcm PKS activity in vitro [183]. While Tcm KLMN and FabD constituted a functional Tcm PKS to synthesize **32** in vitro, omission of FabD completely abolished the production of **32**, conclusively demonstrating the necessity of MAT for a type II PKS [178]. Interestingly, self-malonylation of TcmM ACP in the absence of FabD MAT was noticed but apparently was insufficient to support the Tcm PKS to synthesize **32** in vitro [183].

Revill and co-workers took a similar strategy to investigate the role MAT played in polyketide biosynthesis in *S. coelicolor* [174]. They overexpressed and purified the Act PKS ACP from *E. coli*. Although only 2–3% of the recombinant Act PKS ACP was in the active holo-form, this was enough for them to characterize the Act PKS ACP-dependent MAT from *S. coelicolor* by monitoring the transfer of malonyl CoA to form malonyl-ACP [168, 174]. Deletion of the entire *act* cluster [101] or introduction of the pathway-specific activator of the *act* cluster [220, 221] showed no effect on the MAT activity, suggesting that this MAT was not associated with the *act* cluster [174]. They subsequently purified the MAT enzyme, cloned the *S. coelicolor fabD* gene encoding the MAT protein, and mapped its chromosomal location at approximately 2.8 Mb apart from the *act* cluster [174]. As expected, the *S. coelicolor* FabD MAT was highly homologous to the *E. coli* FabD as well as to other acyltransferase with the -G-H-S- signature motif. Like the *S. glaucescens fabD*, disruption of *fabD* in *S. coelicolor* has not been possible, supporting its essential role in fatty acid biosynthesis [174].

Carreras and Khosla preferred to purify the MTA from *S. coelicolor* by its ability to support the Act PKS activity [184]. They noticed that a reconstituted preparation, consisting of three purified proteins of the minimal Act PKS, failed to catalyze polyketide biosynthesis, but the Act PKS activity can be restored by supplementing it with cell-free extract from *S. coelicolor* CH999, an engineered strain whose *act* cluster has been deleted [101]. They subsequently sought and purified the responsible protein from *S. coelicolor* CH999 by assaying for its ability to complement purified minimal Act PKS in polyketide biosynthesis [184]. N-terminal sequencing of the first 16 residues indeed identified the protein as the FabD MAT, which was purified earlier by Revill and co-workers, based on its ability to charge Act ACP with malonate [174]. Taken together these results from studies for both the Tcm PKS [177, 178, 183] and Act PKS [169, 174, 184], a paradigm is emerging that MAT is essential for bacterial aromatic type II PKSs, which is recruited from the primary metabolism. MAT thus provides another functional connection between fatty acid and polyketide biosynthesis.

Rather unexpected was the observation by Simpson and co-workers that type II PKS ACPs can catalyze self-malonylation upon incubation with malonyl CoA in vitro [172, 173]. These researchers found that the self-malonylation was completed after 30 min when the Act Cys17Ser holo-ACP was incubated with malonyl CoA in phosphate buffer at 30 °C and that no malonylation of the Act Cys17Ser apo-ACP was observed under the identical conditions. (Act holo-ACP behaves similarly to the Cys17Ser mutant that was preferred in these experiments to avoid complication caused by the formation of an intramolecular disulfide bond between the phosphopantetheine thiol and Cys-17.) The formation of a monomalonyl-ACP adduct was established by electron spray mass spectrometry. The site of malonylation was confirmed to be the 4′-phosphopantetheine sulfhydryl by thiol specific reagent that blocked the malonylation completely and by HPLC analysis of proteolytic fragments of the malonyl Act Cys17Ser ACP. Addition of phenylmethylsulfonyl fluoride (PMSF), an effective inhibitor of *E. coli* FAS MAT [22, 213], caused no decrease in the rate of self-malonylation, excluding the possibility that self-malonylation resulted from contamination of trace quantities of the *E. Coli* FAS MAT with the Act Cys17Ser ACP sample. In fact, addition of the recombinant *S. coelicolor* FabD MAT [174] did not accelerate the rate of malonylation. Kinetic characterization showed that the self-malonylation of Act Cys17Ser ACP followed Michaels-Menten kinetics with a K_m of 219 μmol/l for malonyl CoA and a k_{cat} of 0.34/min for the reaction. Complete malonylation was also observed with other type II PKS ACPs but not with FAS ACPs, suggesting that this is a distinct property of type II PKS ACPs. Based on these results, these researchers proposed an alternative model for the initiation of aromatic polyketide biosynthesis by bypassing the MAT activity, suggesting that the KS, CLF, and holo-ACP may constitute a truly minimal PKS in vivo. The latter prediction seems to be inconsistent with the recent study by Carreras and Khosla, who demonstrated that Act KS, CLF, and holo-ACP failed to catalyze polyketide biosynthesis and that MAT was essential for the Act PKS activity in vitro [184]. Bao et al. reached the same conclusion with the reconstituted Tcm PKS. While self-malonylation of TcmM ACP was noticed, the initial rate of TcmM ACP malonylation in the presence of the *S. glaucescens* FAS FabD MAT was 20 times faster than that without MAT. In the absence of MAT, no polyketide was detected, indicating that MAT was essential for the Tcm PKS activity or at least the self-malonylation of TcmM ACP was insufficient to support the minimal Tcm PKS to synthesize polyketide in vitro [183].

The apparent controversy concerning whether or not MAT is essential for polyketide biosynthesis by a PKS complex was finally resolved by Simpson and co-workers [173]. These researchers systematically examined the effect of varying the concentrations of Act holo-ACP to Act KS/CLF proteins on polyketide production in the presence or absence of MAT. They noticed that the concentration of Act holo-ACP appeared to be rate-limiting for the production of polyketide by the minimal Act PKS complex. While the rate of polyketide biosynthesis increased with Act holo-ACP concentration, this rate was significantly lower than expected at low holo-ACP concentration. When Act holo-ACP was present in limiting or in equal molar concentration of Act KS/CLF, MAT was required for the minimal Act PKS complex to synthesize polyketide. When Act holo-ACP was present in excess, efficient polyketide synthesis proceeded without MAT and no difference in the rate of polyketide production was observed. Based on these results, they concluded that the rate of polyketide biosynthesis was dictated by the ratio of holo-ACP to KS/CLF, as well as by the total protein concentration. At low holo-ACP concentration or equimolar concentration of holo-ACP and KS/CLF, holo-ACP was sequestered by the KS/CLF complex, hence reducing the rate of holo-ACP self-malonylation, and the loading of malonyl group to holo-ACP consequently required MAT. At concentrations above the stoichiometric amount of holo-ACP, the excess free ACP was able to undergo self-malonylation, bypassing MAT, to load malonate from CoA to ACP for polyketide biosynthesis. These studies clearly demonstrated two concentration-dependent mechanisms for malonate transfer from CoA to holo-ACP in vitro. However, to what degree each of these two mechanisms contribute to polyketide biosynthesis in vivo remains an open question since the in vivo concentrations of MAT, KS/CLF, holo-ACP, and malonyl CoA are yet to be determined.

3.1.4
β-Ketoacyl Synthase

Functional analysis of the PKS KS (also called KSα, [183]), encoded by *actI-ORF1*, *tcmK*, or their homologs, was primarily based on the extensive knowledge of the FAS KS enzymes, which have been well characterized biochemically and genetically [22–24, 222, 223]. The crystal structure of *E. coli* FAS KS II has been determined recently, revealing the molecular architecture of a KS enzyme [224]. Like fatty acid biosynthesis, KSα catalyzes the decarboxylative condensation between the growing poly-β-ketone intermediates and malonyl-ACP. One could envisage that the acyl group of acyl-ACP is first transfered to the Cys residue at the active site of the KSα, resulting in a thioester; malonyl-ACP is then decarboxylated to generate a carbanion that nucleophilically attacks the thioester to complete one cycle of elongation.

The necessity of Cys in the active site of KSα was first indicated by the specific inhibitory effect of cerulenin on polyketide biosynthesis [181, 182, 225–227]. It has been well established that cerulenin reacts specifically with the Cys residue of the FAS KS enzymes [22, 24, 223, 225–227]. Site-specific mutagenesis of this Cys in ActI-ORF1 to the Cys169Glu mutant [158] and in TcmK to the Cys173Ala or Cys173Ser mutant [163] completely abolished polyketide biosynthesis, confirming that the Cys of KS plays a critical role in the catalytic activity of a type II PKS.

At least the transfer of the acyl group from ACP to the Cys residue of KS prior to further elongation requires, in theory, an AT activity. The observation that Ser-specific inhibitors, like iodoacetamide and PMSF [22, 24], diminished PKS activity [182] supported the presence of such AT activity. KSα indeed contains the highly conserved AT motif of -G-H-S- with the Ser to constitute the active site [36, 38]. However, site-directed mutagenesis of this Ser in ActI-ORF1 to the Ser347Leu mutant [158] reduced but did not abolish polyketide production. Introduction of the Ser351Ala mutation into TcmK [164] also had no noticeable effect on the Tcm PKS activity. These results clearly ruled out this Ser as an essential component of the AT activity of KSα. Intriguingly, the His350Leu-Ser351Ala double mutant of TcmK produced no polyketide metabolite [164]. The idea that His-350 but not Ser-351 may be critical for the AT activity of TcmK was discussed but is still waiting to be proved by experiments [164]. It is sufficient for now that the putative AT activity for KSα has not been identified, if KSα has this second activity.

The function of CLF (also called KSβ [183]), encoded by *actI-ORF2*, *tcmL* or their homologs, was much more elusive. While it is highly homologous to KSα, the KSβ lacks the catalytic site of the Cys residue for condensation. Nontheless, introduction of mutation into either *actI-ORF2* (157,158,165) or *tcmL* [98, 228] completely abolished the production of **11** or **12** in *S. coelicolor* or *S. glaucescens*, respectively, demonstrating the necessity of KSβ for the PKS activity. This led to hypothesis that KSα and KSβ form a functional heterodimer, which was supported by the fact that genes encoding KSα and KSβ are translationally coupled, ensuring equimolar production of the KSα and KSβ proteins [38]. It was the seminal work by Khosla and co-workers that shed light on the function of KSβ [101]. These researchers developed a *Streptomyces* host-vector system to facilitate efficient construction and expression of recombinant PKSs. They constructed various PKS gene cassettes consisting of either a homologous pair of KSα and KSβ, such as ActI-ORF1/ActI-ORF2 or TcmK/L, or a heterologous pair of KSα and KSβ, such as TcmK/ActI-ORF2, along with a few additional genes to convert the nascent poly-β-ketone intermediates into stable metabolites. Although many of the gene cassettes consisting of a heterologous pair of KSα and KSβ were non-functional, in each case where a product could be isolated, the chain length of the polyketide was identical to that of the natural product corresponding to the source of KSβ. Based on these results, they concluded that KSβ plays the role of a chain-length-determining factor; hence, KSβ was named "Chain Length Factor" [101].

Sherman and co-workers discovered that mutations in *actI-ORF1* can be complemented in trans by several heterologous KSα genes, while mutations in *actI-ORF2* were largely not complemented by their heterologous KSβ genes [157, 165]. In contrast, Hutchinson and co-workers found that a *tcmK* mutant cannot be complemented by *actI-ORF1* in trans, but gene cassettes consisting of *tcmK/actI-ORF2/tcmM*, with or without additional CYC genes, restored production of

12 to a *tcmL* mutant of *S. glaucescens* [95]. While the former results reinforced the notion that PKSs consisting of a heterologous pair of KSα and KSβ are often nonfunctional, the latter result clearly demonstrated that KSβ alone is not sufficient to control the chain length. Therefore, these researchers substantiated the functional assignment of KSβ as CLF by emphasizing that it is the joint activities of KSα and KSβ, rather than just KSβ alone, that determine the chain length of the initial poly-β-ketone intermediate [95].

3.1.5
Polyketide Aromatase and Cyclase

Although genetic studies have identified a number of putative polyketide AROs [85–89] and CYCs [36, 49, 84–91], these studies failed to reveal any mechanistic insights for these enzymatic aldol or Claisen condensations. (ARO and CYC are often used interchangeably, and there is little biochemical data to separate them into two subgroups of enzymes.) Biochemical characterizations of these enzymes are difficult because the presumed substrates, whether they are linear or monocyclic poly-β-ketones, are not directly available. Tcm F2 CYC [94] and TcmN [96] are the only polyketide CYCs that have been purified and characterized so far.

Tcm F2 CYC catalyzes the cyclization of **32** to Tcm F1 (**33**) (Fig. 5), a reaction that closely resembles the intramolecular aldol and Claisen condensations involved in the synthesis of polycyclic aromatic polyketides. Unlike the other substrates of polyketide CYCs, **32** is stable enough that it can be isolated [229]. Shen and Hutchinson purified the Tcm F2 CYC from *S. glaucescens* by its ability to catalyze the **32**-to-**33** conversion [94]. The N-terminal sequence of the purified enzyme established that it is encoded by the *tcmI* gene [98], providing direct enzymological evidence for a polyketide CYC. Tcm F2 CYC is a homotrimer in solution and catalyzes the intramolecular cyclization of **32** to **33** at pH 8 and to 9-decarboxy Tcm F1 (**34**) at pH 6.5, requiring no cofactor. A stepwise mechanism was proposed, suggesting that C-C bond formation is a separate event from dehydration (Fig. 5) [94]. Other polyketide CYCs were predicted to operate, like Tcm F2 CYC, by a similar stepwise mechanism [89, 99, 149].

TcmN is a bifunctional protein with the N-terminus catalyzing the cyclization of the Tcm PKS-bound linear decaketide to **32** and the C-terminus catalyzing the O-methylation of Tcm D3 (**35**) to Tcm B3 (**36**) (Fig. 5). Since the linear decaketide cannot be prepared chemically because of its inherent reactivity, Shen and Hutchinson took advantage of the bifunctional nature of TcmN [97] and purified it on the basis of its *O*-methyltransferase activity [96]. They expressed the *tcmN* gene in both *S. lividans* and *E. coli* and purified the TcmN protein to homogeneity. The protein produced in either host was identical, as judged by their enzymological properties as Tcm D3 *O*-methyltransferase. To assay the CYC activity, they used a cell-free preparation of the minimal Tcm PKS that is competent to synthesize the Tcm PKS-bound linear decaketide from acetyl CoA and malonyl CoA in vitro [182]. In the absence of a polyketide cyclase, the linear decaketide underwent aberrant cyclizations to various products, including **32** and SEK15 (**37**) [95, 96]. Remarkably, addition of the purified TcmN to the minimal Tcm PKS preparation subverted all aberrant reactions by regiospecifically cyclizing

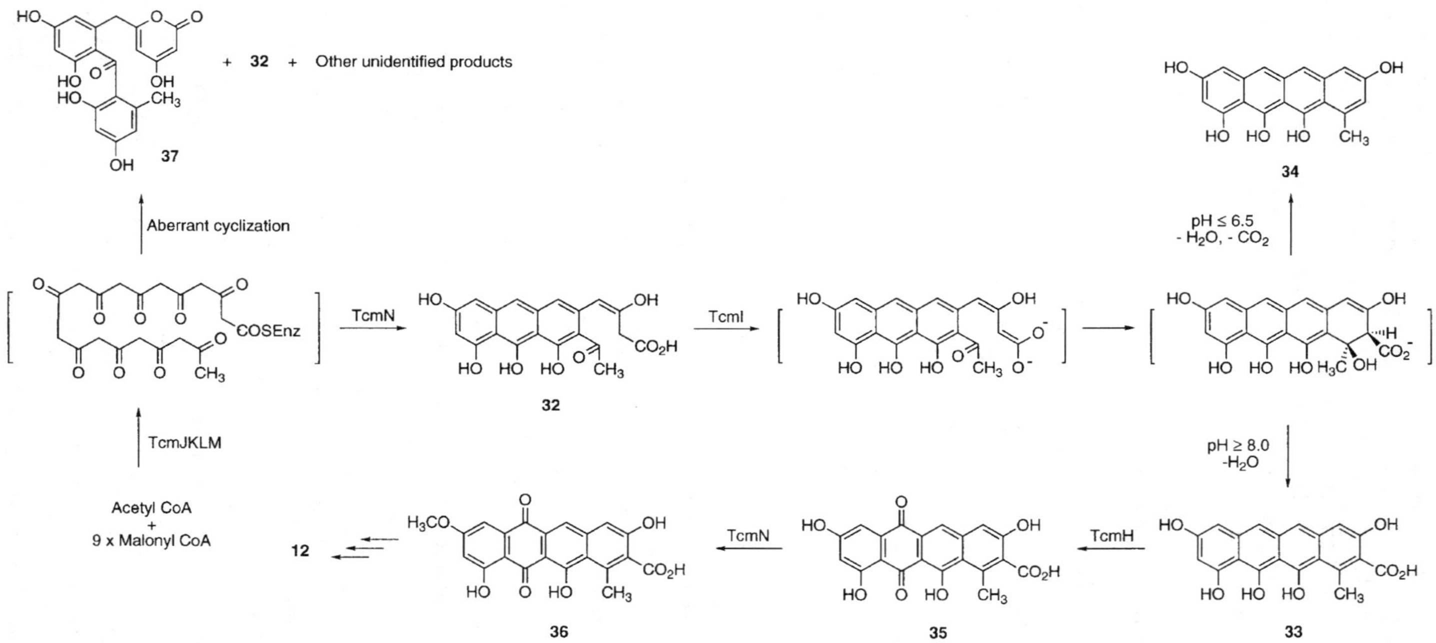

Fig. 5. In vitro synthesis of tetracenomycins and their shunt metabolites from acetyl CoA and malonyl CoA by the TcmKLM polyketide synthase complex, the TcmJ, TcmN, and TcmI cyclases, and the TcmH monoxygenase

the linear decaketide into **32** [96]. These results supported the existence of the putative linear poly-β-ketone intermediates and demonstrated that a CYC such as TcmN is largely responsible for the regiospecific cyclization of these intermediates. The latter is of great significance in formulating hypotheses for studying other polyketide CYCs whose substrates, otherwise unavailable, could be generated in situ by combinations of the minimal PKSs.

Sequence analysis has revealed that polyketide ARO/CYCs occur in two architectural forms: didomain ARO/CYCs, such as ActVII [36] or GrisIV [61], whose N- and C-terminal halves of the protein are similar [41, 89], and monodomain ARO/CYCs, such as TcmN [96, 97] or WhiEVI [45, 85, 100], which shows sequence homology to both the N- and C-terminal halves of the didomain ARO/CYCs. One hypothesis for this architectural difference is that the two halves of the didomain enzymes could fold internally to form the active conformation [96]. For monodomain enzymes to achieve a similar conformation, two molecules of the protein would be required, with each one to provide one-half of the homodimer. Indeed, the purified TcmN is a homodimer in solution [96]. Even the truncated N-terminal half of TcmN, representing the ARO/CYC domain that is encoded by the first 177 codons of *tcmN*, was functional in catalyzing the cyclization of the linear decaketide into **32**, suggesting that the ARO/CYC domain of TcmN can function independently [97]. These results agreed well with the homodimer model for monodomain ARO/CYCs. In contrast, when expressed alone, neither the N-terminal half nor the C-terminal half of the Act or Gris ARO/CYC exhibited any enzyme activity [180]. However, the half proteins were active in the presence of each other, and monodomain ARO/CYCs such as TcmN$^{1-166}$ cannot substitute for the N-terminal half of the didomain ARO/CYCs such as Act ARO/CYC$^{1-145}$ or Gris ARO/CYC$^{1-148}$ despite the high sequence similarity between them [180]. These results agreed well with the model in which didomain enzyme folded internally to form the active conformation. Consequently, neither the N-terminal half or C-terminal half of the didomain ARO/CYCs alone nor a combination of the former with a monodomain ARO/CYC could constitute the active complex.

3.1.6
In Vitro Reconstitution of Type II Polyketide Synthase

Shen and Hutchinson developed the first cell-free system for a bacterial type II PKS in 1993 and succeeded in synthesizing **32** from acetyl and malonyl CoA by the Tcm PKS in vitro [182]. Since then, similar experiments were reported by Khosla and co-workers for the Act PKS [181]. More recently, both the Act PKS [173, 184] and the Tcm PKS [183] have been reconstituted in vitro from individually purified components, making it possible to explore many of the mechanistic details of type II PKSs that have eluded in vivo studies in the past.

Shen and Hutchinson [182] made cell-free preparations from an *S. glaucescens* null mutant bearing plasmids expressing various combinations of the *tcmJKLMN* genes. PKS activity was assayed through the synthesis of **32** from acetyl and malonyl CoA, whose identity was authenticated by its enzymatic conversion to **33** and **35**, catalyzed by the purified Tcm F2 CYC [94, 229] and Tcm F1 monooxygenase, respectively (Fig. 5) [229, 230]. Tcm F2 was also synthesized upon admixture of cell-free extracts prepared from strains bearing the *tcmKMN* and *tcmKLM* genes, suggesting that functional Tcm PKS can be reconstituted in vitro. Thus, they developed methods to prepare TcmM-deficient PKS, which showed diminished Tcm PKS activity, and fully restored the PKS activity to the Tcm-deficient PKS preparation by addition of purified TcmM; addition of either *S. glaucescens* FAS ACP [178] or *E. coli* FAS ACP had no effect. These results not only demonstrated the feasibility of reconstituting a type II PKS in vitro from its individually purified component proteins, but also provided a means to assay each protein biochemically prior to reconstitution.

Bao et al. [183] subsequently constructed gene cassettes consisting of *tcmJΔ-KLMN* and *tcmJKΔLMN*, which carried a deletion at *tcmK* and *tcmL*, respectively. These gene cassettes were introduced into *S. lividans*, from which cell-free extracts, lacking either TcmK or TcmL activity but containing all of the other Tcm PKS components, were prepared. Addition of crude protein extracts, made from recombinant *E. coli* or *Streptomyces* strain expressing either of the *tcmK* or *tcmL* gene, to the *Δ*TcmK- or *Δ*TcmL-PKS preparations, however, failed to restore the synthesis of **32**. Since it has been suggested that TcmK and TcmL are produced in equimolar amounts on the basis that their genes are translationally coupled, they reasoned that functional TcmK and TcmL may not be able to be produced separately. Thus they overexpressed *tcmKL* together in *S. lividans*. The crude preparation of the TcmKL proteins made from the latter strain indeed restored the synthesis of **32** to both the *Δ*TcmK- and *Δ*TcmL-PKS preparation, confirming that both TcmKL were functional. To facilitate the purification, a His_6-tag was fused to TcmK, and the purified [His_6]-TcmKL proteins were identical to the native proteins in complementing either the *Δ*TcmK- or *Δ*TcmL-PKS preparation to synthesize **32**.

Using this functional assay, they purified both the native TcmKL and the [His_6]-TcmKL proteins. Sodium dodecyl sulfate-polyacrylamide gel electrophoresis (SDS-PAGE) analysis showed that TcmK and TcmL were co-purified with 1:1 ratio, and gel filtration confirmed that TcmKL was an $\alpha\beta$ heterodimer in solution. To reconstitute the Tcm PKS, they also overexpressed *tcmM* in *S. lividans* and the *S. glaucescens fabD* in *E. coli*, respectively. Both the purified TcmM ACP, which was confirmed to be fully 4′-phosphopantetheinylated, and the FabD MAT were functional, as judged by the malonyl CoA:ACP transacylase assay. The Tcm PKS was reconstituted in vitro from TcmKL, TcmM, FabD MAT, along with TcmN, catalyzing the synthesis of **32** from malonyl CoA; [His_6]-TcmJ cannot replace TcmN but addition of it greatly increased the production of **32**.

Using a similar approach, Khosla and co-workers (181) made a cell-free preparation from *S. coelicolor* CH999 bearing plasmid expressing the minimal *act* PKS genes of *actI-ORF1,2,3* and demonstrated that this preparation can synthesize **30** [101] and **31** [86] very efficiently from acetyl and malonyl CoA in vitro (Fig. 6). Since it was established in vivo that **30** and **31** resulted from aberrant cyclization of the linear hexaketide intermediate due to the lack of the auxiliary enzymes [86, 101], they examined the interactions of the minimal Act PKS with KR (ActIII), ARO (ActVII), and CYC (ActIV) in vitro [181]. Cell-free extracts were made from *S. coelicolor* CH999 bearing plasmid expressing *actI-ORF1,2,3, actIII, actVII,* and *actIV*. Incubation of this preparation with acetyl CoA,

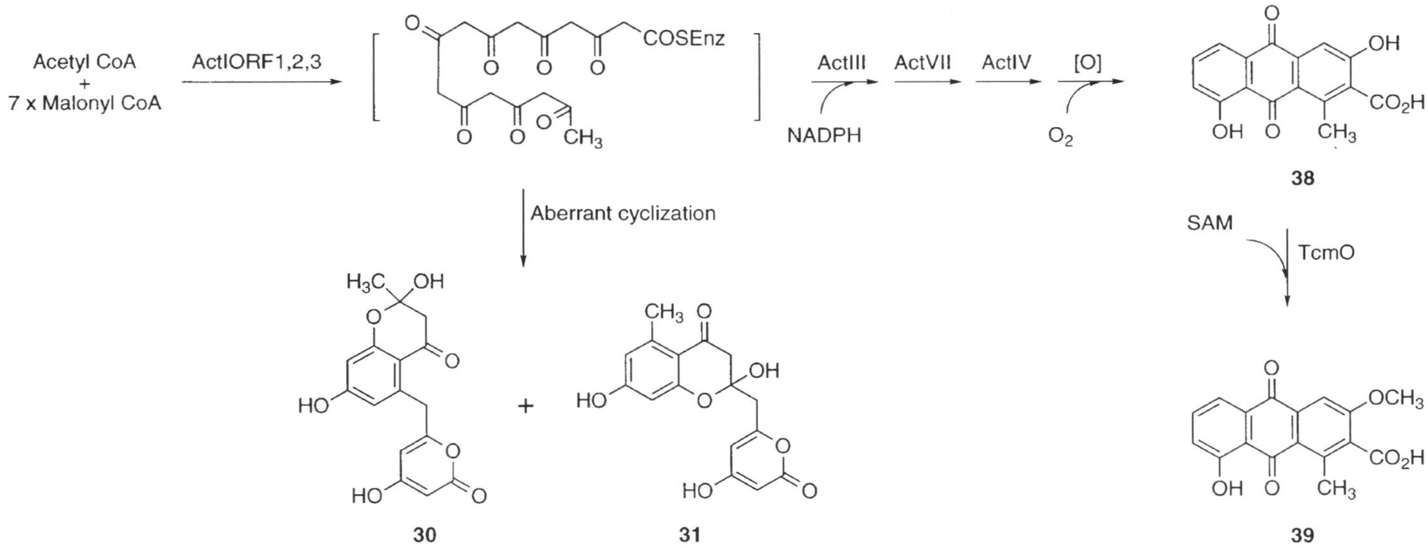

Fig. 6. In vitro synthesis of actinorhodin biosynthetic intermediates and its shunt metabolites from acetyl CoA and malonyl CoA by the ActIORFI,2,3 polyketide synthase complex, the ActIII ketoreductase, the ActVII aromatase, ActIV cyclase, and the TcmO methyltransferase

malonyl CoA, and NADPH indeed resulted in the formation of 3,8-dihydroxy-1-methyl anthraquinone-2-carboxylic acid (DMAC, **38**), the major in vivo product of the complete Act PKS [101]. The identity of **38** was further authenticated by its enzymatic conversion to 8-methoxy DMAC (**39**), catalyzed by the TcmO methyltransferase in the presence of *S*-adenosyl methionine (Fig. 6) [231].

With this cell-free assay for the Act PKS activity, Carreras and Khosla attempted to purify the KSα and KSβ directly from crude extracts of *S. ceolicolor* CH999, overexpressing *actI-ORF1,2,3, actIII, actIV,* and *actVII* [184]. They soon realized that the purification was complicated by the requirement of several proteins for the PKS activity and that separation of these proteins resulted in loss of activity. While a 40% enriched preparation of KSα and KSβ was active in catalyzing polyketide biosynthesis when supplied with purified holo-ACPs, derived from the *fren, gra, oct,* or *tcm* PKS gene clusters, and malonyl CoA, a near homogeneous mixture of KSα and KSβ was inactive under the same assay conditions. KSα and KSβ were co-purified as an equimolar mixture, according to SDS-PAGE analysis, and existed as an $\alpha_2\beta_2$ heterotetramer in solution based on gel filtration data [184]. However, the purified KSα, KSβ, and holo-ACPs can be activated by addition of a crude extract from *S. coelicolor* CH999 host strain that lacks genes encoding these proteins and has no PKS activity of its own. The latter verified the integrity of the purified KSα, KSβ, and holo-ACPs, and suggested an additional protein essential for the PKS activity. By assaying its ability to complement the purified KSα, KSβ, and holo-ACPs in polyketide synthesis, they purified this protein, identified it by N-terminal sequencing as the *S. coelicolor fabD* MAT, and established that KSα, KSβ, holo-ACP, and MAT reconstituted the Act PKS in vitro, synthesizing polyketides from malonyl CoA [184]. Very recently, Simpson and coworkers similarly reconstituted the Act PKS in vitro from individually purified KSα/KSβ and ACP with or without MTA, although their goal was mainly to reexamine if MTA is essential in a reconstituted PKS [173]. As discussed in Sect. 3.1.3, these researchers demonstrated that MTA is not required by the minimal Act PKS for polyketide synthesis in vitro as long as the holo-ACP is in large molar excess over KSα/KSβ and proposed an alternative mechanism of self-malonylation for loading the malonyl group to the ACP of the PKS complex [173].

Reconstitution of the Act PKS [173, 184] and Tcm PKS [183] in vitro from individually purified KSα, KSβ, holo-ACP, and MAT was one of the most exciting advances in the studies of aromatic polyketide biosynthesis, providing excellent model systems for mechanistic studies of bacterial aromatic type II PKS. The methods for Act PKS and Tcm PKS should be applicable to the purification and reconstitution of other aromatic type II PKSs. It will be extremely interesting to examine the effect on polyketide synthesis by swapping functional components of various PKSs in vitro. Summarized here are the highlights that bear fruit directly from the in vitro systems, revealing mechanistic features otherwise inaccessible by in vivo studies.

1. The minimal PKS consists of KSα, KSβ, and holo-ACP. Posttranslational phosphopantetheinylation of apo-ACP into holo-ACP is catalyzed by the endogenous ACPS. When holo-ACP is present in limiting concentration, the minimal PKS requires MAT to synthesize polyketide from malonyl CoA. The

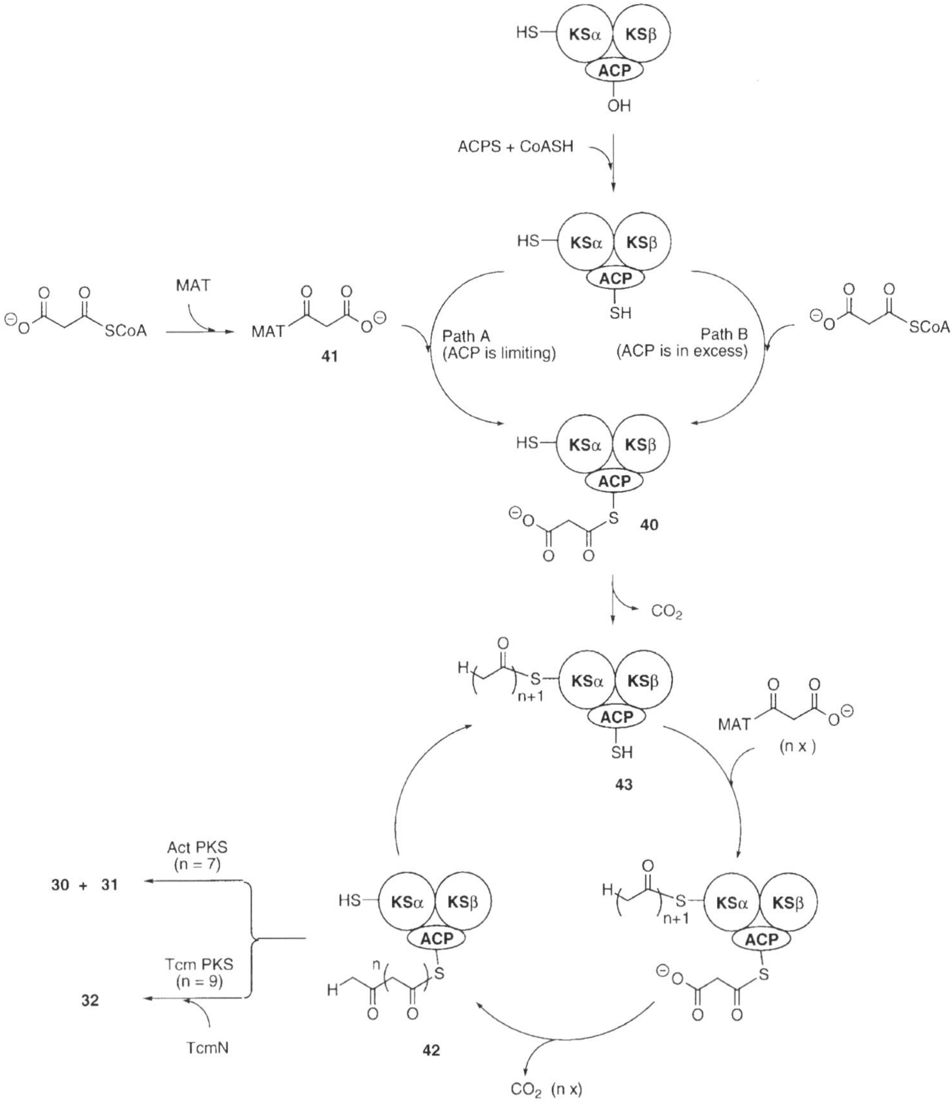

Fig. 7. Stepwise mechanism for type II PKS catalyzed biosynthesis of aromatic polyketides from malonyl CoA

MAT catalyzes the transfer of the malonyl group from malonyl CoA to the holo-ACP to form malonyl-ACP (**40**), involving a transient malonyl-MAT species (**41**) (Fig. 7, path A). ACPS and MAT, both of which are encoded by the fatty acid biosynthetic machinery of the host, provide the functional connections between fatty acid and polyketide biosynthesis.

2. When holo-ACP is present in excess, the minimal PKS is sufficient to support polyketide biosynthesis from malonyl CoA at least in vitro. The loading of

malonyl group from malonyl CoA to holo-ACP is a result of self-malonylation of the holo-ACP (Fig. 7, path B), a property unique to the type II PKS ACP.

3. Acetyl CoA is not required for the synthesis of aromatic polyketides with acetate as an starter unit. Instead, the acetate starter is derived from malonyl CoA by decarboxylation of **40** to acetyl-ACP, which is transferred to KSα to initiate polyketide biosynthesis (Fig. 7). Both of these reactions as well as the transfer of the growing poly-β-ketone intermediates between acyl-ACP (**42**) to acyl-KSα (**43**) are thought to be catalyzed by the KSα/KSβ pair of the PKS. In fact, Carreras and Khosla elegantly demonstrated **41**, **40**, and **43** as covalent intermediates for the Act PKS by malonyl group en route from CoA to polyketide using [^{14}C]malonyl CoA, providing direct evidence for such a mechanism.
4. The full-length poly-β-ketone intermediate, which is tethered to the PKS complex in a thioester linkage at ACP, could spontaneously undergo aberrant cyclization in the absence of additional auxiliary enzymes. Interaction between the PKS complex with additional auxiliary enzymes such as ARO/CYCs dictates subsequent reactions to convert the initial linear poly-β-ketone intermediate into specific aromatic polyketides (Fig. 7).
5. Structural information of the PKSs is emerging. While the stoichiometry of a PKS has yet to be established, and whether or not the difference of the KSα/KSβ complex, being $\alpha_2\beta_2$ for the Act PKS and $\alpha\beta$ for the Tcm PKS, has any particular significance in polyketide synthesis has yet to be determined, it is clear that KSα and KSβ can be co-purified and form a tight complex. The latter in turn sequesters holo-ACP to form a ternary PKS complex. This model is consistent with the observation that KSα and KSβ together control the chain length of the polyketide product and a heterologous pair of KSα and KSβ is often nonfunctional, presumably due to poor interaction between the two subunits or between the KSα/KSβ pair and holo-ACP.

3.2 Fungal Polyketide Synthase

In comparison to bacterial PKSs, in vitro study of the biochemistry and enzymology of the fungal PKSs has been relatively successful. In fact, the complete PKSs that have been purified so far to homogeneity as active enzymes from wild-type organisms (excluding plant PKSs) are all fungal PKSs [105, 232–234]. The iterative type I organization of fungal PKSs certainly provides the inherent stability of a PKS complex by encoding all the enzyme activities on a single protein, facilitating their purification. The most extensively studied fungal PKS is 6MSAS for **17**. Significant progress has also been made recently in studying the FAS/PKS complex for **18**, an early intermediate for AF and ST biosynthesis in *A. parasiticus* and *A. nidulans*, respectively. 6MSAS has been purified to homogeneity [105, 232], and a cell-free extract capable of in vitro synthesis of **18** has been prepared [235, 236]. More recently, functional expressions of the 6MSAS gene in *E. coli* [237], *S. coelicolor* [238], *A. nidulans* [66], and *Saccharomyces cerevisiae* [237] have been realized. These heterologous expression systems will greatly facilitate both mechanistic study and genetic engineering of fungal PKSs.

3.2.1
6-Methylsalicyclic Acid Synthase

Lynen and co-workers first purified 6MSAS from *P. patulum* cells as early as 1970 [105]. Spencer and Jordan improved the original method by optimization of the growth conditions for *P. patulum* to maximize the level of 6MSAS and by addition of PMSF, benzamidine, and 15% glycerol in buffers to extend the half-life of 6MSAS from 20 min to 6 h, greatly facilitating its purification [232]. The purified enzyme exists as a homotetramer of Mr 750,000 as determined by gel filtration, with a subunit Mr of 180,000 as judged by SDS-PAGE analysis [232]. The latter agrees well with an Mr of 190,731 predicted from the 6MSAS gene [65]. The pH optimum for the enzyme is 7.6. 6MSAS catalyzes the synthesis of **17** from one molecule of acetyl CoA (K_m = 10 µmol/l) and three molecules of malonyl CoA (K_m = 7 µmol/l), requiring one molecule of NADPH (K_m = 12 µmol/l). In the absence of NADPH, the 6-methyl-4-hydroxy-2-pyrone (**44**) is the exclusive product, suggesting that the triketide intermediate (**45**) has to be reduced before it can react with the third molecule of malonyl CoA (Fig. 8A). Acetoacetyl CoA

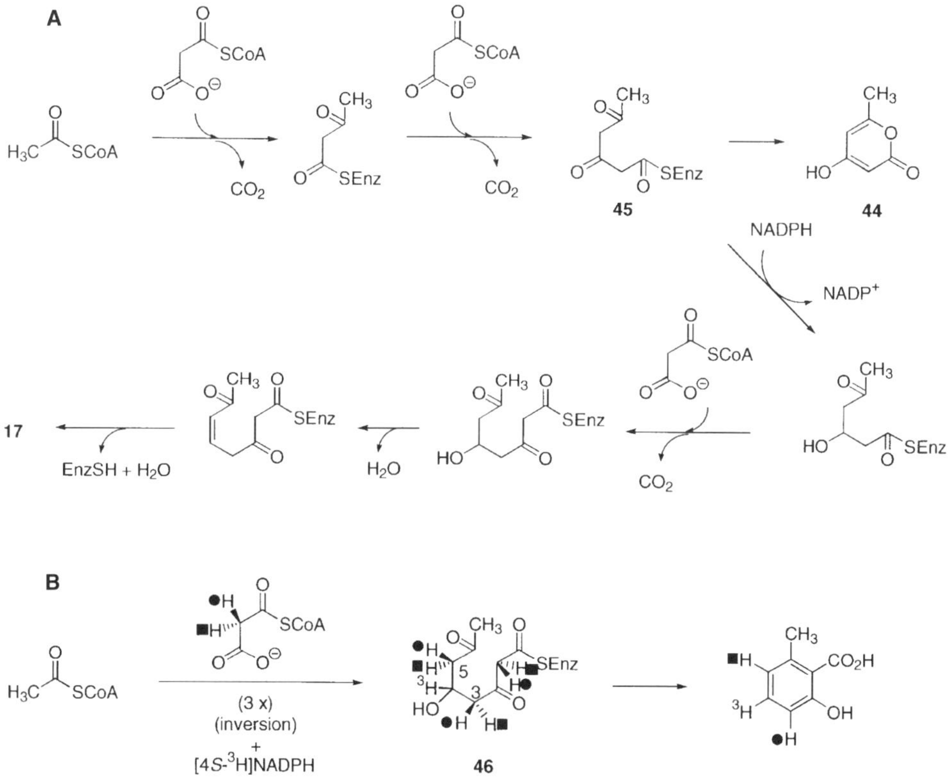

Fig. 8 A, B. 6MSAS catalyzed biosynthesis of 6-methylsalicyclic acid (**17**) and its 2-pyrone shunt metabolite (**44**) from acetyl CoA and malonyl CoA: **A** pathway; **B** stereochemistry

(K_m = 65 µmol/l) could also be used as a substrate, although much less favorably in comparison with acetyl CoA ($[V_{max}/K_m]_{\text{acetyl CoA}}:[V_{max}/K_m]_{\text{acetoacetyl CoA}}$ = 135:1) [232]. The polyketide intermediate is tethered as a thioester to the Cys of the KS domain or the 4′-phosphopantetheine prosthetic group of the ACP domain of 6MSAS at all stages of biosynthesis. Finally, **17** is released from 6MSAS presumably by cleavage of the thioester linkage, although the mechanism of the latter process remains obscure since, like the bacterial type II PKS, no TE activity has been identified for 6MSAS.

Although **17** is achiral, the cryptical stereochemistry of 6MSAS was elegantly established by Spencer and Jordan using chiral malonyl CoA as substrates [239, 240]. They first synthesized (*R*)- and (*S*)-[1-^{13}C,2-^{2}H]malonate, which was converted into malonyl CoA derivatives enzymatically by succinyl CoA transferase. They then used the chiral malonyl CoA as substrates for 6MSAS and determined the absolute stereochemistry of hydrogen atom elimination from the methylene groups originating from malonyl CoA by mass spectrometric analysis of **17**. Since the third malonyl CoA molecule loses both of its methylene hydrogen atoms during the formation of the aromatic ring, these studies revealed only the stereochemistry at C-3 and C-5 (Fig. 8B). Thus, when acetyl CoA and the chiral malonyl CoA were used as substrates, analysis of **17** formed established that the hydrogen atoms at C-3 and C-5 were derived from opposite absolute configurations in malonyl CoA. To determine the absolute stereochemistry at C-3, they replaced acetyl CoA with acetoacetyl CoA as the starter molecule, hence bypassing the first cycle of condensation. Since the latter experiment will only introduce ^{2}H into C-3 of **17**, mass spectrometric analysis indicated that this hydrogen originated from the *pro-R*-H of malonyl CoA. Taking these results into account, they concluded that the hydrogen atom at C-5 of **17** originated from the *pro-S*-H of malonyl CoA and that it is the *pro-S*-H at C-5 and *pro-R*-H at C-3 of the tetraketide intermediate (**46**) that were stereospecifically removed during the biosynthesis of **17** (Fig. 8B). It should be pointed out that these results were interpreted on the assumption that both decarboxylative condensations between acetyl CoA and malonyl CoA and acetoacetyl CoA and malonyl CoA occur with inversion of configuration. While this has been proved for the FAS [241, 242] and for module 2 of the 6-deoxyerythronolide B synthase [243] and may indeed turn out to be the case for 6SMAS, it is far from certain. Finally, the stereochemistry of the NADPH-dependent reduction of **45** was investigated by Schweizer and coworkers using [4*R*-^{3}H]- and [4*S*-^{3}H]NADPH, and only the *pro-S*-H of NADPH was incorporated into **17** (Fig. 8B) [267].

Structural information for 6MSAS is emerging. The fact that the amino acid sequence of 6MSAS and the arrangement of its functional domains are highly homologous to vertebrate type I FAS [65] strongly suggests that there are structural similarities between the two enzymes at the protein level. Jordan and coworkers studied the organization of various functional domains on the subunit and the interactions between subunits of 6MSAS by modification with thio-specific inhibitors and cross-linking reagents and by detailed mapping of the active sites of the KS and ACP domains in the primary sequence [232, 244, 245]. They first demonstrated that 6MSAS was inactivated in a time-dependent process by iodoacetamide and cerulenin and that acetyl CoA but not malonyl CoA can protect 6MSAS against such inactivation. These findings indicated that both iodoacetamide and cerulenin specifically modify the Cys residue of the KS domain, which was subsequently identified as Cys-204 by peptide sequencing. They then showed that di-bromopropan-2-one can cause cross-linking between the thiols of Cys-204 of the KS of one subunit and the 4′-phosphopantetheine (bound at Ser-1733) of the ACP of an adjacent subunit. The latter finding provided unambiguous evidence that the KS and ACP thiols from separate subunits are located close to each other at the condensing site. They further demonstrated that 5,5′-dithiobis(2-nitrobenzoic acid) was able to cross-link the juxtaposed thiols of the adjacent subunits by a direct disulfide linkage and, on this basis, concluded that the two thiol groups are likely to be positioned within approximately 2 Å of each other. From these studies, they proposed that two pairs of functional dimers are present in the 6MSAS homotetramer and that, within each dimer, the KS and ACP domains are juxtaposed to allow the respective Cys and 4′-phosphopantetheine thiols to interact during condensation. Such a structural organization at the condensing site has not only been found in vertebrate FAS [22] but has also been proposed for the 6-deoxyerythronolide B synthase [246].

While very fruitful to study the native enzyme, successful expression of 6MSAS in various heterologous hosts such as bacterial [219, 238], fungal [66], and yeast hosts [237] has made it possible to use genetic tools to decipher its structural-function relationship. Using the 6MSAS gene of *P. patulum* as a probe, Fujii and co-workers cloned the *atX* gene from *A. terrus* that produces the seco-anthraquinone, (+)-geodin [66]. The *atX* gene contains a 70 bp intron near its N-terminus and encodes a protein of 1800 amino acids with an Mr of 190,000 that is highly homologous to 6MSAS (62% identity), in particular at the proposed active sites. They expressed *atX* without removing the intron in *A. nidulans* and confirmed that the recombinant organism indeed produced **17** (330 mg/l isolated yield). This finding not only unambiguously established *atX* as the 6MSAS gene of *A. terrus* but also demonstrated *A. nidulans* as an attractive host for functional analysis of cloned fungal genes and for engineered biosynthesis of fungal polyketides. Since AtX is active, the intron of *atX* must have been correctly processed and the ACP domain of AtX must have been 4′-phosphopantetheinylated, presumably by an endogenous PPTase, in *A. nidulans*. Khosla and co-workers used 6MSAS as a showcase to demonstrate [238] that fungal PKSs can also be efficiently expressed in their *S. coelicolor* CH999-pRM5 system, which was developed originally for functional expression of bacterial PKSs [101]. To allow optimal expression in *S. coelicolor*, the 6MSAS gene was modified by excising the intron and by altering four of the first seven codons into ones most frequently used in streptomycetes. Significant amount of **17** (67 mg/l on solid medium) was produced, suggesting that fungal PKSs like 6MSAS can be expressed, folded, and posttranslationally modified into functional proteins in this actinomycete model host. Kealey and co-workers [237] examined expressions of 6MSAS of *P. patulum* in *E. coli* and *S. cerevisiae* to take advantage of the extensive molecular biology and recombinant technology available to these organisms. While 6MSAS was very well overexpressed in *E. coli* in ATCC medium 765 (at ~5% of the total protein), most of it was neither soluble, nor 4′-phosphopantetheinylated. To circumvent these problems, they lowered the incubation temperature to enhance protein folding, added glycerol to the medium to increase the levels of intracellular malonyl CoA, and co-expressed the *sfp* PPTase to ensure 4′-phosphopantetheinylation of the apo-6MSAS, resulting in the production of **17** (75 mg/l). Similarly, expression of 6MSAS in *S. cerevisiae* produced only a trace quantity of **17** (<0.75 μg/l), presumably due to the lack of modification of the apo-6MSAS by the endogenous PPTase. Remarkably, co-expression of 6MSAS and *sfp* in *S. cerevisiae* resulted in high-level production of **17** up to 1.7 g/l, which is more than twice as much as that produced by the natural host *P. patulum*. These findings agreed well with the relaxed substrate specificity of the Sfp PPTase and clearly demonstrated the feasibility of expressing functional PKSs in these organisms, and by implication in other organisms as well.

3.2.2
The Aflatoxin Polyketide Synthase/Fatty Acid Synthase Complex

The enzymology of AF biosynthesis has received great attention over the years, and enzyme activities have been detected in cell-free preparations for most of the steps of the AF biosynthetic pathway [68, 107]. Most recently, a new protocol for cell-free enzyme preparation from *A. parasiticus* was developed by Watanabe and Townsend [236] that involved rapid concentration and efficient dialysis by membrane filtration to remove primary and secondary metabolites, cofactors, and small biomolecules (MW$<$10,000). All enzymes of the AF biosynthetic pathway have been dramatically stabilized by this procedure, and the effects of added substrate and cofactors can be assayed against virtually no background reactions. This significant methodological advance has enabled them to demonstrate for the first time in a cell-free system all biosynthetic steps from the initial construction of **18** by a pair of specialized FAS (Fas-1 A/Fas-2 A) and a simple type I PKS (PksA) to the conversion of **18** to AF. It is noteworthy that these authors emphasized the general applicability of this improved method of cell-free enzyme preparation and stabilization to a wide range of biological systems.

Although the in vitro synthesis of **18** by Fas-1 A/Fas-2 A/PksA from acetyl CoA and malonyl CoA has been achieved, details of this study are yet to be published [236]. However, evidence is emerging to support the physical association of Fas-1 A/Fas-2 A and PksA in initiating **18** biosynthesis. Feeding experiments [247–249] with ^{13}C labeled C-4, C-6, and C-8 carboxylic acids and their corresponding *N*-acetylcysteamine (NAC) thioesters suggested that the biosynthesis of **18** utilizes a C-6 starter unit. Genetic analysis confirmed that the synthesis of the C-6 starter unit requires a dedicated FAS such as *stcJ* and *stcK* in *A. nidulans* (115) and *fas-1 A* and *fas-2A* in *A. parasiticus* [117]. Insertional inactivation of these FAS genes resulted in *A. nidulans* [115] or *A. parasiticus* mutants that no longer produce ST [115] or AF [117], but these mutants can be complemented by addition of the NAC thioester of hexanoate. Townsend and co-workers examined how the C-6 starter unit was channeled between FAS and PKS [235]. They found that although *A. parasiticus* mutants, Dis-1 and Dis-2, which were generated by insertional mutagenesis at *fas-1A*, can be complemented by the addition of hexanoyl-NAC, the production of **18** and AF is much lower than that observed in the parent strain. Similar results have also been observed

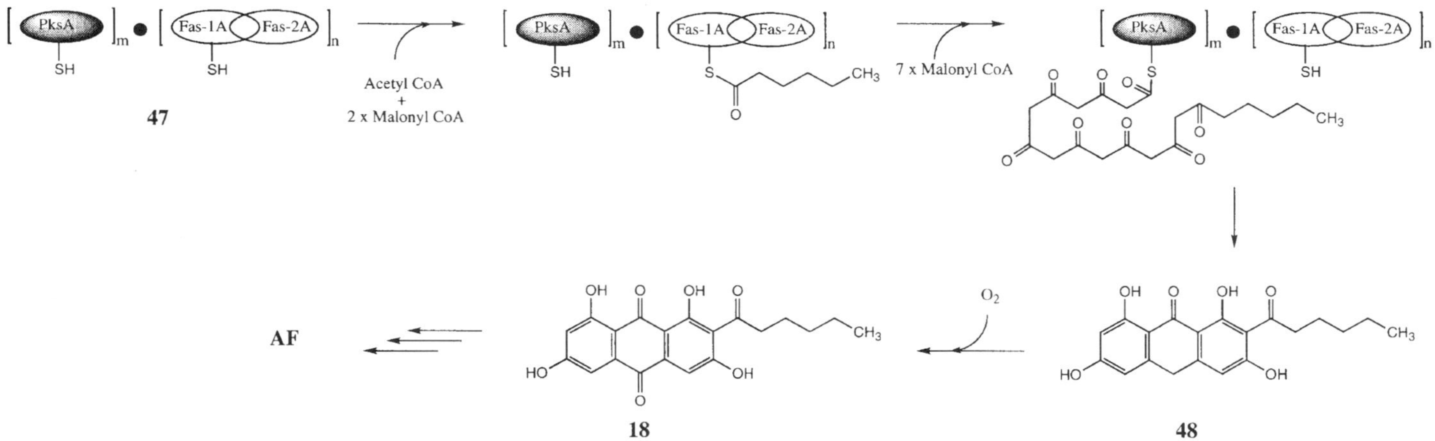

Fig. 9. Model for the FAS/PKS complex catalyzed biosynthesis of norsolorinic acid (**18**) from acetyl CoA and malonyl CoA

with analogous *A. nidulans* FAS mutants; the overall yield of ST produced by the latter grown on hexanoic acid-enriched medium was 20 times lower than that by the wide-type strain [115]. On the basis of a series of control experiments, they concluded that the reduced efficiency of **18** synthesis in Dis-1 and Dis-2, and by implication in the *stcJ*- and *stcK*-mutants too, might result from impaired loading of the hexanoyl-NAC onto PksA. They proposed that FAS and PKS form a complex such as **47** and that Fas-1 A (and possibly also Fas-2 A) is required to mediate transfer of the C-6 starter unit to the PksA (perhaps by direct transfer without the need for conversion to the CoA ester) to initiate AF biosynthesis (Fig. 9) [235]. The mutated form of Fas-1 A would fail to complex with the PksA, reducing the effectiveness of the starter unit binding and transfer. Finally, it is quite puzzling how the hypothetical linear poly-β-ketone is cyclized into the anthrone intermediate (**48**) since no CYC/ARO, such as those associated with bacterial type II PKSs [36, 49, 84–91], has been identified within fungal PKS. The distinctive TE of PksA presumably is responsible for the release of **48** from the FAS/PKS complex, as in bacterial type I PKS [118–121]. Oxidation of **48** to **18** is likely catalyzed by a monooxygenase, and similar enzymes have been identified from several polyketide biosynthetic pathways [230, 250].

3.3 Plant Polyketide Synthase

The biochemistry and enzymology of CHS and its closely related plant PKSs have been extensively studied [72, 75]. The native enzymes have been isolated from various plants [70, 71, 73, 74, 124, 125, 127, 128], and the recombinant proteins have been produced in *E. coli* and purified [71, 76, 124, 125, 132–135, 139]. The active site for the condensation reaction has been mapped to a single Cys residue [132]. The dimeric nature of CHS and STS has been established with each homodimer to constitute two active sites [76, 127, 133]. The subtle differences among the CHS and STS proteins have been compared, which seems to have a profound effect on substrate specificity and regiospecificity for the condensation and cyclization reactions, respectively [74, 127, 128, 133, 134]. The interactions between CHS and other proteins, such as KR [133, 136–139] and methylmalonyl CoA-specific condensing enzyme [76], have been examined to account for the biosynthesis of deoxychalcones and methylchalcones.

3.3.1 *Chalcone Synthase and Stilbene Synthase*

CHS and STS are closely related plant PKSs with about 70% amino acid identity. Both enzymes use phenylpropanoid CoAs, bound to the active site (thiol group of Cys-169) prior to condensation, as starter units and catalyze three sequential decarboxylative condensations with malonyl CoA in the synthesis of a tetraketide intermediate. The latter undergoes regiospecific cyclization into **19** and **22**, respectively (Fig. 10 A) [72, 75]. It has been suggested that the regiospecificity is likely to have resulted from a different spatial stabilization and folding of the tetraketide intermediate prior to cyclization [74, 127, 128, 133, 134]. Although

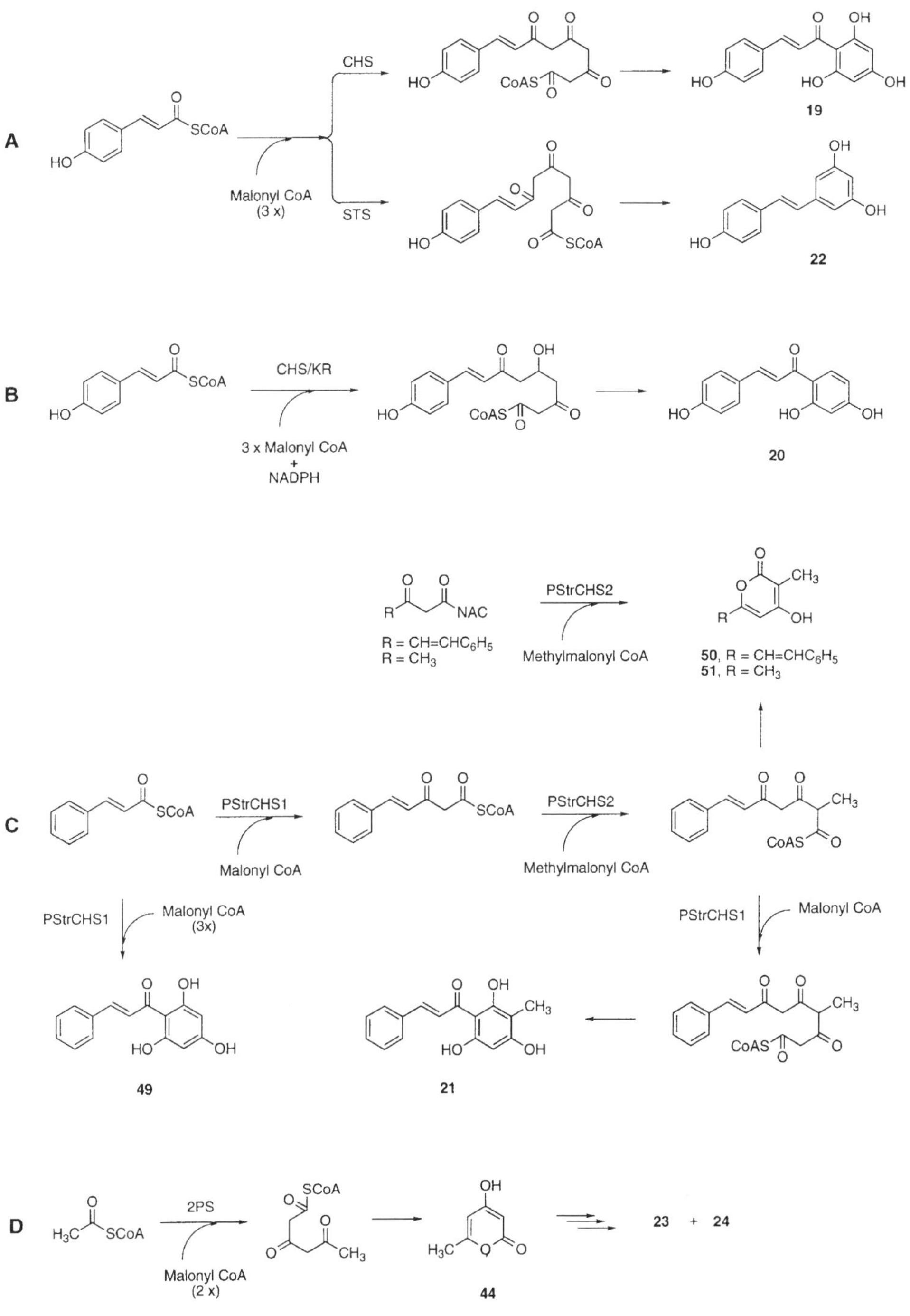

Fig. 10 A–D. Pathways for plant polyketide synthase catalyzed biosynthesis of: **A** chalcone and stilbene; **B** 6′-deoxychalcone; **C** methylchalcone; **D** 2-pyrone

differences between the consensus sequences of CHS and STS have been identified that might be important in determining the folding pattern of the condensation product, attempts to change STS into CHS and vice versa by either mutagenesis or formation of STS/CHS heterodimer were unsuccessful [128, 132–134].

Schröder and co-workers examined the role of Cys in CHS and STS by site-directed mutagenesis and tested the mutants after expression of the proteins in *E. coli* [132]. Of all of the six conserved Cys, only Cys-169 was essential for both enzymes. Inhibitor studies also suggested that Cys-169 is the main target of thiol-specific reagents such as cerulenin and iodoacetamide. These results exclude a role for other cysteines in malonyl binding and support the model that CHS and STS use malonyl CoA directly as a substrate for the condensation reactions. Interestingly, while the amino acids surrounding Cys-169 are highly conserved in CHS and STS, as would be expected for a closely related group of enzymes, no similarity was found at the active sites of KSs in both FAS and PKS [132]. It is tempting to suggest, based on these data, that CHS and STS could have evolved independently from the latter enzymes.

Although no structural information is available yet, it has been well established that both CHS and STS comprise a homodimer of subunits consisting of 388–400 amino acids [76, 127, 133]. To distinguish between independent or co-operative action of the active sites in the CHS and STS dimer, Schröder and co-workers prepared heterodimers in which one of the subunits is inactive (133). This was accomplished by co-expression of the native and mutant subunits under identical promoter-translation start configuration, since in vitro reconstitution of active CHS or STS from subunits has so far been impossible. The heterodimers were indeed fully functional and displayed close to 50% activity of the parent enzyme, as would be expected for a dimer possessing only one active site. These data clearly established that each subunit performs all the three condensation reactions, with two products being synthesized per dimer and reaction cycle.

3.3.2
Deoxychalcone Synthase

In the presence of a KR [133, 136–139], CHS also catalyzes the synthesis of **20**, in addition to **19**, from phenylpropanoid CoA and malonyl CoA (Fig. 10B). This co-action between a condensing enzyme and a KR is reminiscent of bacterial type II PKS. Furuya and co-workers first detected the KR activity with crude extracts from *Glycyrrhiza echinata* [136]. Welle and Grisebach subsequently purified the enzyme from soybean, demonstrating that the KR was a discrete enzyme consisting of a single peptide of 34 KDa [138]. Using the soybean KR antiserum to screen a cDNA library, Schröder and co-workers cloned the KR gene from soybean *Glycine max* L., overexpressed it in *E. coli*, and demonstrated by Southern analysis that KR-homologs are widely distributed among other plants capable of deoxychalcone biosynthesis [137, 139].

The KR is monomeric and co-acts with CHS in 1:2 molar ratio at the optimal pH of 6.0. The enzyme has a high affinity for NADPH (K_m = 17 μmol/l) and replacement of NADPH by NADH in the enzyme assay decreased the yield of **20** by 92%. It is the *pro-R*-H of NADPH that is transferred to the substrate [138]. Although the mechanism of the complex interaction between the dimeric CHS and the monomeric KR remains unknown, the fact that high salt concentration inhibited the reaction suggested an ionic interaction [138]. Schröder and co-workers further noticed that **20** represented at most 35–50% of the products under all conditions tested, including the one with a large excess of KR; the other product was always the non-reduced **19**. Since they have expressed in *E. coli* the functional KR, CHS dimer, and CHS heterodimer (which possesses only one active site), they examined the biosynthesis of **20** by co-incubation of the recombinant KR with either the CHS dimer or the CHS heterodimer. In all experiments, both **20** and **19** were synthesized in a constant ratio that was the same as that observed from the native enzymes. Based on these results, they concluded that the synthesis of **20** required no other plant factor and that the formation of two products may be an intrinsic property of the interaction between dimeric CHS and monomeric KR [133]. Finally, it was remarkable that the combination of the CHS from parsley, a plant that does not contain deoxyflavonoids, with the soybean KR also resulted in the synthesis of **20** [133, 138]. The latter data support the notion that deoxychalcone biosynthesis might depend only on the presence of the KR rather than on the nature of CHS or the specific interaction between KR and CHS.

3.3.3
Methylchalcone Synthase

The biosynthesis of C-methylated chalcones involves a specific CHS that performs the condensing reaction with methylmalonyl CoA as the extender unit. Schröder and co-workers cloned two CHS-type genes from *Pinus strobus*, *pstrCHS1* and *pstrCHS2*, whose deduced proteins were 87.6% identical [76]. Upon expression in *E. coli*, however, only PstrCHS1 was capable of synthesizing chalcone **49** with substrate specificity very similar to those of typical CHSs; PstrCHS2 was inactive with all of these substrates. In contrast, incubation of PstrCHS2 with a NAC derivative of the diketide intermediate and methylmalonyl CoA resulted in the synthesis of the substituted 3-methyl-2-pyrone (**50**), and the PstrCHS2 protein preferred methylmalonyl CoA over malonyl CoA, catalyzing only one condensation. While the diketide intermediate cannot be substituted by a NAC derivative of cinnamic acid, replacing it with acetoacyl-NAC yielded an analogous 2-pyrone (**51**), reinforcing the notion that PstrCHS2 specifically accepted a diketide as substrate (Fig. 10C). The latter result is remarkable because acetyl CoA is known to be a very poor substrate for typical CHS [72, 75]. Based on these results, these authors concluded that CHSs are not capable of performing sequential condensation using two different extender units. Instead, the two dimers of PstrCHS1 and PstrCHS2 likely formed a complex, with the former to catalyze the first and third condensations and with the latter to catalyze the second condensation, using malonyl CoA and methylmalonyl CoA as extender units, respectively, in **21** biosynthesis (Fig. 10C). The growing diketide and triketide intermediates are likely channeled between the two homodimers, depending on the choice of the extender units [76]. It is noteworthy that such strategy for a CHS to select a different extender unit is remarkably similar to the bacterial noniterative type I PKS and could be of general implication for the biosynthesis of other C-methylated flavonoids. Since many C-methylated flavonoids are known, including some dimethylated chalcone derivatives, it would be very interesting to investigate if the CHSs for these methylated chalcones adopt similar mechanisms for substrate selection and intermediate channeling [76].

3.3.4
2-Pyrone Synthase

In contrast to most CHS-like plant PKSs that prefer phenylpropanoid CoAs as starter units, 2PS is one of the few exceptions that prefers acetyl CoA as a starter unit. It catalyzes two sequential condensation reactions between acetyl CoA and malonyl CoA to yield **44**, a precursor for the biosynthesis of **23** and **24** in *Gerbera* (Fig. 10D) [135]. (6-Hydroxymellein synthase is another example that prefers acetyl CoA as a starter unit. Although Kurosaki and co-workers have extensively studied this enzyme from carrot cell extracts, it has yet to be established if it belongs to the CHS superfamily, since neither has its amino acid sequence been determined, nor has its corresponding gene been cloned [251–255].) 2PS, encoded by the *gchs2* gene that was cloned from *G. hybrida* as a CHS, is 73% identical to CHS enzymes [71, 123, 135]. Schröder and co-workers expressed the *gchs2* gene in *E. coli* and purified the recombinant protein [135]. GCHS2 differs from CHS in its specialization for only two condensation reactions, its complete lack of activity with phenylpropanoid CoAs, and its high activity with acetyl CoA. Acetoacetyl CoA, the predicted intermediate, was ~2.5 times more efficient as a substrate than acetyl CoA. In the absence of acetyl CoA, GCHS2 could synthesize **44** by decarboxylation of malonyl CoA to produce acetyl CoA, a property that is well known for FAS [208] and other PKSs [83, 173, 184, 209]. Although the enzyme also accepts other small hydrophobic CoAs as starter units in vitro, leading to other 4-hydroxy-2-pyrones, neither these products nor their derivatives are known from *Gerbera* [135].

The establishment of GCHS2 as 2PS [135], as well as the identification of PstrCHS2 as a methylmalonyl CoA-specific CHS [76], demonstrated that CHS-like proteins contain plant PKSs with surprisingly diverse functions. The fact that relatively small differences in these proteins could have a profound effect on substrate specificity, intermediate channeling, and condensation progressivity should aid genetic engineering of plant PKSs for the production of new metabolites. Equally noteworthy are the synthesis of **44** from acetyl CoA and malonyl CoA by 6MSMS [232] and 6-hydroxymellein synthase [251–255] and the synthesis of an analog of **44** from propionyl CoA and methylmalonyl CoA by a truncated 6-deoxyerythrolide B synthase fused with a TE domain [256]. Since the plant PKSs of GCHS2 and 6-hydroxymellein synthase, the iterative fungal PKS of 6MSAS, and the bacterial noniterative type I PKS are structurally distinct, the fact that **44** or its analog could be synthesized by all four enzymes from similar substrates underscores the great flexibility and diversity in protein catalysis.

4
Genetic Engineering of Polyketide Synthase for Novel Aromatic Polyketides

Production of novel aromatic polyketides by genetic engineering of PKSs depends critically on the intrinsic substrate specificity of individual PKS components, on the organizational tolerance of these components to form a functional PKS complex, and on the repertoire of PKS genes encoding diverse aromatic polyketide structures. Equally essential are the molecular biological techniques and tools needed for rapid and convenient construction of engineered PKSs for combinatorial biosynthesis. Although significant progress has been made in all of the above-mentioned aspects and the concept of generating structural diversity by metabolic pathway engineering has been proven beyond doubt, biosynthesis of novel aromatic polyketides by engineered PKSs remains at the stage governed primarily by intuition and serendipity. Any so-called "design-rules" for engineered biosynthesis have to be treated with great caution because they are derived exclusively from empirical observations rather than based on a mechanistic understanding of the PKS complex. Excellent reviews have appeared recently summarizing the principle and practice of engineered biosynthesis for various systems [9–15, 25–34, 257]. The following discussion highlights the recent progress in synthesizing novel aromatic polyketides by engineered type II bacterial PKSs. Genetic engineering of either fungal PKS or plant PKS is yet to be explored.

4.1
Expression System

Genetic engineering of PKS involves heterologous expression of a defined set of PKS genes in a heterologous host, and screening and characterization of the resulting metabolites. In general, an ideal host should be genetically well characterized so that recombinant DNA can be easily introduced, recovered, and maintained. Heterologous genes should be efficiently expressed in this host, and the gene products should be correctly folded and posttranslationally modified. The host itself should produce relatively few endogenous metabolites so that their interference with the detection and characterization of novel compounds can be reduced to a minimum. An ideal vector should be based on a stable replicon with a marker, often an antibiotic resistance gene, for selection in the desired host. An *E. coli* replicon and selection marker are often incorporated into the vector as well, so that rapid genetic manipulation can be carried out in *E. coli* before introduction of the final construct into the desired host. In addition, a strong promoter is often preferred to ensure efficient expression of the recombinant PKS, and an inducible or regulated promoter should have the added benefit of minimizing the potential toxicity of heterologous proteins to the host. Since almost all of the known type II bacterial PKS are from *Streptomyces* and related actinomycetes, various *Streptomyces* species have become the host of choice for engineered biosynthesis of aromatic polyketides, despite the fact that these organisms are not well characterized genetically. The *S. coelicolor*-pRM5-CH999-based host-vector system, developed by Hopwood and co-workers [101], is one of the best systems used in engineered biosynthesis. *S. coelicolor* CH999 is derived from *S. coelicolor* A3(2) by deleting the entire *act* gene cluster to reduce endogenous polyketide biosynthesis. pRM5, a *Streptomyces-E. coli* shuttle vector, contains a divergently paired *actI* and *actIII* promoter whose activity is developmentally regulated by the ActII-ORF4 protein. More recently, Ziermann and Betlach have engineered several *S. lividans* equivalents of *S. coelicolor* CH999, such as *S. lividans* K4–114 and K4–155 [258], which can be transformed with pRM5-based plasmids with superior frequency. Alternatively, Simpson and co-workers demonstrated the utility of the thiostrepton induction system, designed for *Streptomyces* by Bibb and co-workers [259], in *S. coelicolor* CH999 for engineered biosynthesis [173]. This system provides a convenient method of induction by the *tipA* promoter, and functional PKS was efficiently expressed upon induction, resulting in aromatic polyketide biosynthesis [173]. Hutchinson and co-workers took advantage of the promoter mutant of *S. glaucescens* WHM1077 [260, 261], in which the *tcm* PKS and most of the auxiliary genes were not expressed, and used it as host for PKS expression [95, 96, 182]. They have chosen the constitutive *ermE** promoter and constructed the expression vectors initially in the *Streptomyces* plasmid of pIJ486 [97] and lately in the *Streptomyces-E. coli* shuttle plasmid pWHM3 [162–164] to facilitate genetic manipulations. In all these host-vector systems, the endogenous ACPS and MTA are apparently sufficient to support the plasmid-borne PKS for polyketide synthesis. Lack of posttranslational modifications such ACP phosphopantetheinylation could be easily circumvented by co-expressing an ACPS gene, should the need arise [237].

4.2 Chain Length

The carbon chain length of an aromatic polyketide is determined by the KSα/KSβ pair of the minimal PKS complex, and substitution of the ACP subunit of a minimal PKS complex with a heterologous ACP will not alter the chain length specificity [95, 101, 157, 165]. So far production of aromatic polyketides with chain length of C-10 [100], C-14 [100], C-16, C-18, C-20, C-22, and C-24 have all been demonstrated by engineered PKSs. By implication, engineered PKSs should soon provide access to aromatic polyketides with other chain lengths as the total number of cloned PKS genes increases exponentially. Thus, for engineered synthesis of octaketides, one could choose the KSα/KSβ pair from the minimal PKS of *act* [80, 95, 101], *fren* [80, 161], or *gra* [53]. The KSα/KSβ from the minimal PKS of *fren* is the only one known so far to produce a nonaketide [80, 159, 161]. For the production of decaketides, one has the choice of KSα/KSβ from the minimal PKS of *dps* [153, 162, 163], *jad* [162], *mtm* [150, 154, 160], *oct* [152], or *tcm* [80, 95, 162]. Engineered biosynthesis of both undecaketides and dodecaketides has been demonstrated with KSα/KSβ from the *cur*, *sch*, and *whiE* PKS gene clusters [100]. The minimal *pms* [44] should be another candidate for engineered dodecaketide biosynthesis.

It is noteworthy that exceptions have been observed in which the KSα/KSβ pair of an engineered PKS displayed a relaxed chain length specificity. For example, both octaketides and nonaketides have been synthesized by engineered PKS constructs consisting of the *fren* KSα/KSβ pair [80, 159, 161]. Similarly, several engineered PKS constructs consisting of the *whi*E KSα/KSβ pair synthesized a mixture of undecaketides, dodecaketides, as well as other short polyketides [100]. One could argue that the *fren* PKS has a relaxed intrinsic chain length specificity because *Streptomyces roseofulvus*, from which the *fren* genes were cloned, produces both the octaketide nanaomycin and the nonaketide frenolicin [41]. However, it is believed that the *whiE* cluster encodes for a dodecaketide spore pigment in *S. coelicolor* [100]. Consequently, the production of a mixture of polyketides with variable chain length by the minimal WhiE PKS or other engineered PKSs consisting of the *whiE* KSα/KSβ pair suggests that subunits other than KSα/KSβ could also modulate the chain length specificity [100]. These exceptions underscore the uncertainty for rational design of aromatic polyketide by engineering artificial PKS without considering the interactions among individual proteins [80, 100, 153, 159, 162].

4.3 Starter Unit

Most of the type II PKSs studied so far used acetate as a starter unit. The minimal PKS appears to have intrinsic starter unit specificity for acetyl CoA, which could also be derived by decarboxylation from malonyl CoA, at least in vitro [173, 183, 184]. Little is known about how a type II PKS chooses a starter unit other than acetate. The biosynthesis of **13** and **14** in *S. peucetius* and *Streptomyces* sp. C5, and oxytetracycline in *Streptomyces rimosus* are known examples in which a type II PKS used propionyl CoA [84, 153, 163] and malonamyl CoA [152] as a starter unit, respectively. However, engineered PKSs consisting of the KSα/KSβ pair from either the *dps* [163] or *otc* [152] cluster produced polyketides exclusively with an acetate starter, reinforcing the notion that the minimal PKS has an intrinsic specificity for an acetate starter. At least for the biosynthesis of **13** and **14** the choice of the propionate starter unit has been shown to depend on dedicated subunits, in addition to the minimal *dps* PKS, and on the protein/protein interactions among all the subunits of the resulting PKS complex [84, 153, 163].

4.4 Ketoreduction

Among the KRs that have been cloned from various pathways for aromatic polyketide biosynthesis, the KR of ActIII [50, 80, 83, 100–102, 150, 159–162], DpsE [84, 153, 162, 163], and JadE [51, 162] have been studied for their utility in engineered polyketide biosynthesis. The three KRs carry out the same regiospecific reduction at the C-9 carbonyl group of a linear poly-β-ketone intermediate, although the natural substrate for ActIII is presumably a linear octaketide [50] and that for DpsE [49] and JadE [51] is a linear decaketide. With two exceptions in which the C-7 carbonyl group of a linear octaketide [161] and a linear undecaketide [100] was reduced, respectively, ActIII regiospecifically reduced the C-9 carbonyl group of the linear poly-β-ketone intermediates with chain length ranging from C-16 [80, 83, 101, 161], C-18 [80, 161], C-20 [80, 83, 101, 102, 150], C-22 [100], to C-24 [100] (Fig. 11, path C). Similarly, JadE was also functional in regiospecific reduction of the C-9 carbonyl group for both the linear octaketide and decaketide intermediate [162]. In contrast, DpsE only reduced a linear decaketide intermediate and was nonfunctional to an octaketide [162], suggesting that KRs might exhibit different chain length specificity. Again, these results should be interpreted with great caution because they were derived empirically from a few examples. Other possibilities, such as inefficient channeling of the nascent octaketide intermediate between the minimal PKS and KR, could also result in the apparent lack of the DpsE function.

4.5 Cyclization

Despite the fact that the type II PKSs studied so far synthesize polycyclic aromatic structures such as anthraquinones, benzo[*a*]anthraquinones, or naphthaquinones in the wild-type organisms, most of the aromatic polyketides generated by engineered PKSs are structurally distinct from the natural products with common structural moieties such as benzene, 2-pyrone, or naphthalene [15, 29, 80, 86, 88, 162]. With the exception of the folding and the first and second ring cyclizations of the nascent linear poly-β-ketone intermediate, current understanding of the role of CYC/ARO in engineered biosynthesis remains rudimentary. In fact, many of the common structural moieties observed in engineered biosynthesis most likely resulted from spontaneous cyclizations of the highly active poly-β-ketone precursors, whose chemo- and regioselectivity were mainly governed by thermodynamics. The following discussion concerns mainly the folding and first and second ring cyclizations of a linear poly-β-ketone in engineered aromatic polyketide biosynthesis.

The minimal PKSs alone fold the nascent linear poly-β-ketone at C-9 position predominantly and initiate C-7/C-12 condensation for the first ring cyclization (Fig. 11, path A). Such regiospecific folding and cyclization have been demonstrated for the linear polyketide intermediates with chain length ranging from C-14 [100], C-16 [80, 86–88, 95, 101, 159, 161], C-18 [88, 159, 161], C-20 [80, 88, 95, 96, 101, 154, 160], C-22 [100], to C-24 [100]. Whether or not these processes are catalyzed by the minimal PKS is still an open question. Spontaneous cyclizations have been proposed to account for metabolites resulting from other folding and cyclization patterns [80, 86, 88, 95, 96, 100]. Since Tcm PKS [95, 96], Mtm PKS [150, 154, 160], and possibly WhiE PKS [100] fold the linear poly-β-ketone at C-11 and catalyze the first ring cyclization by C-9/C-14 condensation in the wild-type organisms, it is very unlikely for their minimal PKSs to have an intrinsic regiospecificity for C-9 folding and C-7/C-12 first ring cyclization.

Specific CYCs and their interaction with the minimal PKS determine the regiospecificity for the folding and first ring cyclization of the nascent linear poly-β-ketone intermediates. Thus, upon co-action between the minimal PKS and the TcmN or TcmJN CYCs, the nascent linear poly-β-ketone intermediates

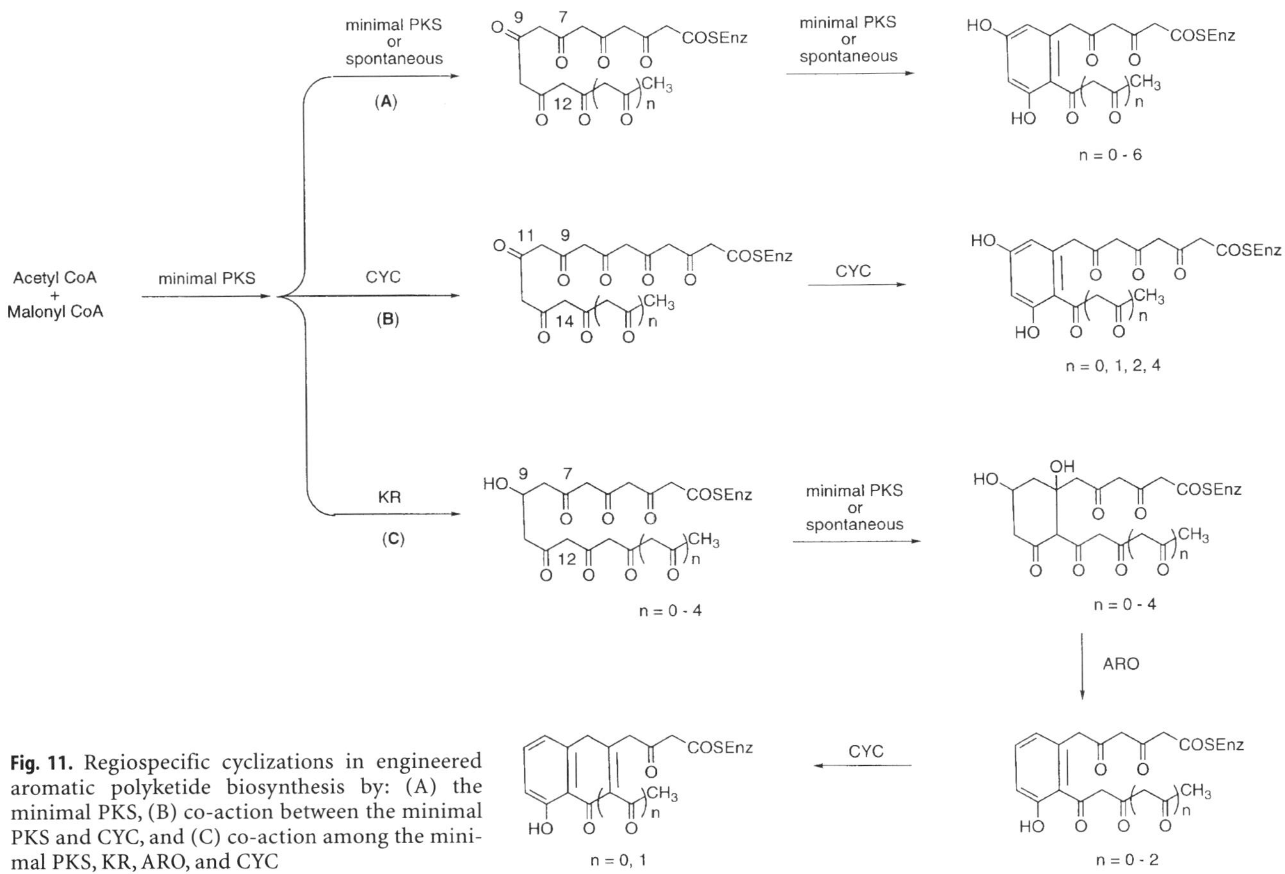

Fig. 11. Regiospecific cyclizations in engineered aromatic polyketide biosynthesis by: (A) the minimal PKS, (B) co-action between the minimal PKS and CYC, and (C) co-action among the minimal PKS, KR, ARO, and CYC

with chain length ranging from C-16 [92, 95], C-18 [159], to C-20 [92, 95, 96, 162] can be regiospecifically folded at C-11 to proceed the C-9/C-14 first ring cyclization. Co-action between the minimal PKS and the WhiE ORFVI CYC resulted in similar regiospecificity for a linear decaketide [85] and dodecaketide [100], as does the CurFG CYCs for a linear dodecaketide [100] and the putative Mtm CYC for a linear decaketide [154, 160] (Fig. 11, path B). Other CYCs such as ActIV [80, 88], JadI [162], and DpsF [162] or AROs such as Gris ORFIV [88], ActVII [88], Fren ORFIV [88] were nonfunctional towards non-reduced linear poly-β-ketone intermediates.

A C-9 reduced linear polyketide synthesized by a minimal PKS and KR, on the other hand, can undergo regiospecific cyclization between C-7/C-12, but subsequent dehydration to form the first aromatic ring requires a specific ARO. In the presence of an additional CYC, the one-ring intermediates can be further cyclized to form the second ring (Fig. 11, path C). CYCs such as TcmN and TcmJN, which act on non-reduced linear poly-β-ketone intermediates, are nonfunctional towards the C-9 reduced intermediates [92, 162]. Thus, ActVII can aromatize C-16 intermediates [80, 83, 87, 88], Fren ORFIV can aromatize both C-16 and C-18 intermediate, and both the Gris ORFIV [88] and WhiE ORFVI [85] can aromatize intermediates ranging from C-16, C-18, to C-20. For the second ring cyclization, ActIV can act on both C-16 and C-18 intermediates [80, 83, 87, 88]. WhiE ORFVI appears to be unique among all known CYCs or AROs. To a non-reduced linear poly-β-ketone intermediate, it acts as a CYC for regiospecific C-11 folding and C-9/C-14 first ring cyclization [100]. To a C-9 reduced linear nonaketide or decaketide, it acts as ARO for the first ring aromatization and as CYC for the second ring cyclization [85].

5 Perspectives

The study of aromatic polyketide biosynthesis is at a very exciting new stage, and genetics and molecular biology have served us well beyond the call of duty to the development of knowledge of the field. We have witnessed the exponential growth of cloned PKS genes, the conceptual breakthroughs in developing aromatic PKS models, the technological advances in rapid manipulation of PKS genes and aromatic polyketide biosynthetic pathways, the development of various systems for functional expression of PKS enzymes, and the successful in vitro reconstitutions of aromatic PKSs. While it is clear that mechanistic understanding of aromatic PKSs is still lacking, such insights eventually have to come from in vitro studies. Current lines of inquiry have opened up so many new opportunities to such studies. There is little doubt that biochemical and enzymological characterization of the aromatic PKSs will reveal the basic catalytic and molecular recognition features and structure-function relationships of these remarkable biosynthetic systems over the next few years. The potential of combinatorial biosynthesis for the production of novel polyketides will not be fully realized until we can engineer polyketide biosynthesis rationally. With all these successful precedents in bacteria, we could easily envisage similar prosperity for engineered polyketide biosynthesis in fungi and plants.

Addendum. Several new reports on aromatic polyketide biosynthesis have recently appeared. I should at least mention the reports on JadI [268], DnrD [269], and DpsC [270, 271], by Hutchinson and co-workers; ectopic expression of the minimal *whiE* PKS genes in *S. coelicolor* [272], by Moore and co-workers; and the ability of a PKS ACP to catalyze malonylation of a FAS ACP [273], by Reynolds and co-workers. JadI is a CYC that is required for the biosynthesis of angucyclines but is nonfunctional when combined with the minimal PKS genes of other clusters [268]. Overexpression in *E. coli* and purification and characterization of the recombinant enzyme confirmed DnrD as a CYC that catalyzes the cyclization of aklanonic acid methyl ester to aklaviketone in the biosynthesis of **13** and **14** [269], a reaction very similar to those catalyzed by TcmI [94] and TcmN [96]. DpsC was found to exhibit KS III activity, which specifies the propionate starter unit for the Dps PKS, shedding light on the mechanism how type II PKS chooses a starter unit other than acetate [270, 271]. It is remarkable that a single recombinant expressing the *whiE* minimal PKS in *S. coelicolor* is capable of generating a library of more than 30 polyketides that differ in chain length and folding/cyclization regiospecificity, underscoring once again the great promise of creating novel molecules by combinatorial biosynthesis [272]. Finally, the finding that, in addition to the self-malonylation process, TcmM can catalyzes the malonylation of AcpP was quite unexpected, demonstrating for the first time that PKS ACPs and FabD can catalyze the same reaction [273]. However, TcmM is a relatively poor catalyst when compared with FabD. The differences in the catalytic efficiency of these two proteins finally provided a satisfactory explanation for the early observations that FabD-independent polyketide biosynthesis in vitro proceeds only at high concentration of a PKS ACP [173, 273].

6
References

1. O'Hagan D (1991) The polyketide metabolites. Ellis Horwood, Chichester, UK
2. (a) Simpson TJ (1984) Nat Prod Rep 1:28; (b) Simpson TJ (1985) Nat Prod Rep 2:321; (c) Simpson TJ (1987) Nat Prod Rep 4:339; (d) Simpson TJ (1991) Nat Prod Rep 8:573
3. (a) O'Hagan D (1992) Nat Prod Rep 9:447; (b) O'Hagan D (1993) Nat Prod Rep 10:593; (c) O'Hagan D (1995) Nat Prod Rep 12:1
4. (a) Rawlings BJ (1997) Nat Prod Rep 14:335; (b) Rawlings BJ (1997) Nat Prod Rep 14:523; (c) Rawlings BJ (1998) Nat Prod Rep 15:275
5. Monaghan RL, Tkacz JS (1990) Annu Rev Microbiol 44:271
6. Vining LC, Stuttard C (1995) Genetics and biochemistry of antibiotic production. Butterworth-Heinemann, Boston
7. Strohl WR (1997) Biotechnology of antibiotics, 2nd edn. Marcel Deckker, New York
8. Cragg GM, Newman DJ, Snader KM (1997) J Nat Prod 60:52
9. Hutchinson CR (1988) Med Res Rev 8:557
10. Katz L, Donadio S (1993) Ann Rev Microbiol 47:875
11. Hopwood DA (1993) Curr Opinion Biotechnol 4:531
12. Hutchinson CR (1994) Bio/Technol 12:375
13. Piepersberg W (1994) Crit Rev Biotechnol 14:251
14. Tsoi CJ, Khosla C (1995) Chem & Biol 2:355
15. Hutchinson CR, Fujii I (1995) Ann Rev Microbiol 49:201
16. Ikeda H, Omura S (1995) J Antibiot 48:549
17. Khosla C, Zawada RJX (1996) TIBTECH 14:335
18. Lal R, Khanna R, Kaur H, Khanna M, Dhingra N, Lal S, Gartemann KH, Eichenlaub R, Ghosh PK (1996) Crit Rev Microbiol 22:201
19. Nakagawa A, Omura S (1996) J Antibiot 49:717
20. Fu H, Khosla C (1996) Mol. Divers 1:121
21. Baltz RH, McHenney MA, Cantwell CA, Queener SW, Solenberg PJ (1997) Antonie van Leeuwenhock 71:179
22. Wakil SJ (1989) Biochemistry 28:4523

23. Boom TV, Cronan JE (1989) Ann Rev Microbiol 43:317
24. Magnuson K, Jackowski S, Rock CO, Cronan JE (1993) Microbiol Rev 57:522
25. Carreras CW, Pieper R, Khosla C (1997) Top Cur Chem 188:85
26. Hopwood DA, Sherman DH (1990) Ann Rev Genet 24:37–66
27. Hopwood DA, Khosla C (1992) Ciba Foundation Sym 171:88
28. Pieper R, Kao C, Khosla C, Lou G, Cane DE (1996) Chem Soc Rev 297
29. Hopwood DA (1997) Chem Rev 97:2465
30. Ikeda H, Omura S (1997) Chem Rev 97:2591
31. Katz L (1997) Chem Rev 97:2557
32. Khosla C (1997) Chem Rev 97:2577
33. Staunton J, Wilkinson B (1997) Chem Rev 97:2611
34. Staunton J, Wilkinson B (1998) Top Curr Chem 195:49
35. Malpartida F, Hopwood DA (1994) Nature 309:462
36. Fernandez-Moreno MA, Martinez E, Boto L, Hopwood DA, Malpartida F (1992) J Biol Chem 267:19,278
37. Motamedi H, Hutchinson CR (1987) Proc Natl Acad Sci USA 84:4445
38. Bibb MJ, Biro S, Motamedi H, Collins JF, Hutchinson CR (1989) EMBO J 8:2727
39. Malpartida F, Hallam SE, Kieser HM, Motamedi H, Hutchinson CR, Butler MJ, Sugden DA, Warren M, McKillop C, Bailey CR, Humphreys GO, Hopwood DA (1987) Nature 325:818
40. Arrowsmith TJ, Malpartida F, Sherman DH, Birch A, Hopwood DA, Robinson JA (1992) Mol Gen Genet 234:254
41. Bibb MJ, Sherman DH, Omura S, Hopwood DA (1994) Gene 142:31–39
42. Blanco G, Brian P, Pereda A, Mendez C, Salas JA, Chater KF (1993) Gene 130:107
43. Butler MJ, Friend EJ, Hunter IS, Kaczmarek FS, Sugden DA, Warren M (1989) Mol Gen Genet 215:231
44. Dairi T, Hamano Y, Igarashi Y, Furumai T, Oki T (1997) Biosci Biotech Biochem 61:1445
45. Davis NK, Chater KF (1990) Mol Microbiol 4:1679
46. Decker H, Hagg S (1995) J Bacteriol 177:6126
47. Decker H, Rhor J, Motamedi H, Zahner H, Hutchinson CR (1995) Gene 166:121
48. Gould SJ, Hong ST, Carney JR (1998) J Antibiot 51:50
49. Grim A, Madduri K, Ali A, Hutchinson CR (1994) Gene 151:1
50. Hallam SE, Malaprtida F, Hopwood DA (1988) Gene 74:305
51. Han L, Yang K, Ramalingam E, Mosher RH, Vining LC (1994) Microbiology 140:3379
52. Hong ST, Carney JR, Gould SJ (1997) J Bacteriol 179:470
53. Ichinose K, Bedford DJ, Tornus D, Bechthold A, Bibb MJ, Revill WP, Floss HG, Hopwood DA (1998) Chem Biol 5:647
54. Kakinuma S, Ikeda H, Omura S (1991) Tetrahedron 47:6059
55. Kim ES, Bibb MJ, Butler MJ, Hopwood DA, Sherman DH (1994) Gene 141:141
56. Lombo F, Blanco G, Fernandez E, Mendez C, Salas JA (1996) Gene 172:87
57. Niemi J, Ylihonko K, Hakala J, Parssinen R, Kopio A, Mantsala P (1994) Microbiology 140:1351
58. Sherman DH, Malpartida F, Bibb MJ, Kieser HM, Bibb MJ, Hopwood DA (1989) EMBO J 8:2717
59. Ye J, Dickens ML, Plater R, Li Y, Lawrence J, Strohl WR (1994) J Bacteriol 176:6270
60. Ylihonko K, Tuikkanen J, Jussila S, Cong L, Mantsala P (1996) Mol Gen Genet 251:113
61. Yu TW, Bibb MJ, Revill PW, Hopwood DA (1994) J Bacteriol 176:2627
62. Seow KT, Meurer G, Gerlitz M, Wendt-Pienkowski E, Hutchinson CR, Davies J (1997) J Bacteriol 179:7360
63. Gaisser S, Trefzer A, Stockert S, Kirschning A, Bchthold A (1997) J Bacteriol 179:6271
64. Nowak-Thompson B, Gould SJ, Loper JE (1997) Gene 204:17
65. Beck J, Ripka S, Siegner A, Schiltz E, Schweizer E (1990) Eur J Biochem 192:487
66. Fujii I, Ono Y, Tada H, Gomi K, Ebizuka Y, Sankawa U (1996) Mol Gen Genet 253:1
67. Kimura N, Tsuge T (1995) J Bacteriol 175:4427
68. Minto RE, Townsend CA (1997) Chem Rev 97:2537

69. Takano Y, Kubo Y, Shimizu K, Mise K, Okuno T, Furusawa I (1995) Mol Gen Genet 249:162
70. Borejsza-Wysocki W, Hrazdina G (1996) Plant Physiol 110:791
71. Helariutta Y, Kotilainen M, Elomaa P, Kalkkinen N, Bremer K, Teeri TH, Albert VA (1996) Proc Natl Acad Sci USA 93:9033
72. Martin CR (1993) Internal Rev Cytol 147:233
73. Reimold U, Kroger M, Kreuzaler F, Hahlbrock K (1983) EMBL J 2:1801
74. Schroder G, Brown JWS, Schroder J (1988) Eur J Biochem 172:161
75. Schroder J (1997) Trends Plant Sci 2:373
76. Schroder J, Raiber S, Berger T, Schmidt A, Schmidt J, Soares-Sello AM, Bardshiri E, Strack D, Simpson TJ, Veit M, Schroder G (1998) Biochemistry 37:8417
77. Gould SJ (1997) Chem Rev 97:2499
78. Fujii I, Ebizuka Y (1997) Chem Rev 97:2511
79. Hutchinson CR (1997) Chem Rev 97:2525
80. McDaniel R, Ebert-Khosla S, Fu H, Hopwood DA, Khosla C (1994) Proc Natl Acad Sci USA 91:11,542
81. Bartel PL, Connors NC, Strohl WR (1990) J Gen Microbiol 136:1877
82. Dickens ML, Ye J, Strohl WR (1996) J Bacteriol 178:3384
83. Fu H, Ebert-Khosla S, Hopwood DA, Khosla C (1994) J Am Chem Soc 116:4166
84. Rajgarhia VB, Sttrohl WR (1997) J Bacteriol 179:2690
85. Alvarez MA, Fu H, Khosla C, Hopwood DA, Bailey J (1996) Nature Biotechnol 14:335
86. Fu H, Hopwood DA, Khosla C (1994) Chem Biol 1:205
87. McDaniel R, Ebert-Khosla S, Hopwood DA, Khosla C (1994) J Am Chem Soc 116:10,855
88. McDaniel R, Ebert-Khosla S, Hopwood DA, Khosla C (1995) Nature 375:549
89. Sherman DH, Bibb MJ, Simpson TJ, Johnson D, Malpartida F, Frenandez-Moreno M, Martinez E, Hutchinson CR, Hopwood DA (1991) Tetrahedron 47:6029
90. Dickens ML, Ye J, Strohl WR (1995) J Bacteriol 177:536
91. Lomovskaya N, Doi-Katayama Y, Filippini S, Nastro C, Fonstein L, Colombo AL, Hutchinson CR (1998) J Bacteriol 180:2379
92. McDaniel R, Hutchinson CR, Khosla C (1995) J Am Chem Soc 117:6805
93. Madduri K, Hutchinson CR (1995) J Bacteriol 177:3879
94. Shen B, Hutchinson CR (1993) Biochemistry 32:11,149
95. Shen B, Summers RG, Wendt-Pienkowski E, Hutchinson CR (1995) J Am Chem Soc 117:6811
96. Shen B, Hutchinson CR (1996) Proc Natl Acad Sci USA 93:6600
97. Summers RG, Wendt-Pienkowski E, Motamedi H, Hutchinson CR (1993) J Bacteriol 174:1810
98. Summers RG, Wendt-Pienkowski E, Motamedi H, Hutchinson CR (1993) J Bacteriol 175:7571
99. Zhang HI, He XG, Adefarati A, Gallucci J, Cole SP, Beale JM, Keller PJ, Chang CJ, Floss HG (1990) J Org Chem 55:1682
100. Yu T-W, Shen Y, McDaniel R, Floss HG, Khosla C, Hopwood DA, Moore B (1998) J Am Chem Soc 120:7749
101. McDaniel R, Ebert-Khosla S, Hopwood DA, Khosla C (1993) Science 262:1546
102. Fu H, Ebert-Khosla S, Hopwood DA, Khosla C (1994) Biochemistry 33:9321
103. Decker H, Gaisser S, Pelzer S, Schneider P, Westrich L, Wohlleben W, Bechthold A (1996) FEMS Microbiol Lett 141:195
104. Kraus J, Loper J (1995) Appl Environ Microbiol 61:849
105. Dimroth P, Ringelmann E, Lynen F (1971) Eur J Biochem 13:98
106. Wang IK, Reeves C, Gaucher GM (1991) Can J Microbiol 37:86
107. Dutton MF (1988) Microbiol Rev 52:274
108. Trail F, Mahanti N, Linz J (1995) Microbiology 141:755
109. Watanabe CMH, Townsend CA (1998) 120:6231
110. Chang PK, Cary JW, Yu JH, Bhatnagar D, Cleveland TE (1995) Mol Gen Genet 248:270
111. Feng GH, Leonard TJ (1995) J Bacteriol 177:6264

112. Yu JH, Chang PK, Cary JW, Wright M, Bhatnagar D, Cleveland TS, Payne GA, Linz JE (1995) Appl Environ Microbiol 61:2365
113. Yu JH, Leonard TJ (1995) J Bacteriol 177:4792
114. Brown DW, Yu JH, Kelkar HS, Fernandes M, Nesbitt TC, Keller NP, Adams TH, Leonard TJ (1996) Proc Natl Acad Sci 93:1418
115. Brown DW, Adams TH, Keller NP (1996) Proc Natl Acad Sci USA 93:14,873
116. Kottig H, Rottner G, Beck K-F, Schweizer M, Schweizer E (1991) Mol Gen Genet 226:310
117. Mahanti N, Bhatnagar D, Cary JW, Joubran J, Linz JE (1996) Appl Environ Microbiol 62:191
118. Aggarwal R, Caffrey P, Leadlay PF, Smith CJ, Staunton J (1995) J Chem Soc Chem Commun 1519
119. Brown MJ, Cortes J, Cutter AL, Leadlay PF, Staunton J (1995) J Chem Soc Chem Commun 1517
120. Cortes J, Wiesmann KEH, Roberts GA, Brown MJB, Staunton J, Leadlay PF (1995) Science 268:1487
121. Kao CM, Lou G, Katz L, Cane DE, Khosla C (1995) J Am Chem Soc 117:9105
122. Ehmann B, Schafer E (1988) Plant Mol Biol 11:869
123. Helariutta Y, Kotilainen M, Elomaa P, Teeri TH, Albert VA (1995) Plant Mol Biol 28:935
124. Preisig-Muller R, Gehlert R, Melchior F, Stietz U, Kindl H (1997) Biochemistry 36:8349
125. Raiber S, Schroder G, Schroder J (1995) FEBS Lett 361:299
126. Ryder TB, Hedrick SA, Bell JN, Liang X, Clouse SD, Lamb CJ (1987) Mol Gen Genet 210:219
127. Schoppner A, Kindl H (1984) J Biol Chem 259:6806
128. Schuz R, Heller W, Hahlbrock K (1983) J Biol Chem 258:6730
129. (a) Akada S, Kung SD, Dube SK (1990) Nucl Acids Res 18:3398; (b) Akada S, Kung SD, Dube SK (1990) Nucl Acids Res 18:5899
130. Akada S, Kung SD, Dube SK (1991) Plant Mol Biol 16:751
131. (a) Akada S, Kung SD, Dube SK (1993) Plant Physiol 102:317; (b) Akada S, Kung SD, Dube SK (1993) Plant Physiol 102:321; (c) Akada S, Kung SD, Dube SK (1993) Plant Physiol 102:699
132. Lanz T, Tropf S, Marner FJ, Schroder J, Schroder G (1991) J Biol Chem 266:9971
133. Tropf S, Karcher B, Schroder G, Schroder J (1995) J Biol Chem 270:7922
134. Schroder G, Schroder J (1992) J Biol Chem 267:20,558
135. Eckermann S, Schroder G, Schmidt J, Strack D, Edrada RA, Helariutta Y, Elomaa P, Kotilainen M, Kilpelainen I, Proksch P, Terri TH, Schroder J (1998) Nature 396:387
136. Ayabe SI, Udagawa A, Furuya T (1988) Arch Biochem Biophys 261:458
137. Harano K, Okada N, Furuno T, Takahashi T, Ayabe SI, Welle R (1993) Plant Cell Rep 12:66
138. Welle R, Grisebach H (1988) FEBS Lett 236:221
139. Welle R, Schroder G, Schiltz E, Grisebach H, Schroder J (1991) Eur J Biochem 196:423
140. Carper E, Nishimura C, Shinohara T, Dietzchold B, Wistow G, Craft C, Kador P, Kinoshita JH (1987) FEBS Lett 220:209
141. Anderson S, Marks CB, Lazarus R, Miller J, Stafford K, Seymour J, Light D, Rastetter W, Estell D (1985) Science 230:144
142. Kouzarides T, Ziff E (1988) Nature 336:646
143. Ibrahim RK (1997) Trends Plant Sci 2:249
144. Haydock SF, Aparicio JF, Molnar I, Schwecke T, Khaw LE, Kong A, Marsden AFA, Galloway IS, Staunton J, Leadlay PF (1995) FEBS Lett 374:246
145. Marsden AFA, Caffrey P, Aparicio JF, Loughran MS, Staunton J, Leadlay PF (1994) Science 263:378
146. Ruan X, Pereda A, Stassi DL, Zeidner D, Summers RG, Jackson M, Shivakumar A, Kakavas S, Staver MJ, Donadio S, Katz L (1997) J Bacteriol 179:6416
147. Stassi DL, Kakavas SJ, Reynold KA, Gunwardana G, Swanson S, Zeidner D, Jackson M, Liu H, Buko A, Katz L (1998) Proc Natl Acad Sci USA 95:7305
148. Harris TM, Harris CM (1986) Pure Appl Chem 58:283

149. Bartel PL, Zhu CB, Lampel JS, Dosch DC, Connors NC, Strohl WR, Beale JM, Floss HG (1990) J Bacteriol 172:4816
150. Blanco G, Fu H, Mendez C, Khosla C, Salas J (1996) Chem Biol 3:193
151. Decker H, Summers RG, Hutchinson CR (1994) J Antibiot 47:54
152. Fu H, Ebert-Khosla S, Hopwood DA, Khosla C (1994) J Am Chem Soc 116:6443
153. Gerlitz M, Meurer G, Wendt-Pienkowski E, Madduri K, Hutchinson CR (1997) J Am Chem Soc 119:7392
154. Kantola J, Blanco G, Hautala A, Kunnari T, Hakala J, Mendez C, Ylihonko K, Mantsala P, Salas J (1997) Chem Biol 4:751
155. Khosla C, McDaniel R, Ebert-Khosla S, Torres R, Sherman DH, Bibb MJ, Hopwood DA (1993) J Bacteriol 175:2197
156. Khosla C, Ebert-Khosla S, Hopwood DA (1992) Mol Microbiol 6:3237
157. Kim ES, Hopwood DA, Sherman DH (1994) J Bacteriol 176:1801
158. Kim ES, Cramer KD, Shreve AL, Sherman DH (1995) J Bacteriol 177:1202
159. Kramer PJ, Zawada RJX, McDaniel R, Hutchinson CR, Hopwood DA, Khosla C (1997) J Am Chem Soc 119:635
160. Kunzel E, Wohlert SE, Beninga C, Haag S, Decker H, Hutchinson CR, Blanco G, Mendez C, Salas JA, Rohr J (1997) Chem Eur J 3:1675
161. McDaniel R, Ebert-Khosla S, Hopwood DA, Khosla C (1993) J Am Chem Soc 115:11,671
162. Meurer G, Gerlitz M, Wendt-Pienkowski E, Vining LC, Rhor J, Hutchinson CR (1997) Chem Biol 4:433
163. Meurrer G, Hutchinson CR (1995) J Am Chem Soc 117:5899
164. Meurrer G, Hutchinson CR (1995) J Bacteriol 177:477
165. Sherman DH, Kim ES, Bibb MJ, Hopwood DA (1992) J Bacteriol 174:6184
166. Yu TW, Hopwood DA (1995) Microbiology 141:2779
167. Cox RJ, Hitchman TS, Byrom KJ, Findlow SC, Tanner JA, Crosby J, Simpson TJ (1997) FEBS Lett 405:267
168. Crosby J, Sherman DH, Bibb MJ, Revill WP, Hopwood DA, Simpson TJ (1995) Biochim Biophy Acta 1251:32
169. Crump MP, Crosby J, Dempsey CE, Murray M, Hopwood DA, Simpson TJ (1996) FEBS Lett 391:302
170. Crump MP, Crosby J, Dempsey CE, Parkinson JA, Murray M, Hopwood DA, Simpson (1997) Biochemistry 36:6000
171. Gramajo H, White J, Hutchinson CR, Bibb MJ (1991) J Bacteriol 173:6475
172. Hitchman TS, Crosby J, Byrom KJ, Cox RJ, Simpson TJ (1998) Chem Biol 5:35
173. Matharu A-L, Cox RJ, Crosby J, Byrom KJ, Simpson TJ (1998) Chem Biol 5:699
174. Revill WP, Bibb MJ, Hopwood DA (1995) J Bacteriol 177:3946
175. Revill WP, Bibb MJ, Hopwood DA (1996) J Bacteriol 178:5660
176. Ritsema T, Gehring AM, Stuitje AR, van der Drift KMGM, Dandal I, Lambalot EH, Walsh CT, Thomas-Oates JE, Lugtenberg BJJ, Spaink HP (1998) Mol Gen Genet 257:641
177. Shen B, Summers RG, Gramajo H, Bibb MJ, Hutchinson CR (1992) J Bacteriol 174:3818
178. Summers RG, Ali A, Shen B, Wessel WA, Hutchinson CR (1995) Biochemistry 34:9389
179. Tropf S, Revill WP, Bibb MJ, Hopwood DA, Schweizer M (1998) Chem Biol 5:135
180. Zawada RJX, Khosla C (1997) J Biol Chem 272:16,184
181. Carreras CW, Pieper R, Khosla C (1996) J Am Chem Soc 118:5158
182. Shen B, Hutchinson CR (1993) Science 262:1535
183. Bao W, Wendt-Pienkowski E, Hutchinson CR (1998) Biochemistry 37:8132
184. Carreras CW, Khosla C (1998) Biochemistry 37:2084
185. Revill WP, Leadlay PF (1991) J Bacteriol 173:4379
186. Caffrey P, Green B, Packman LC, Rawlings BJ, Staunton J, Leadlay PF (1991) Eur J Biochem 195:823
187. Lambalot RH, Walsh CT (1995) J Biol Chem 270:24,658
188. Gehring AM, Lambalot RH, Vogel KW, Drueckhammer DG, Walsh CT (1997) Chem Biol 4:17
189. Lambalot RH, Gehring AM, Flugel RS, Zuber P, LaCelle M, Marahiel MA, Reid R, Khosla C, Walsh CT (1993) Chem Biol 3:923

190. Stuible HP, Meier S, Wagner C, Hannappel E, Schweizer E (1998) 273:22,334
191. Walsh CT, Gehring AM, Weinreb PH, Quadri LEN, Flugel RS (1997) Curr Opin Chem Biol 1:309
192. Carreras CW, Gehring AM, Walsh CT, Khosla C (1997) Biochemistry 36:11,757
193. Marahiel MA, Stachelhaus T, Mootz HD (1997) Chem Rev 97:2651
194. von Dohren H, Keller U, Vater J, Zocher R (1997) Chem Rev 97:2675
195. Kleinkauf H, von Dohren H (1996) Eur J Biochem 236:335
196. Stein T, Vater J (1996) Amino Acids 10:201
197. Kleinkauf H, von Dohren H (1990) Eur J Biochem 192:1
198. Stachelhaus T, Huser A, Marahiel MA (1996) Chem Biol 3:913
199. Gehring AM, Bradley KA, Walsh CT (1997) Biochemistry 36:8495
200. Gehring AM, Mori I, Walsh CT (1998) Biochemistry 37:2648
201. Gehring AM, Mori I, Perry RD, Walsh CT (1998) Biochemistry 37:11,637
202. Nakano MM, Corbell N, Besson J, Zuber P (1992) Mol Gen Genet 232:313
203. Weinreb PH, Quadri LEN, Walsh CT, Zuber P (1998) Biochemistry 37:1575
204. Quadri LEN, Weinreb PH, Lei M, Nakano MM, Zuber P, Walsh CT (1998) Biochemistry 37:1585
205. Borchert S, Stachelhaus T, Marahiel MA (1994) J Bacteriol 176:2458
206. Ku J, Mirmira RG, Liu L, Santi DV (1997) 4:203
207. Jaworski JG, Post-Beittenmiller MA, Ohlrogge JB (1989) Eur J Biochem 184:603
208. Kresze GB, Steber L, Oseterhelt D, Lynen F (1977) Eur J Biochem 79:191
209. Pieper R, Ebert-Khosla S, Cane D, Khosla C (1996) Biochemistry 35:2045
210. Holak TA, Nilges M, Prestegard JH, Gronenborn AM, Clore GM (1988) Eur J Biochem 175:9
211. Kim Y, Prestegard JH (1989) Biochemistry 28:8792
212. Andrec M, Hill RB, Prestegard JH (1995) Protein Sci 4:983
213. Joshi VC, Wakil SJ (1971) Arch Biochem Biophys 143:493
214. Ruch FE, Vagelos PR (1973) J Biol Chem 248:8086
215. Ruch FE, Vagelos PR (1973) J Biol Chem 248:8095
216. Clark DP, Cronan JE (1981) Methods Enzymol 72:693
217. Magnuson K, Oh W, Larson TJ, Cronan JE (1992) FEBS Lett 299:262
218. Verwoert IIGS, Verbree EC, van der Linden KH, Nijkamp HJJ, Stuitje AR (1992) J Bacteriol 174:2851
219. Harder ME, Ladenson RC, Schimmel SD, Silbert DF (1974) J Biol Chem 249:7468
220. Fernandez-Moreno MA, Caballero JL, Hopwood DA, Malpartida F (1991) Cell 66:769
221. Gramajo HC, Takano E, Bibb MJ (1993) Mol Microbiol 837:845
222. Garwin JL, Klages AL, Cronan JE (1980) J Biol Chem 255:11,949
223. Funabashi H, Kawaguchi A, Tomoda H, Omura S, Okuda S, Iwasaki S (1989) J Biochem 105:751
224. Huang W, Jia J, Edwards P, Dehesh K, Schneider G, Lindqvist Y (1998) EMBO J 17:1183
225. Omura S (1981) Methods Enzymol 72:520
226. Robert G, Leadlay PF (1983) FEBS Lett 159:13
227. Robert G, Leadlay PF (1984) Biochem Soc Trans 12:642
228. Motamedi H, Wendt-Pienkowski E, Hutchinson CR (1986) J Bacteriol 167:575
229. Shen B, Nakayama H, Hutchinson CR (1993) J Nat Prod 56:1288
230. Shen B, Hutchinson CR (1993) Biochemistry 32:6656
231. Fu H, Alvarez MA, Khosla C, Bailey J (1996) Biochemistry 35:6527
232. Spencer JB, Jordan PM (1992) Biochem J 288:839
233. Spencer JB, Jordan PM (1992) J Chem Soc Chem Commun 646
234. Woo E-R, Fujii I, EbizukaY, Sankawa U, Kawaguchi A, Huang S, Beale JM, Shibuya M, Mocek U, Floss HG (1989) J Am Chem Soc 111:5498
235. Watanabe CMH, Wilson D, Linz JE, Townsend CA (1996) Chem Biol 3:463
236. Watanabe CMH, Townsend CA (1998) J Am Chem Soc 120:6231
237. Kealey JT, Liu L, Santi DV, Betlach MC, Barr PJ (1998) Proc Natl Acad Sci USA 95:505
238. Bedford DJ, Schweizer E, Hopwood DA, Khosla C (1995) J Bacteriol 177:4544

239. Jordan PM, Spencer JB (1991) Tetrahedron 47:6051
240. Spencer JB, Jordan PM (1992) Biochemistry 31:9107
241. Hanson KR, Rose IA (1975) Acc Chem Res 8:1
242. Sedgwick B, Morris D, French SJ (1978) J Chem Soc Chem Commun 1704–1706
243. Weissman KJ, Timoney M, Bycroft M, Grice P, Hanefeld U, Staunton J, Leadlay PF (1997) Biochemistry 36:13,849
244. Child CJ, Spencer JB, Bhogal P, Shooling-Jordan PM (1996) Biochemistry 35:12,267
245. Child CJ, Shoolingin-Jordan PM (1998) Biochem J 330:933
246. Staunton J, Caffrey P, Aparicio JF, Roberts GA, Bethell SS, Leadlay PF (1996) Nat Struct Biol 3:188
247. Brobst SW, Townsend CA (1994) Can J Chem 72:200
248. Townsend CA, Christensen SB, Trautwein K (1984) J Am Chem Soc 106:3868
249. Townsend CA, Christensen SB (1983) Tetrahedron 21:3575
250. Kendrew SG, Hopwood DA, Marsh ENG (1997) J Bacteriol 179:4305
251. Kurosaki F, Itoh M, Kizawa Y, Nishi A (1993) Arch Biochem Biophys 163:157
252. Kurosaki F, Itoh M, Kizawa Y, Nishi A (1994) Phytochem 35:297
253. Kurosaki F (1995) Arch Biochem Biophys 321:239
254. Kurosaki F (1996) FEBS Lett 379:97
255. Kurosaki F (1996) Arch Biochem Biophys 328:213
256. Pieper R, Luo G, Cane DE, Khosla C (1995) J Am Chem Soc 117:11,373
257. Cane DE, Walsh CT, Khosla C (1998) Science 282:63
258. Ziermann R, Betlach MC (1999) BioTechniques 26:106
259. Takano E, White J, Thompson CJ, Bibb MJ (1995) Gene 166:133
260. Decker H, Motamedi H, Hutchinson CR (1993) J Bacteriol 175:3876
261. Decker H, Hutchinson CR (1993) J Bacteriol 175:3887
262. Cole ST, Brosch R, Parkhill J, Garnier T, Churcher C, and additional 37 authors (1998) Nature 393:537
263. GeneBank AL031317
264. Quadri LEN, Sello J, Keating TA, Weinreb PH, Walsh CT (1998) Chem Biol 5:631
265. Watanabe A, Ono Y, Fujii I, Sankawa U, Mayorga ME, Timberlake WE, Ebizuka Y (1998) Tetrahedron Lett 39:7733
266. Watanabe A, Fujii I, Sankawa U, Mayorga ME, Timberlake WE, Ebizuka Y (1999) Tetrahedron Lett 40:91
267. Schorr R, Mittag M, Müller G, Schweizer E (1994) J Plant Physiol 143:407
268. Kulowski K, Wendt-Pienkowski E, Han L, Yang K, Vining LC, Hutchinson CR (1999) J Am Chem Soc 121:1786
269. Kendrew SG, Katayama K, Deutsch E, Madduri K, Hutchinson CR (1999) Biochemistry 38:4794
270. Bao W, Sheldon PJ, Hutchinson CR (1999) Biochemistry 38:9752
271. Bao W, Sheldon PJ, Wendt-Pienkowski E, Hutchinson CR (1999) J Bacteriol 181:4690
272. Shen Y, Yoon P, Yu T-W, Floss HG, Hopwood DA, Moore BS (1999) Proc Natl Acad Sci USA 96:3622
273. Zhou P, Florova G, Reynolds KA (1999) Chem Biol 6:577

Cyclization Enzymes in the Biosynthesis of Monoterpenes, Sesquiterpenes, and Diterpenes

Edward M. Davis · Rodney Croteau

Institute of Biological Chemistry, Washington State University, Pullman, WA 99164-6340, USA
E-mail: croteau@mail.wsu.edu

Terpene synthases catalyze the first committed steps in the biosynthesis of monoterpenes, sesquiterpenes, and diterpenes. An overview is presented of the enzymology and mechanism of these terpene synthases, and their molecular cloning, expression, and sequence analysis. Detailed structural and functional evaluation of four representative monoterpene, sesquiterpene, and diterpene synthases is also presented.

Keywords. Terpene, Monoterpene, Sesquiterpene, Diterpene, Cyclase, Geranyl diphosphate, Farnesyl diphosphate, Geranylgeranyl diphosphate

Abbreviations

CPP	copalyl diphosphate
DMAPP	dimethylallyl diphosphate
FPP	farnesyl diphosphate
GPP	geranyl diphosphate
GGPP	geranylgeranyl diphosphate
Hm-VS	*Hyoscyamus muticus* vetispiradiene synthase
IPP	isopentenyl diphosphate
KS	kaurene synthase
LPP	linalyl diphosphate
NPP	nerolidyl diphosphate
Nt-EAS	*Nicotiana tabacum* 5-*epi*-aristolochene synthase
tps	terpene synthase
WM	Wagner-Meerwein rearrangement

1 Introduction

The terpenes are a structurally diverse and widely distributed family of natural products containing well over 30,000 defined compounds identified from all kingdoms of life [1]. The majority of terpenes have been isolated from plants where they serve a broad range of roles in primary metabolism (including several plant hormones and the most abundant plant terpenoid, phytol, the side chain of the photosynthetic pigment chlorophyll) and in ecological interactions (as chemical defenses against herbivores and pathogens, pollinator attractants, allelopathic agents, etc.) (Table 1). Many terpenes are of economic importance, including the essential oils, carotenoid pigments and natural rubber. Terpenes are constructed by the repetitive joining of branched C_5 isoprenoid units and are categorized according to size with the C_{10} family, historically considered the smallest, named the monoterpenes. The nomenclature thus evolved with the sesquiterpenes (C_{15}), diterpenes (C_{20}), sesterterpenes (C_{25}), triterpenes (C_{30}), tetraterpenes (C_{40}), and polyterpenes ($>C_{40}$). Isoprene, discovered as a natural product in the 1960's, was thus considered a hemiterpene (C_5). The meroterpenes consist of a terpene-derived unit attached to a non-terpene moiety; examples include chlorophyll, plastoquinone, certain indole alkaloids, and prenylated proteins.

All terpenoids are derived from the universal precursor isopentenyl diphosphate (IPP) which, following isomerization to dimethylallyl diphosphate (DMAPP) by IPP isomerase, is sequentially elongated by prenyltransferases to geranyl diphosphate (GPP, C_{10}), farnesyl diphosphate (FPP, C_{15}), and geranylgeranyl diphosphate (GGPP, C_{20}) (Scheme 1). These acyclic intermediates function at the branch points of isoprenoid metabolism from which the monoterpenes (from GPP), sesquiterpenes (from FPP), and diterpenes (from GGPP) diverge through the action of the terpenoid synthases. The terpenoid synthases, often called cyclases since the products of the reaction are most often cyclic,

Table 1. Classification, structures, and physiological and ecological roles of selected terpenoids

Monoterpenes	medicinal	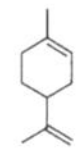	limonene
		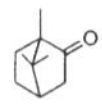	camphor
	flavoring	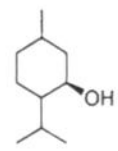	menthol
Sesquiterpenes	perfumary raw material	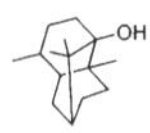	patchoulol
	phytohormone	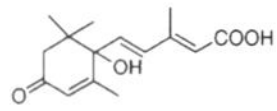	abscisic acid
	antibiotic	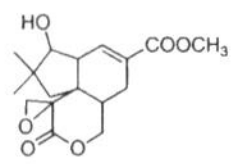	pentalenolactone
Diterpenes	anti-cancer drug	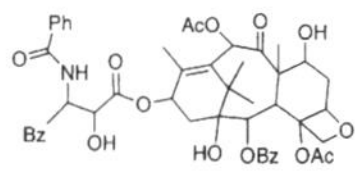	taxol
	phytoalexin	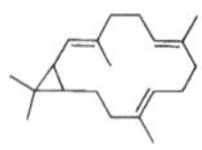	casbene
	phytohormone	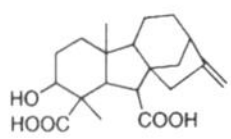	gibberellin
Triterpene	membrane component	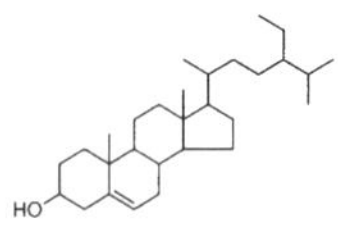	β-sitosterol
Tetraterpene	plant pigment (provitamin A)	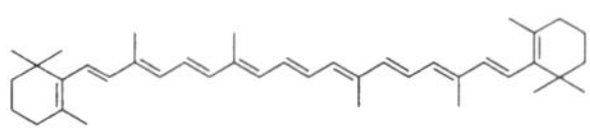	β-carotene
Meroterpenes	photosynthesis	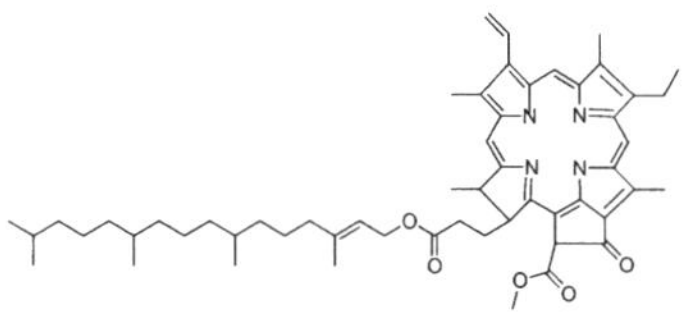	chlorophyll
	electron transport	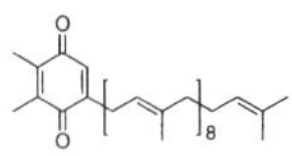	plastoquinone
Polyterpene	commercial (industrial raw material)	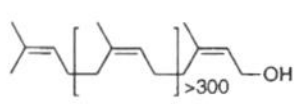	rubber

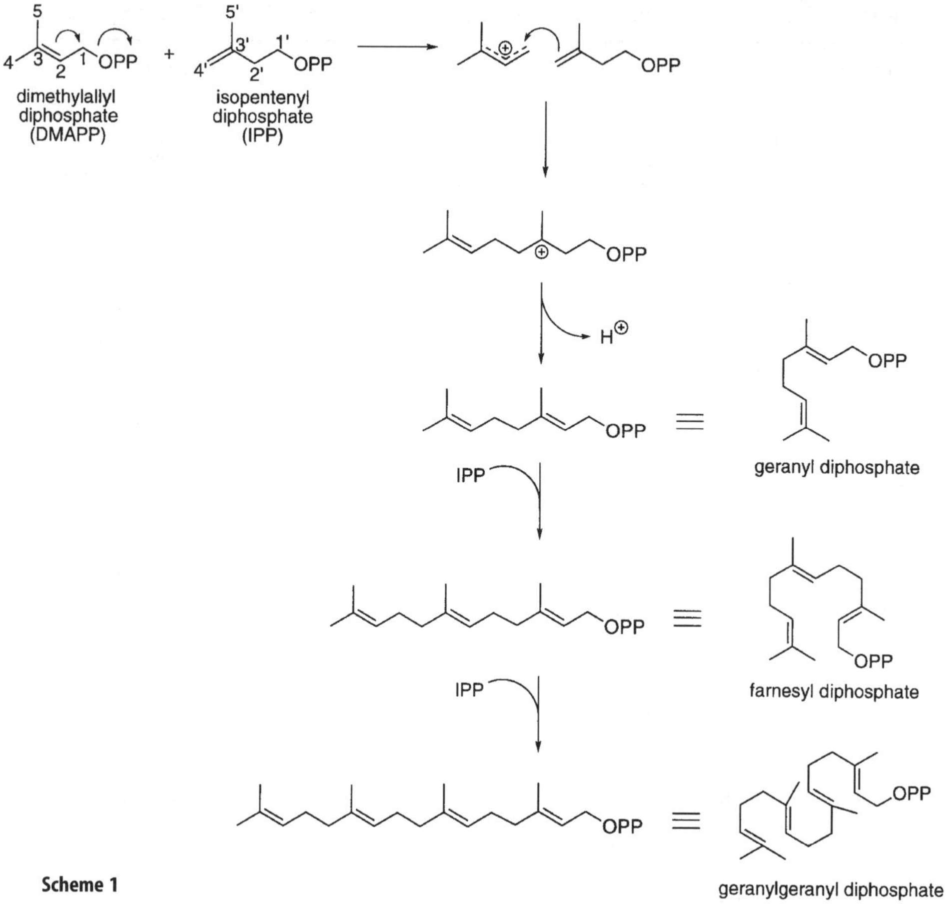

Scheme 1

serve to convert the three acyclic branch point intermediates to the parent skeletons of the various monoterpene, sesquiterpene, and diterpene types. This family of enzymes, which forms the focus of this review, not only catalyzes the committed step in the biosynthesis of most terpenoids but is also largely responsible for the great structural diversity encountered in this class of natural products.

The past five years have witnessed a major reevaluation of the biogenetic origin of IPP as the universal precursor of terpenoids. Since its elucidation in the late 1950's [2], the classical acetate-mevalonate pathway (Scheme 2A) has been considered the only biosynthetic route to terpenoids. However, a mevalonate-independent pathway, with its origins in the condensation of pyruvate (via hydroxyethyl thiamine pyrophosphate) and glyceraldehyde-3-phosphate to form 1-deoxyxylose-5-phosphate (Scheme 2B), has been discovered to operate in certain bacteria [3] and, more recently, in the plastids of both lower and higher plants [4–6]. It is now clear that terpenoid biosynthesis in plants is compartmentalized, with the sesquiterpenes (and triterpenes) arising

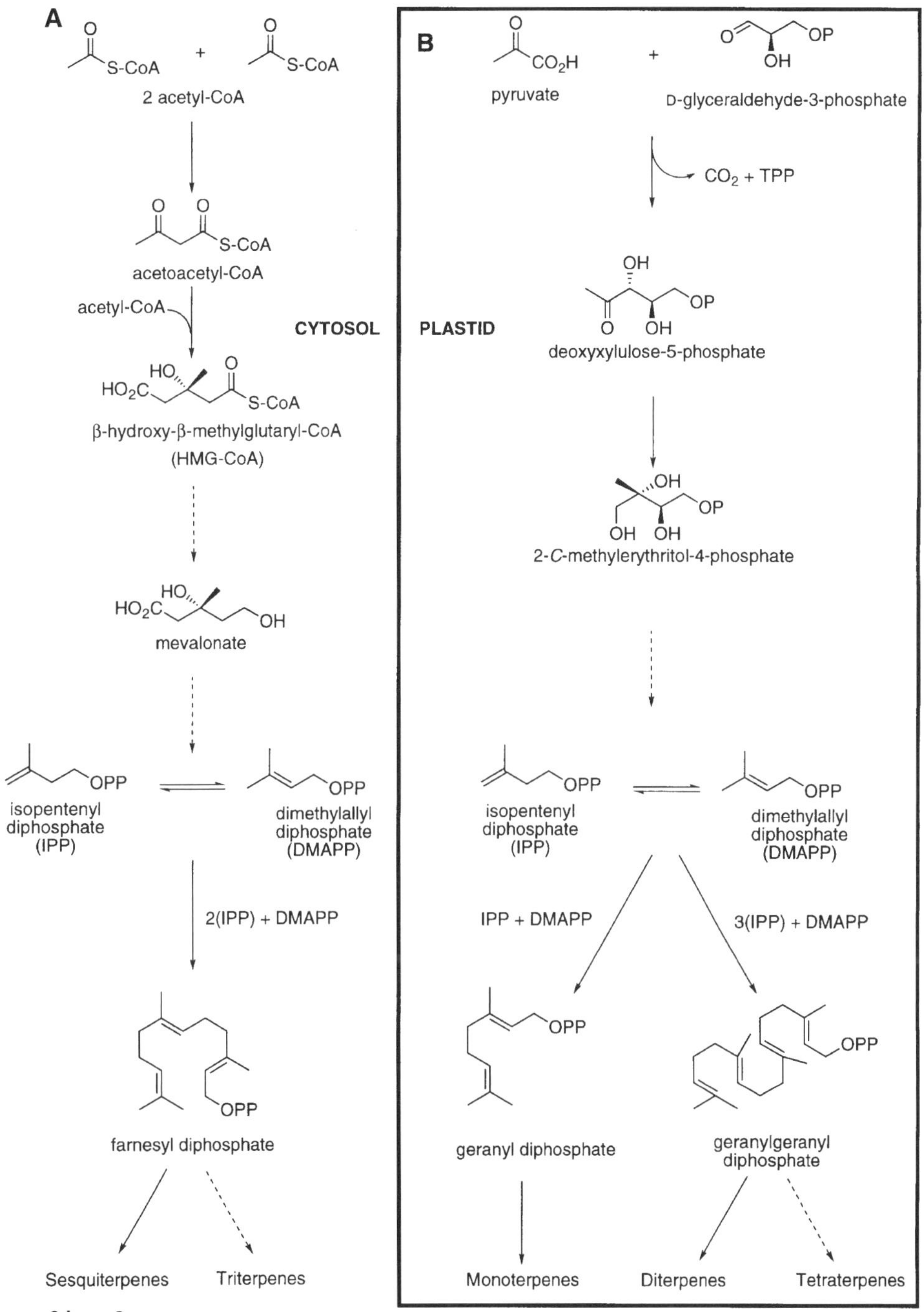
A
2 acetyl-CoA
acetoacetyl-CoA
acetyl-CoA
CYTOSOL
β-hydroxy-β-methylglutaryl-CoA
(HMG-CoA)
mevalonate
isopentenyl diphosphate (IPP)
dimethylallyl diphosphate (DMAPP)
2(IPP) + DMAPP
farnesyl diphosphate
Sesquiterpenes
Triterpenes
B
pyruvate
D-glyceraldehyde-3-phosphate
CO2 + TPP
PLASTID
deoxyxylulose-5-phosphate
2-C-methylerythritol-4-phosphate
isopentenyl diphosphate (IPP)
dimethylallyl diphosphate (DMAPP)
IPP + DMAPP
3(IPP) + DMAPP
geranyl diphosphate
geranylgeranyl diphosphate
Monoterpenes
Diterpenes
Tetraterpenes

Scheme 2

in the cytosol [7] via the acetate-mevalonate pathway, and the monoterpenes and diterpenes (and tetraterpenes) arising in the plastids [6–9] via the still incompletely understood mevalonate-independent pathway of bacterial endosymbiotic origin. Each compartment contains IPP isomerase [10] and the requisite prenyltransferases for construction of the appropriate precursors.

2
Prenyltransferases and Terpenoid Synthases

The prenyltransferases catalyze the head-to-tail condensation of IPP with either DMAPP (geranyl diphosphate synthase), GPP (farnesyl diphosphate synthase), or FPP (geranylgeranyl diphosphate synthase), although FPP synthase and GGPP synthase can also utilize only IPP and DMAPP as the initial substrates in multistep elongation sequences via bound intermediates [11]. The electrophilic reaction mechanisms of the prenyltransferases involves initial ionization of the allylic diphosphate ester substrate to form a charge-delocalized carbocation which undergoes C4′-C1 coupling via the terminal double bond of IPP to generate a tertiary carbocation followed by deprotonation to complete the reaction (Scheme 1). The terpenoid synthase reaction may be regarded as the intramolecular analogue of the intermolecular electrophilic coupling reaction catalyzed by the prenyltransferases, although the detailed reaction variants encountered in the former case are far greater than in the latter [11, 12]. More than a common mechanistic link was established between the prenyltransferases and the terpenoid synthases (cyclases) with the discovery that avian FPP synthase is capable of generating cyclic C_{15} olefins as products with FPP as substrate [12, 13]. The two enzyme types also exhibit notable conservation in primary sequence and physical characteristics (see below).

As might be expected based on the similarities in reaction mechanism, the prenyltransferases and terpenoid synthases share a number of common properties. Both are functionally soluble, acidic enzymes, with native sizes in the 35–80 kDa range, require only a divalent metal ion for catalysis, and exhibit Michaelis constants for the prenyl substrate that rarely exceed 10 μM and turnover numbers typically ranging from 0.03 to 0.3 s^{-1} [14]. The similarities between the prenyltransferases and the terpenoid synthases are clearly of evolutionary significance; however, it is equally clear that the diversity in type and mechanistic complexity set the terpenoid synthases apart from the more conservative prenyltransferases. The typical cyclization reaction involves ionization of the prenyl diphosphate substrate with intramolecular attack by a double bond of the prenyl chain at C1 to yield a cyclic carbocationic intermediate. Subsequent steps may involve further internal additions via the remaining double bonds to generate additional rings, and may be accompanied by hydride shifts, methyl migrations and Wagner-Meerwein rearrangements (WM) before termination of the reaction by deprotonation or nucleophile (often water) capture. The terpenoid synthases commonly produce multiple products, although high fidelity cyclizations yielding a single product are also known.

The reaction mechanisms have been well established for several terpenoid synthases, and general stereochemical models for the reaction types now allow sound rationalization for the biosynthesis of most monoterpenes, sesquiterpenes, and diterpenes; comprehensive reviews of this area are available [15–20]. Additionally, the cDNAs encoding more than 30 terpenoid synthases of plant origin (and eight of microbial origin) have been isolated, allowing comparative study of the primary structures of these enzymes. Interestingly, the plant terpenoid synthases bear little resemblance in primary structure to their microbial counterparts, although in three-dimensional structure their similarity is remarkable. Sufficient information is now available to permit inference of structural and phylogenetic relationships among the terpenoid synthases, which indicates structural evolution of the gene type based upon taxonomic origin rather than substrate utilization and mechanistic similarity of the encoded catalysts [21]. This ability of very similar enzymes to catalyze different reactions, and quite different enzymes to catalyze ostensibly identical reactions, coupled to the overall diversity and complexity of the reaction types often leading to multiple end-products from the same active site, dictate that no single enzyme is adequate as a model for all terpenoid cyclases. Nevertheless, there are certain synthases whose physical and chemical characteristics allow the more general mechanistic and stereochemical aspects of cyclization to be probed and which can serve as a foundation for understanding cyclase catalysis. Toward this end, the present review focuses on four terpenoid synthases as models possessing certain structural and mechanistic features which, although not universal to all cyclases, allow unifying principles to be described.

3
Monoterpene Synthases

Monoterpene synthases capable of generating acyclic, monocyclic and bicyclic products as olefins, alcohols and diphosphate esters (Scheme 3) have been isolated from several lower plants (liverworts) and higher plants, including angiosperms and gymnosperms. The gymnosperm monoterpene synthases differ from their angiosperm counterparts in several respects. All gymnosperm monoterpene cyclases have an absolute requirement for a monovalent cation such as K^+, utilize Mn^{2+} or Fe^{2+} as the required divalent cation and possess alkaline pH optima [22–24], whereas the angiosperm cyclases prefer Mg^{2+} or Mn^{2+}, do not require a monovalent cation, and possess pH optima closer to neutrality [25–31].

More than a dozen cDNAs encoding monoterpene synthases from a range of plant sources have now been isolated [22, 32–36]. The corresponding recombinant enzymes generate a diverse set of products and provide an exceptional set of catalysts for the detailed study of structure-function relationships in monoterpene cyclization. However, it was earlier work with native synthases that first allowed definition of the cyclization reaction [15]. Thus, the stereochemical model for monoterpene cyclization was developed largely from studies with the bicyclic monoterpene cyclases (+)- and (–)-bornyl diphosphate synthases, (+)- and (–)-pinene synthases, and (–)-*endo*-fenchol synthase. It was also the purification of several native monoterpene synthases that gave the first indication that

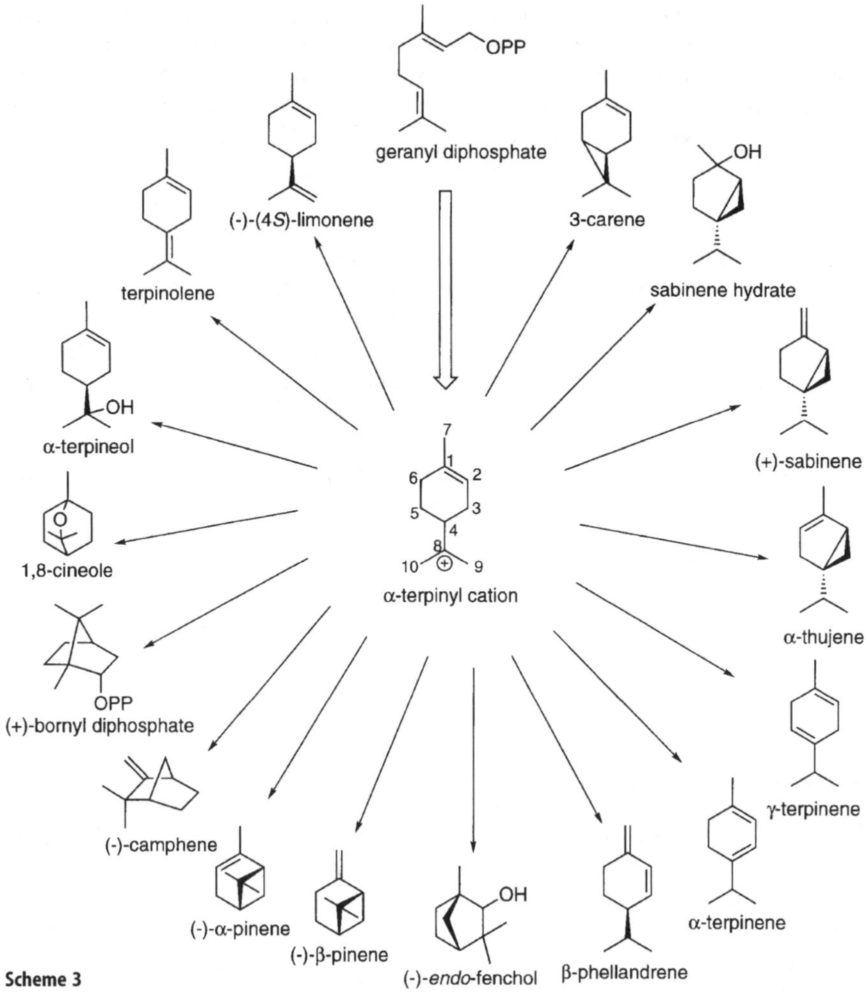

Scheme 3

these enzymes catalyzed the formation of multiple products in fixed ratios [37]. This observation was subsequently supported by studies that exploited isotopically sensitive branching in the multiple-product reaction channel [38–40], and was unambiguously confirmed by demonstrating the identical product profiles of the native enzymes and their recombinant, heterologously expressed counterparts [33, 34]. With the cloning and characterization of the monoterpene synthases, each of which generates multiple products at the same active site, it is likely that this seemingly unusual property is, in fact, common. Multiple product formation can be rationalized as a consequence of the electrophilic cyclization mechanism in which several highly reactive carbocationic intermediates are generated in the course of the reaction, any of which may be stabilized by deprotonation or other means. Stabilization of such intermediates in the course of the often complex cyclization sequence is a common theme that will be developed throughout.

3.1 Mechanism and Stereochemistry

The monoterpene cyclization reaction (Scheme 4) consists of the stereoselective binding of GPP at the active site in either a right-handed or left-handed helical conformer followed by metal ion-dependent ionization of the diphosphate ester and *syn* migration of the diphosphate group to C3 to generate enzyme-bound (3*R*)- or (3*S*)-linalyl diphosphate (LPP), respectively, thereby allowing the requisite rotation about the newly generated C2-C3 single bond. Subsequent C6-C1 cyclization of the *anti,endo* conformer of LPP is initiated by a second ionization of this bound intermediate to form the universal monocyclic intermediate, the α-terpinyl cation. Whether the (4*R*)- or (4*S*)- antipode is formed depends on the configuration at C3 of the linalyl intermediate which, in turn, depends on the initial folding of GPP at the active site. Termination of the reaction via deprotonation or nucleophile capture may be preceded by myriad intervening cyclizations to the remaining double bond, hydride shifts, and other rearrangements shown in Table 2.

Apparent from Scheme 4 is the necessity of the ionization-isomerization step to the cisoid linalyl intermediate for cyclization to the α-terpinyl cation, a reaction not directly possible from the geranyl substrate because of the *trans*-C2,C3 bond. A notable feature of all monoterpene cyclases is the ability to utilize LPP

Table 2. Summation of mechanisms involved in the biosynthesis of selected monoterpenes derived from the α-terpinyl cation. Numbering is based on the α-terpinyl cation shown in Scheme 3

Product	Mechanism	Reference
limonene	deprotonation	25
terpinolene	deprotonation	32
α-terpineol	capture of C8 cation by water	15
1,8-cineole	capture of C8 cation by water, C8-C1 ether linkage	41, 42
bornyl diphosphate	C1-C8 ring closure, C2 cation capture by diphosphate	43
camphene	C1-C8 ring closure, WM rearrangement, deprotonation	32, 44
α-pinene	C2-C8 ring closure, deprotonation	31
β-pinene	C2-C8 ring closure, deprotonation	31
endo-fenchol	C2-C8 ring closure, WM rearrangement, C2 cation capture by water	29
β-phellandrene	1,3 hydride shift, deprotonation	32, 45
α-terpinene	1,2 hydride shift, deprotonation	45
γ-terpinene	1,2 hydride shift, deprotonation	45
α-thujene	1,2 hydride shift, C2-C4 ring closure, deprotonation	46
sabinene	1,2 hydride shift, C2-C4 ring closure, deprotonation	47
sabinene hydrate	1,2 hydride shift, C2-C4 ring closure, C1 capture by water	30
3-carene	deprotonation-mediated C5-C8 ring closure	48

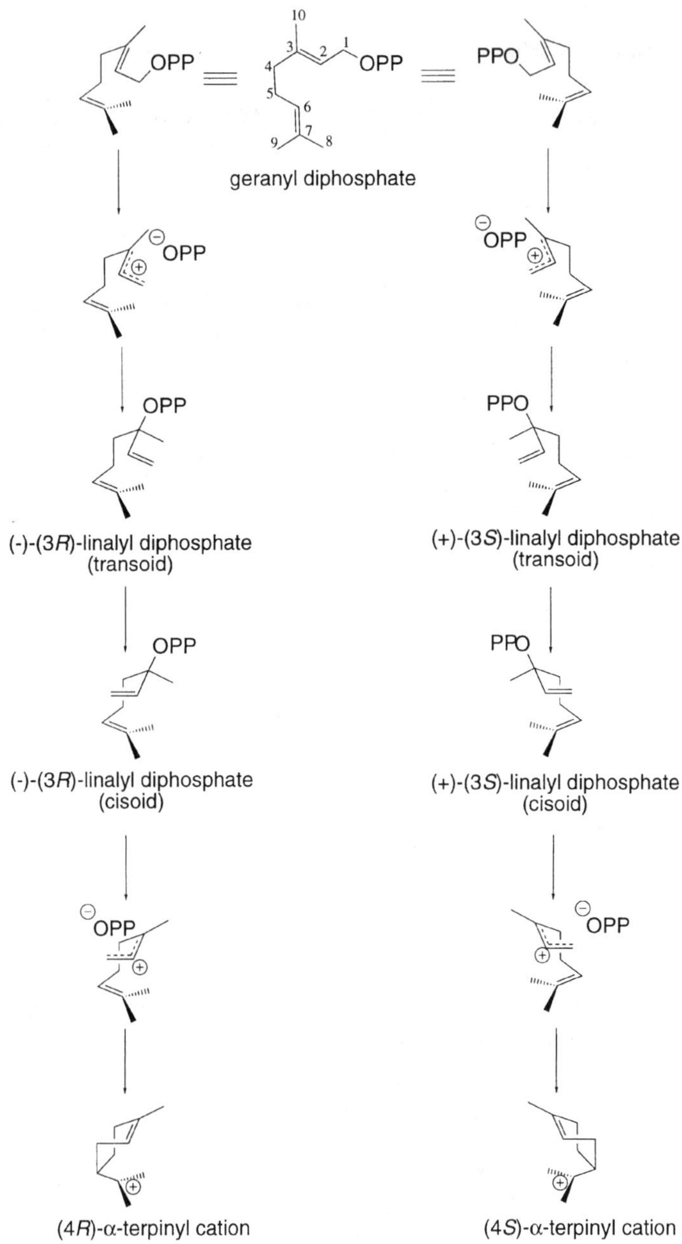
OPP
OPP
PPO
geranyl diphosphate
OPP
OPP
OPP
PPO
(-)-(3R)-linalyl diphosphate (transoid)
(+)-(3S)-linalyl diphosphate (transoid)
OPP
PPO
(-)-(3R)-linalyl diphosphate (cisoid)
(+)-(3S)-linalyl diphosphate (cisoid)
OPP
OPP
(4R)-α-terpinyl cation
(4S)-α-terpinyl cation

Scheme 4

as an alternate substrate, effectively uncoupling the normally linked cyclization from the isomerization step [15, 43–50]. Stereochemical deductions from the utilization of (3*R*)- or (3*S*)-LPP as an alternate substrate for cyclization are described below. Kinetic analyses show that V_{rel} and the catalytic efficiency (V_{rel}/K_m) for LPP are substantially higher than for GPP or the *cis*-isomer neryl diphosphate [30, 31, 43, 48–50]. These results support the intermediacy of LPP and implicate the isomerization component of the reaction as the slow step of the coupled sequence. The absence of detectable free intermediates, and the lack of exchange between LPP generated at the active site and that supplied exogenously, confirm that the isomerization and cyclization are tightly coupled [31, 44, 47, 48]. The mutually competitive nature of LPP and GPP as alternate substrates [30, 42, 48, 51] confirms that the two steps occur at the same active site. Although the isomerization step of the coupled sequence has never been directly observed, indirect evidence for the required isomerization has been obtained using the non-cyclizable substrate analogs 6,7-dihydro-GPP [52] and 2,3-cyclopropyl(methano)-GPP (Fig. 1) [53]. In the former instance, the

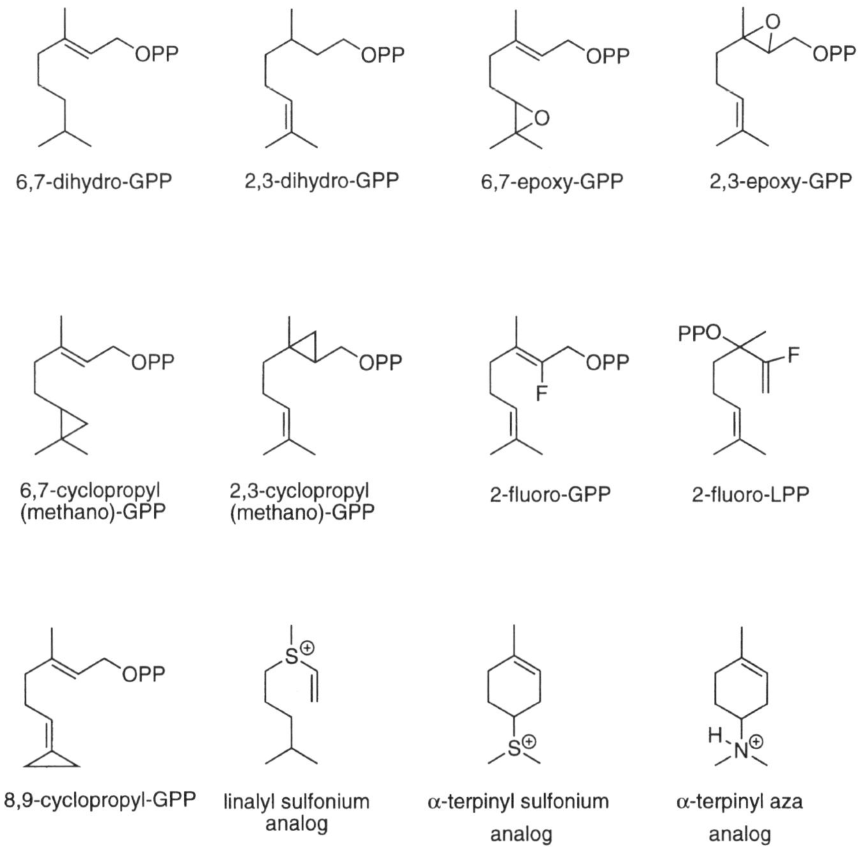

Fig. 1. Selected monoterpene synthase substrate analogs

enzymatic products generated were shown to resemble more closely the solvolytic products of 6,7-dihydro-LPP rather than those derived from 6,7-dihydro-GPP with the latter substrate, the analog of LPP, i.e., the homoallylic diphosphate ester, could be detected as an aborted reaction product consistent with the isomerization step.

With the clear indication that the isomerization of GPP was slower than the cyclization of LPP, evidence for the ionization steps of these reaction components was sought using the competitive inhibitors 2-fluoro-GPP and 2-fluoro-LPP (Fig. 1) [54]. A similar approach employing substrate analogs bearing electron-withdrawing substituents has been used to examine the ionization steps of prenyltransferase [55] and sesquiterpene cyclase catalysis [56]. With the monoterpene cyclases, significant rate suppressions were observed for the enzymatic reactions with each analog that paralleled the rate suppressions of model solvolytic (but not S_N2 displacement) reactions [57]. These results indicated that both the isomerization of GPP and the cyclization of LPP are initiated by ionization of the primary and tertiary allylic diphosphates, respectively, and furthermore implicated the ionization of GPP as the rate-limiting step of the overall, coupled reaction sequence.

The absolute dependency of the monoterpene cyclases on the divalent metal ion prompted inquiries into the role of this cofactor in the reaction. Solvolytic decomposition of GPP occurs under acidic conditions [58], or at neutrality in the presence of divalent cation [59, 60], indicating that neutralization of the negatively charged diphosphate is required to promote ionization. The divalent cation:GPP stoichiometry in solution is 2:1 [59, 60] suggesting that the bis-metal ion:diphosphate complex is the probable leaving group in enzyme catalysis. The crystal structures of FPP synthase [61] and two sesquiterpene synthases [62, 63] clearly identify at least two Mg^{2+} coordination sites in the enzyme-substrate complex, consistent with the divalent cation-assisted ionization of the substrate. A monovalent cation requirement is observed for a range of enzymes [64], and a mechanistic role for the monovalent cation in gymnosperm monoterpene synthase catalysis has been suggested [22] based on the possible similarity in function to the heat-shock cognates which utilize the cation to assist in phosphate ester hydrolysis [65, 66].

Additional evidence for the electrophilic nature of the monoterpene synthase reaction is provided by the mechanism-based inhibitor 8,9-cyclopropyl-GPP (Fig. 1). Metal ion-dependent isomerization and cyclization to the stabilized analog of the α-terpinyl cation promotes alkylation and inactivation of several synthases [67]. By using the inhibitor in labeled form, active site residues may be identified [68]. Substrate protection of monoterpene synthases against inactivation by classical amino acid-directed reactions provides another approach to the identification of active site residues. By this means, essential cysteines, histidines, and arginines were identified in all monoterpene cyclases examined, although substrate protection is afforded only to histidine and cysteine residues in the angiosperm cyclases [25, 69], and only to arginine and cysteine residues in the gymnosperm enzymes [23, 70].

Studies aimed at evaluating the determinants of GPP binding at the active site were conducted with several bicyclic monoterpene synthases using analogs which differed minimally in structural features of the substrate (Fig. 1) [71]. To uncover the nature of the interactions of the C2-C3 and C6-C7 olefinic domains, inhibition constants were determined for the non-cyclizable analogs of GPP in which each double bond was reduced or converted to the epoxide derivative. The results indicated that binding of the C2-C3 double bond is largely a consequence of geometry, whereas it is the electronic nature of the C6-C7 double bond that influences binding [71]. Inhibition of cyclases by pyrophosphate analogs demonstrated that three key features are required to promote tight binding of the diphosphate ester substrate: the ionization state at the enzymatic pH optimum, the structural flexibility of the diphosphate, and the ability to chelate divalent metal ion [72]. The inability of monoterpene cyclases to utilize geraniol and geranyl monophosphate, coupled to the inhibition of cyclization of GPP by these analogs in the presence of inorganic diphosphate, clearly establish the diphosphate moiety of the substrate as the primary binding determinant [72–75].

The aza and sulfonium analogs of the linalyl and α-terpinyl carbocation intermediates of the isomerization-cyclization reaction (Fig. 1) were utilized to determine the nature of active site interactions with several monoterpene synthases [49, 75, 76]. Synergistic inhibition with each analog in the presence of inorganic diphosphate was demonstrated, suggesting ion-pairing of the diphosphate anion with various carbocationic intermediates generated in the course of the reaction sequence.

The configuration and conformation of intermediates in the isomerization-cyclization reaction can be determined by establishing the stereochemical alterations at C1 and C3 of the geranyl substrate and linalyl intermediate in the course of the coupled transformation. Several bicyclic monoterpene synthases were chosen for this purpose, since the hydrogens at C1 of the labeled precursor can be unambiguously located by conversion of the product to a ketone followed by stereoselective, base-catalyzed exchange, for example in the conversion of (+)-bornyl diphosphate to (+)-camphor (Scheme 5) [43, 77]. Retention of configuration at C1 in the isomerization and cyclization of 1*R*- and 1*S*-labeled GPP was observed for each cyclase, consistent with the predicted *syn*-migration of the diphosphate moiety from C1 to C3 followed by rotation about C2-C3 of the linalyl intermediate, and *anti*-C1-C6 cyclization (i.e., carbon-carbon bond formation to the same face of C1 from which the diphosphate departed). Retention of configuration at C1 of GPP is universally observed among the monoterpene synthases [15], and additionally eliminates the *cis*-isomer neryl diphosphate as a possible reaction intermediate. A consequence of the *syn* diphosphate migration is that the 1pro*R*- and 1pro*S*-hydrogens of GPP will become the 1*Z*-hydrogen of (3*R*)-LPP and (3*S*)-LPP, respectively. This prediction was confirmed in a set of experiments analogous to those described above but utilizing (3*R*)- and (3*S*)-[1*Z*-^{3}H]-LPP as substrates [50, 77, 78].

The ability of the monoterpene synthases to utilize LPP as an alternate substrate allows ready determination of the stereochemistry at C3 resulting from the isomerization component of the coupled reaction sequence. Thus, depletion of the preferred enantiomer of racemic LPP can be monitored during the reac-

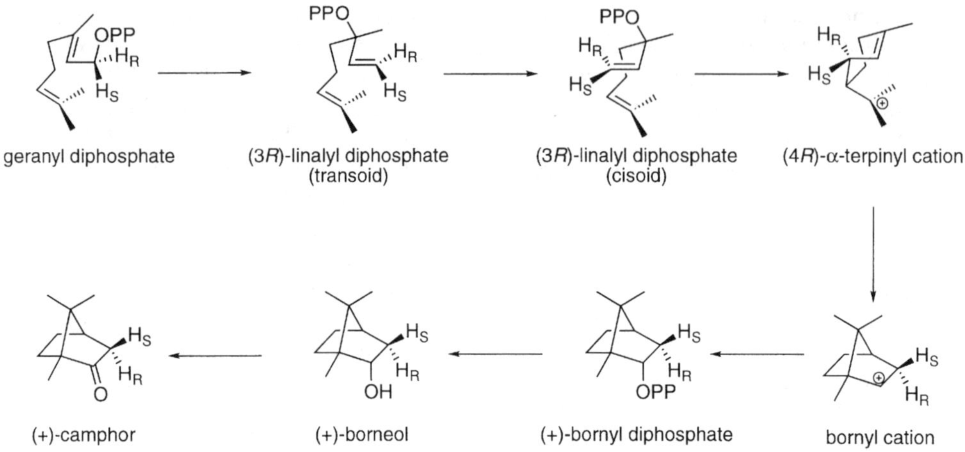

Scheme 5

tion, or the enantiomerically pure antipodes of LPP can be tested directly by comparison of relative velocities, the K_m values and any alteration of product outcome. By this means, it was established that the preferential conversion of (3*R*)- and (3*S*)-LPP by a given cyclase to the stereochemically appropriate products was consistent with a *syn* isomerization-*anti*, *endo*-cyclization sequence (Scheme 4). Although all monoterpene synthases examined to date appear to be completely stereospecific in the transformation of GPP, and highly stereoselective in the utilization of LPP, most lack complete enantio-discrimination in generating antipodal products from the unnatural LPP enantiomer [50, 77, 78]. In spite of the inefficient cyclization of the abnormal antipode, relative to the preferred enantiomer of LPP, or to GPP, these reactions are largely in accord with the proposed stereochemical model (Scheme 4). Nevertheless, an unusually high proportion of aberrant products have been observed from these unnatural cyclizations which are consistent with the suggestion that diphosphate binding and ionization occur before the proper *anti,endo*-positioning of the olefinic chain, such that cyclization occurs from the extended (*exo*) conformation from which the opposite configuration at C4 of the α-terpinyl cation results [15]. Thus, a preassociation mechanism has been proposed in which the binding and ionization of the substrate diphosphate are slow enough to allow appropriate positioning of the olefinic chain before the reaction is initiated.

3.2
Limonene Synthase

Interest in limonene synthase stems in part from the fact that (–)-limonene is the common precursor of menthol and carvone, respectively, of the essential oils of peppermint and spearmint species (Scheme 6) [79]. The cyclization leading to limonene is the simplest of all terpenoid cyclizations and the reaction has ample precedent in solvolytic model studies [58, 59, 80]. Thus, it is not

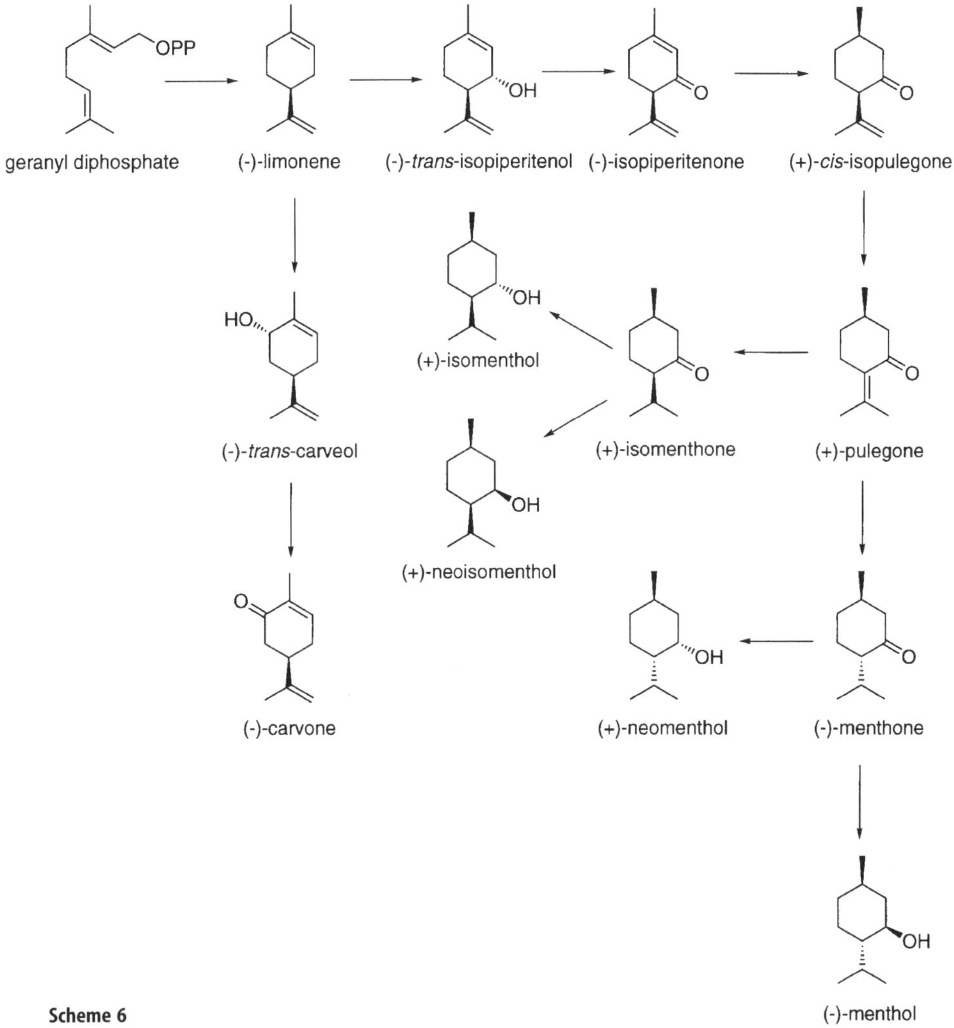

Scheme 6

surprising that limonene synthase has become an archetype for this class of enzyme [18]. The overall stereochemistry of the cyclization to limonene has been examined using 1*R*- and 1*S*-^{2}H labeled GPP as substrates with the (4*S*)-(−)-limonene synthase from spearmint and the (4*R*)-(+)-limonene synthase from *Citrus unshiu* [81]. NMR analysis of the derived olefins to locate the position of deuterium indicated a process involving *syn*-migration of the diphosphate to the intermediate (3*S*)-LPP in the first instance and to (3*R*)-LPP in the second, with *anti,endo*-cyclization in each case, to yield the respective (4*S*)-(−)- and (4*R*)-(+)-antipodes of limonene [81]. Related experiments with isotopically labeled GPP established that both the (+)- and (−)-limonene synthases from *Citrus* and mint regiospecifically eliminate a proton from the same methyl group of the (4*R*)- and (4*S*)-terpinyl cation intermediate (corresponding to the *Z*-methyl group of LPP) in the formation of the methylene group of limonene [82,83]. These results confirmed the same conclusion derived by natural abundance deuterium NMR analysis of (+)-limonene from *Citrus limonum* [84], and together indicate the same relative positioning, with respect to the substrate diphosphate moiety, of the enzyme base involved in the terminating deprotonation step of these antipodal reactions.

(4*S*)-(−)-Limonene synthase was first purified from mint species [85] following the preliminary isolation of the leaf oil glands as a highly enriched starting material [86]. The enzyme has been thoroughly characterized and exhibits properties typical of those of other terpenoid synthases from angiosperms [25,85]. The native enzyme is *N*-terminally blocked; however, internal microsequencing provided sufficient amino acid sequence information to permit an oligonucleotide-based screen of a leaf library from which the corresponding cDNA was isolated. The recombinant enzyme, heterologously expressed in *E. coli*, and the native enzyme from both peppermint and spearmint produce the same olefin mixture from GPP comprised of (4*S*)-(−)-limonene as the principal product (95%), with roughly equal amounts of myrcene, (−)-α-pinene, and (−)-β-pinene (Scheme 7) [33]. The reaction mechanism can be rationalized by postulating the direct deprotonation of an isopropyl methyl of the α-terpinyl cation to yield limonene for most reaction cycles. Approximately 4% of the time, internal addition of the α-terpinyl cation proceeds to the pinyl cation from which two alternate deprotonations of equal probability generate (−)-α-pinene and (−)-β-pinene. Roughly 2% of the reaction cycles result in the failure of either the geranyl or linalyl cation to proceed to cyclization, leading to either one or both of these acyclic species suffering direct deprotonation to myrcene.

cDNAs encoding (−)-limonene synthase have been subsequently isolated from other plants. Comparison of deduced amino-acid sequence for the enzyme from the closely related species *Mentha candicans* and *Perilla frutescens* [35] shows greater than 98% and 70% identity, respectively, whereas the (−)-limonene synthase from the phylogenetically-distant gymnosperm grand fir exhibits only 31% sequence identity at the amino acid level when compared to the enzyme from spearmint [22]. Although the data set is still limited, angiosperm monoterpene synthases exhibit greater than 50% sequence identity among themselves while within the gymnosperm synthases, even though of more ancient lineage, even higher levels of sequence identity are found [21].

Consistent with the plastidial location of monoterpene biosynthesis [8], immunocytochemical studies with limonene synthase confirmed the localization of this enzyme to the leucoplasts of peppermint oil glands [87]. Plastid targeting requires the translation of a preprotein bearing an *N*-terminal transit peptide that directs the newly synthesized enzyme to the plastid for proteolytic processing to the mature form [88]. The limonene synthase cDNA encodes such a preprotein, with an *N*-terminal sequence that exhibits the typical properties of a transit peptide (rich in serine and threonine (25–30%), rich in small hydrophobic amino acids with few acidic residues, and a propensity to form amphiphilic helices [89,90]). A precise cleavage site between the transit peptide and mature protein is not obvious in the limonene synthase preprotein, and mass spectral

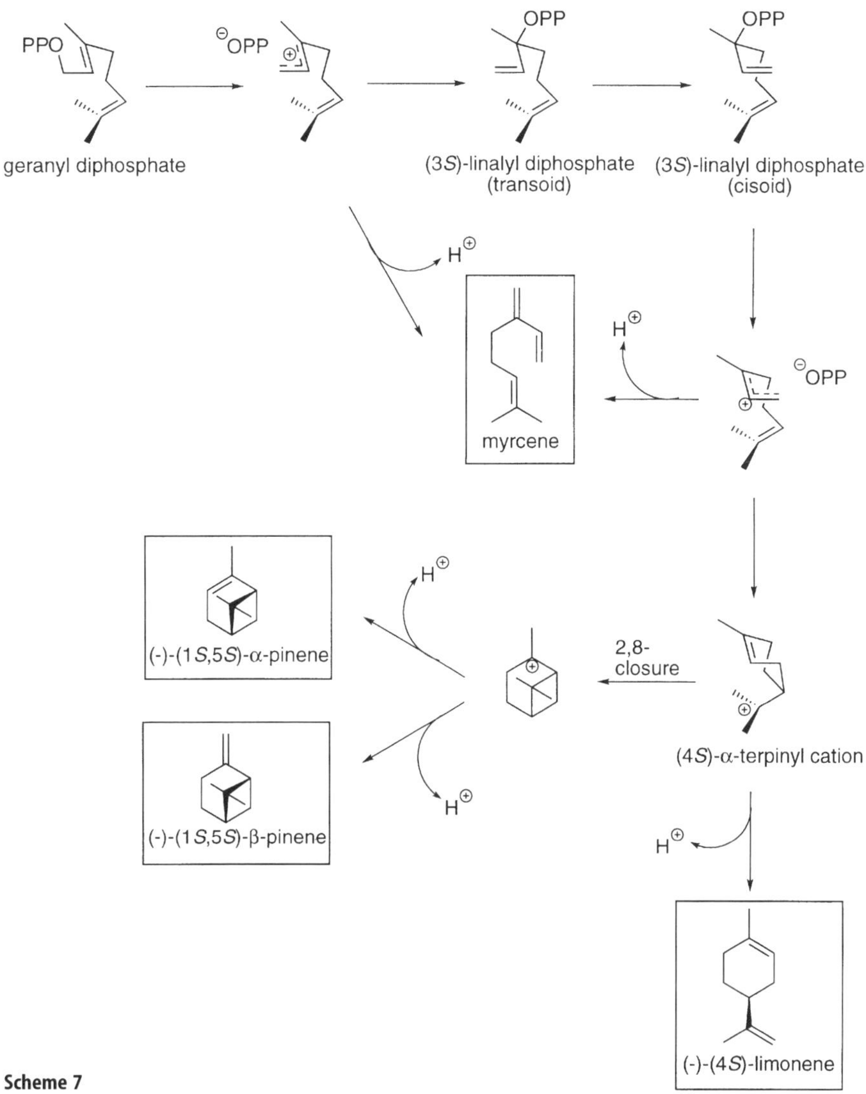

Scheme 7

analysis of the native enzyme, as well as plastid import studies [88], indicate that proteolytic processing is imprecise and *N*-terminal modification is variable, such that a family of enzymes of 65 ± 0.8 kDa are produced.

A series of *N*-terminal truncations of limonene synthase focused on a tandem pair of arginine residues (R58R59) that are conserved in all monoterpene synthases, and revealed that truncation immediately upstream of this element yielded a highly competent "pseudomature" form of the enzyme [91]. Truncation at or beyond the first arginine residue or substitution of the second arginine residue yielded enzymes completely incapable of utilizing GPP. Interestingly, these constructs (including a truncation at S89) yielded enzymes still capable of cyclizing (3*S*)-LPP with kinetics comparable to the unmodified form, suggesting a critical role for the R58R59 pair in the ionization and isomerization of GPP to (3*S*)-LPP [91]. The localization of the isomerization step toward the amino terminus and the cyclization active site toward the carboxy terminus (see below) may allow dissection of limonene synthase into separable isomerization and cyclization components. The conservation of the arginine pair among the monoterpene cyclases and the plant IPP isomerases is intriguing [92], since only in the former instance is the migration of the diphosphate required as a reaction step. The conservation of the first arginine of the motif in all terpene synthases as well as in prenyltransferases, in which ionization but not isomerization is required, suggests a role for this residue in the diphosphate ester ionization step.

Substrate protection against inactivation by amino acid-specific modifying reagents have implicated histidine [69], cysteine [25], and methionine [25, 93] at or near the active site of limonene synthase. Consistent with other angiosperm monoterpene synthases, arginine-directed reagents inhibit limonene synthase activity but the enzyme is not protected by substrate [70]. This interesting observation implies that the *N*-terminal arginine pair of the monoterpene synthases, that is required for the isomerization step, must remain exposed on the surface of the active site during catalysis. While such substrate protection studies have suggested a number of residues potentially involved in substrate binding and catalysis, the cloning of (–)-limonene synthase permits sequence alignment with related enzymes from which highly conserved residues can be identified as targets for testing of function by site-directed mutagenesis.

The absolute conservation of a DDxxD motif among the terpene cyclases and the distantly related prenyltransferases suggests a common role for this element in binding the common prenyl diphosphate substrates. The prenyltransferases in fact bear two such elements, domain I for binding the allylic diphosphate cosubstrate and domain II for binding the homoallylic (isopentenyl) diphosphate substrate [61, 94]. The crystal structures of avian FPP synthase bound to the allylic substrates [61, 94], the sesquiterpene cyclase 5-*epi*-aristolochene synthase from tobacco with bound farnesyl substrate analog [62], and a fungal sesquiterpene synthase pentalenene synthase [63] all identify divalent metal ion bridges between the first and third aspartate carboxylates and the allylic substrate diphosphate, clearly implicating coordination of the metal ion chelated diphosphate moiety and assistance in the critical ionization step of the reaction. Consistent with this catalytic role, FPP synthase (yeast enzyme) is intolerant of alanine for aspartate substitutions [95], and glutamate for aspartate replacements in rat FPP synthase had little effect on Michaelis constants for the allylic and homoallylic substrates but depressed k_{cat} values were observed for substitutions at the second and third aspartates of domain I (for allylic diphosphate binding) [94–96]. Glutamate substitution at each position of the DDxxD motif in the fungal sesquiterpene cyclase, trichodiene synthase, yielded mutant enzymes only marginally altered in kinetic constants; however, product distributions were modified, including generation of a novel sesquiterpene [97–99]. Similar conservative substitutions in the DDxxD motif in a plant diterpene cyclase, casbene synthase, also revealed minimal kinetic influence in the mutant enzymes [100].

Since the role of substrate binding and ionization is different for the monoterpene synthases, because of the requisite migration of the diphosphate from C1 to C3 in the isomerization of GPP to LPP, site-directed mutants of the conserved aspartates of (4*S*)-(–)-limonene synthase were examined. Each aspartate of the DDxxD motif (residues 352–356) of the pseudomature form of the mint (4*S*)-(–)-limonene synthase (R58 truncation) was changed to alanine and glutamate, and the mutant enzymes were expressed in *E. coli*, purified and evaluated for kinetic constants and product distribution. Four of the six mutants generated (D352A, D352E, D353E, and D356A) were so severely compromised in activity that neither K_m nor k_{cat} could be measured. Thus, although it was clear that these aspartate residues are important, their involvement in binding, catalysis, product release, or indirect structural effects could not be determined. While the K_m value for GPP was unchanged for the D353A mutant, a k_{cat} value some three orders of magnitude lower than the wild-type pseudomature enzyme suggests a primary role in catalysis. Since the D353E mutant did not turnover substrate detectably, it would appear that the lack of a carboxyl function at position 353 (D353A) is of less significance than its misplacement (D353E). Mutant synthase D356E also supported catalysis but at levels too low for adequate kinetic determination. Nevertheless, a normal product distribution of olefins indicates that, although catalysis was seriously impaired, substrate positioning at the active site that directs product outcome, had not been influenced.

The nearly complete catalytic intolerance for glutamyl and alanyl substitutions in the DDxxD motif of limonene synthase is novel and unlike the much less pronounced effects of comparable substitutions in the sesquiterpene cyclase trichodiene synthase [97, 98]. However, pre-steady state kinetic analysis of trichodiene synthase [101] and several other sesquiterpene synthases [102] has recently shown that product release is rate limiting in these cases, and thus can mask the kinetic influence of the aspartate mutations on earlier steps in the catalytic cycle. In the instance of monoterpene cyclase catalysis, product release is not the slow step since comparison of k_{cat} values with GPP and LPP as substrate clearly reveals the initial ionization-isomerization to be rate limiting. Thus, perturbations that influence the first ionization step will be fully reflected in overall rate suppression for limonene synthase. This kinetic sensitivity at the initial steps of the reaction cycle does not, however, explain the near complete intolerance of limonene synthase to aspartate substitution in the DDxxD motif and it is thus tempting to speculate a more specific, but presently unidentified, influence on the requisite isomerization of GPP.

Evaluation of the aforementioned crystal structures [61–63, 94], has revealed that the second aspartate of the DDxxD motifs of these sesquiterpene synthases and prenyltransferase (corresponding to D353 of (–)-limonene synthase) is positioned such that the carboxylate is oriented away from the substrate diphosphate and may form a salt bridge with a spatially adjacent, conserved arginine. This arginine corresponds to R315 in limonene synthase, and is the first of a highly conserved RxR element found about 35 amino acids upstream of the DDxxD motif of most terpene cyclases. Together, the aspartate and arginine pair are thought to assist in directing the diphosphate anion away from the reacting carbocation formed upon ionization [61–63]. It seems plausible that the monoterpene synthases utilize the aspartate-arginine pair in the secondary ionization of LPP to promote cyclization, whereas the initial ionization and isomerization of GPP involve the amino terminal tandem arginine pair.

4 Sesquiterpene Synthases

The diversity of cyclic sesquiterpene skeletal types is greater than the monoterpene types due to the presence in FPP of a third double bond and the longer, more flexible chain. The cyclizations of FPP (Scheme 8) may occur from the distal double bond to generate either the ten- (*E,E*-germacradienyl cation) or eleven- (*trans*-humulyl cation) membered ring but topological constraints prevent cyclization from the central double bond without prior isomerization of FPP to cisoid nerolidyl diphosphate (NPP), thereby permitting the formation of the six- (bisabolyl cation) or seven- (cycloheptanyl cation), as well as ten- (*Z,E*-germacradienyl cation) and eleven- (*cis*-humulyl cation) membered rings [16]. cDNAs encoding sesquiterpene cyclases that catalyze both the direct cyclization of FPP and the coupled isomerization-cyclization have been obtained (Scheme 9). It is apparent from the skeletal diversity of this class that further cyclizations, hydride shifts, methyl migrations, and/or Wagner-Meerwein rearrangements may occur following the generation of the initial cyclic carbocation, and prior to termination of the reaction by deprotonation or nucleophile capture. Comparison of the deduced amino-acid sequences of the cloned plant sesquiterpene cyclases indicates significant conservation among them, with the highest levels of homology based upon taxonomic relationships rather than mechanistic considerations [21]. Consistent with the cytosolic localization of sesquiterpene biosynthesis, none of these sesquiterpene cyclases are encoded with a transit peptide. While the majority of sesquiterpene synthases generate multiple products, presumably as a consequence of the premature quenching of the highly unstable carbocationic intermediates comprising the reaction cascade, several of these enzymes display considerable fidelity in generating one principal product. Enzyme structural differences leading to multiproduct formation, or to a single product, are considered below.

In general properties, the sesquiterpene cyclases prefer Mg^{2+} to Mn^{2+} as cofactor, with Mn^{2+} being inhibitory at higher concentrations [103–109]; they exhibit no K^{+} requirement regardless of taxonomic origin, in contrast to this distinction between the gymnosperm and angiosperm monoterpene cyclases. Contrary to the monoterpene synthases, few studies with amino acid-directed reagents have been carried out with the sesquiterpene synthases; however, early work with humulene synthase from *Salvia officianalis* [107] identified the presence of substrate-protectable histidine and cysteine residues, and essential (but not protectable) arginine residues, reminiscent of the results with angiosperm monoterpene cyclases.

A refined model for sesquiterpene synthase catalysis [16, 19, 110] proposes ionization of the diphosphate and, where necessary, *syn* migration of the diphosphate to C3 allowing C2-C3 bond rotation and reionization of the tran-

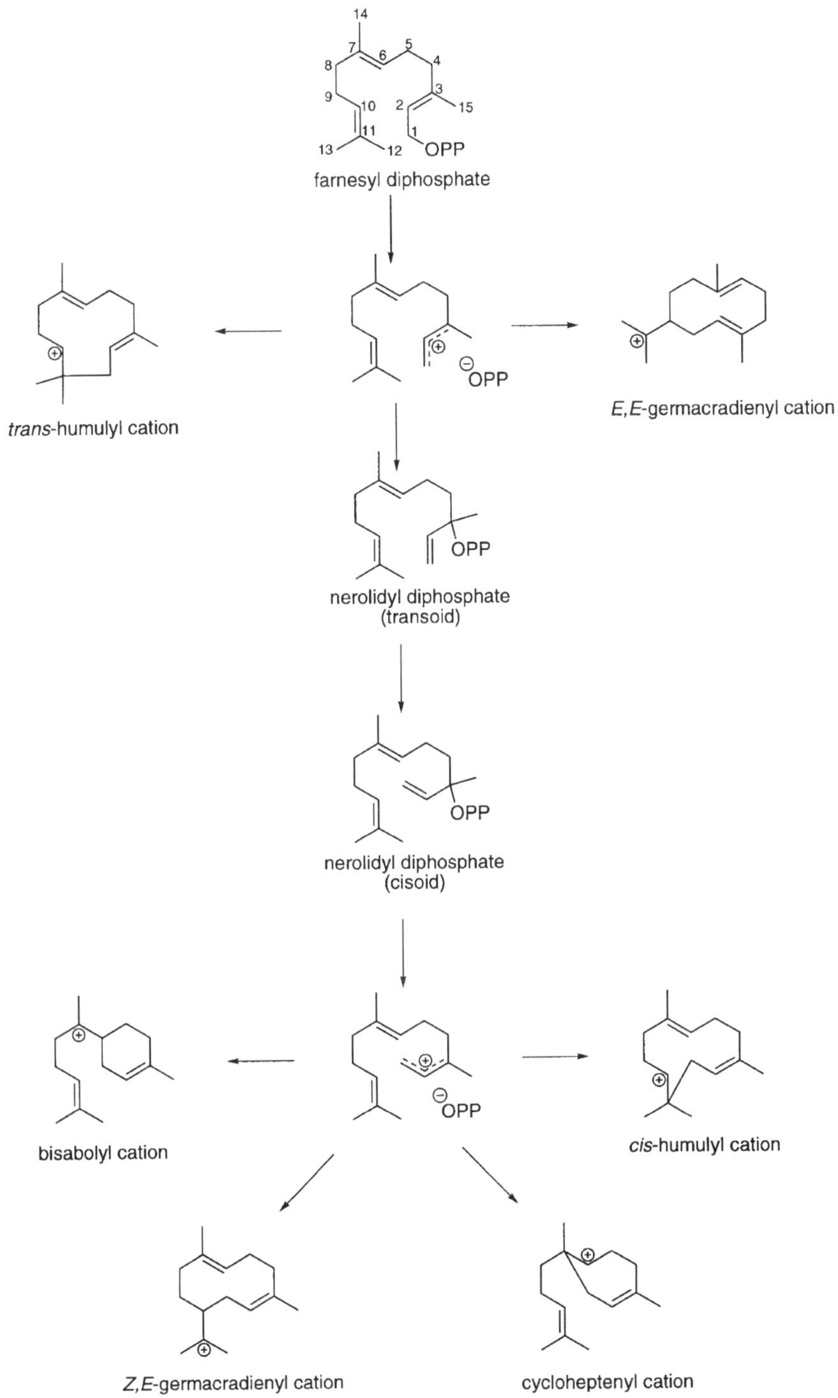
14
7
6
5
4
8
9
10
2
3
15
11
1
13
12
OPP
farnesyl diphosphate
OPP
trans-humulyl cation
E,E-germacradienyl cation
OPP
nerolidyl diphosphate (transoid)
OPP
nerolidyl diphosphate (cisoid)
OPP
bisabolyl cation
cis-humulyl cation
Z,E-germacradienyl cation
cycloheptenyl cation

Scheme 8

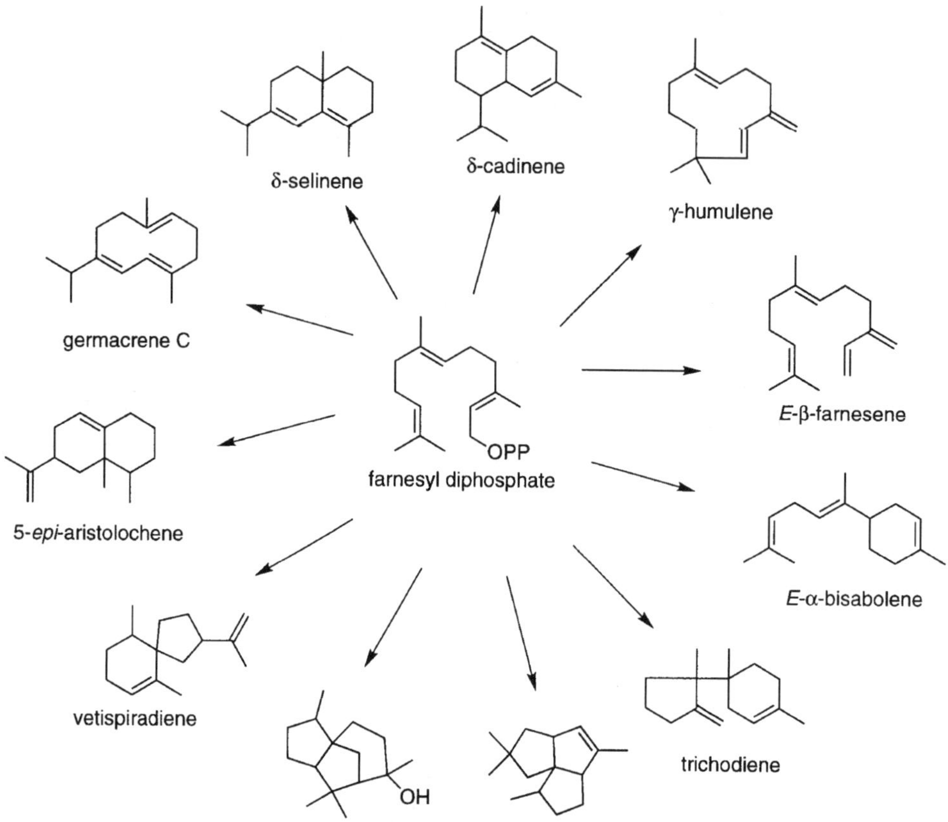

Scheme 9

sient NPP intermediate thus formed, followed by an *anti* intramolecular attack of the central or distal double bond on C1 to generate the first cyclic carbocation which may react further. As with the monoterpene synthases, the stereochemical alteration at C1 of the substrate can be used to determine whether the reaction proceeds via direct cyclization of FPP (resulting in inversion of configuration) or via the intermediacy of NPP (resulting in retention of configuration). Several microbial sesquiterpene cyclization reactions that were predicted to proceed via the bisabolyl cation were shown to occur with retention of configuration at C1 of FPP [111–113], whereas cyclizations thought to proceed via the *E*,*E*-germacradienyl cation and humulyl cation were shown to occur with inversion of configuration at C1 of the substrate [114–116]. These results are entirely consistent with the proposed model for sesquiterpene cyclization.

Although the primary structures of the microbial sesquiterpene cyclases share little with their plant counterparts, similar tertiary structures, common electrophilic reaction mechanisms, and conservation of the DDxxD substrate binding motif make it instructive to describe trichodiene synthase from *Fusarium sporotrichioides*, about which considerable information is available.

The recombinant enzyme has been functionally expressed to greatly facilitate the understanding of structure-function relationships. This cyclase catalyzes initial *syn* isomerization of FPP to NPP, with rotation about the C2-C3 bond and anti-allylic attack leading to C6-C1 bond closure to generate the bisabolyl cation. This sequence is followed by C11-C7 closure, a 1,4-hydride shift, two suc-

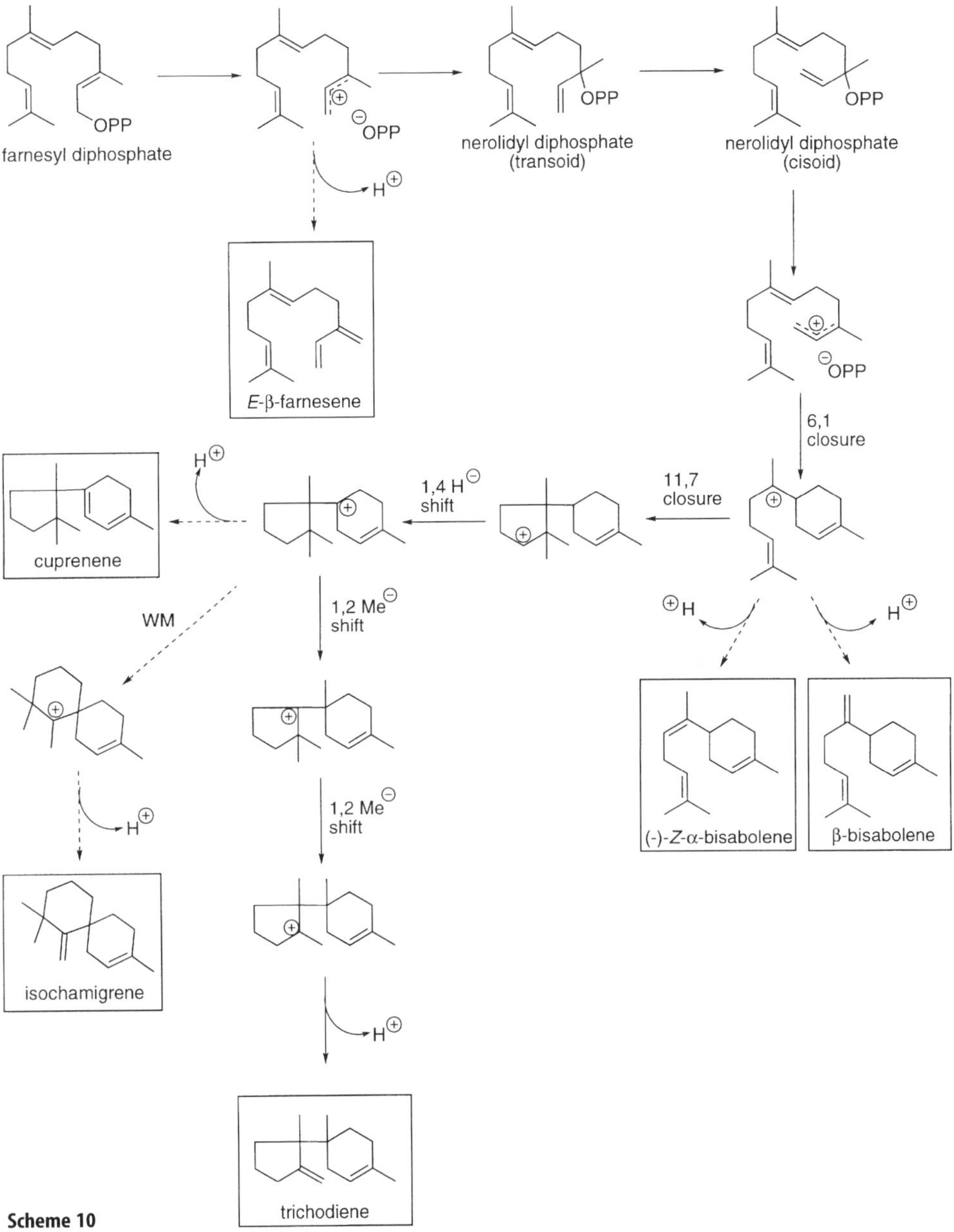

Scheme 10

cessive 1,2-methyl migrations, and a final deprotonation step (Scheme 10, solid arrows) [19].

The role of the DDxxD element of trichodiene synthase in coordinating the divalent cation and in positioning the substrate via the diphosphate moiety was examined for a series of aspartate to glutamate substitutions. Each of these replacements corrupted this normally single product cyclase into a multiproduct enzyme which additionally generated (–)-Z-α-bisabolene, β-bisabolene, E-β-farnesene, cuprenene, and isochamigrene (Scheme 10, dashed arrows) [98]. The aberrant products generated by these site-directed mutants can be formed by alternate deprotonations of normal cationic intermediates of the reaction cascade with the exception of isochamigrene, which requires a ring expansion via a Wagner-Meerwein rearrangement prior to the terminating deprotonation [98]. These results suggest that changes in the binding orientation of the substrate diphosphate moiety affect positioning of the olefinic chain and can thereby alter product outcome.

The participation of arginine residues in the ionization of the substrate diphosphate ester is likely for both plant and fungal terpene cyclases, and mutagenesis of a conserved DRRYR motif (residues 302–306 in trichodiene synthase) identified critical functions for R304 and Y305 in maintaining product fidelity of this enzyme [117]. Again, multiproduct formation via alternate deprotonations of cationic intermediates suggested a role for these residues in positioning the diphosphate moiety [117] and further implicated the importance of substrate orientation in determining product disposition.

The role of cyclase structural elements in defining product outcome was examined using chimeras formed from two plant sesquiterpene cyclases, 5-*epi*-aristolochene synthase from tobacco (Nt-EAS) and vetispiradiene synthase from *Hyoscyamus muticus* (Hm-VS) [118]. These phylogenetically-related cyclases catalyze several common reaction steps leading to their respective products (Scheme 11), including the transient formation of enzyme-bound germacrene A via deprotonation of the *E,E*-germacradienyl cation, reprotonation at C6 of germacrene A, C2-C7 closure, and a 1,2-hydride shift. The resulting eudesmyl cation formed by this shared reaction sequence may undergo a 1,2-methyl migration or a Wagner-Meerwein rearrangement leading to 5-*epi*-aristolochene or vetispiradiene, respectively. Three critical regions were identified from domain-swapping experiments, a Nt-EAS-specific domain (residues 261–379), an Hm-VS-specific domain (residues 379–442), and a ratio-determining domain (residues 342–379) which clearly established that product specificity is determined by the carboxyl terminal region of the protein [118]. A comparison of the active site cavity of Nt-EAS, deduced from the crystal structure, with a model of the Hm-VS active site identified highly conserved residues lining the substrate binding cavity. The domain swapping experiments therefore led to the hypothesis that product specificity is determined by the layers surrounding the active site rather than by the highly conserved residues which line the active site itself [62], an intriguing hypothesis which has not yet been validated with other terpenoid synthases.

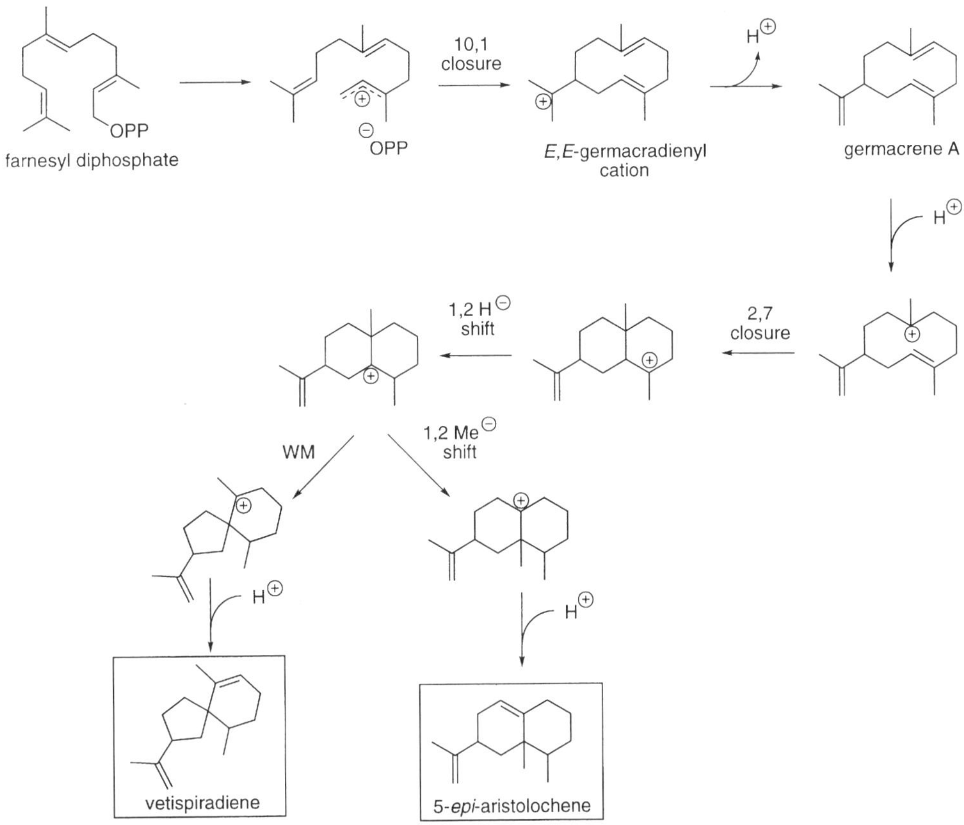

Scheme 11

4.1 Γ-Humulene Synthase and δ-Selinene Synthase

A similarity-based PCR cloning approach [119] led to the isolation and identification of two cDNA clones from grand fir encoding δ-selinene synthase and γ-humulene synthase [120]. Heterologous overexpression of these cDNA clones produced catalytically active enzymes capable of generating an extraordinary array of olefins from FPP, the product compositions of which are consistent with the mixture of sesquiterpenes found in grand fir primary oleoresin [120]. The deduced amino acid sequences of these two cyclases are 65% identical and 85% similar yet each makes a complex mixture of mostly unique products, probably resulting from subtle differences in substrate binding orientations. The recombinant proteins possess properties typical of other sesquiterpene cyclases, although γ-humulene synthase has an exceptional divalent cation requirement in showing no apparent preference for Mg^{2+} or Mn^{2+}.

δ-Selinene synthase and γ-humulene synthase generate 34 and 52 sesquiterpene olefin products, respectively, with roughly half of these of unidentified structures corresponding to less than 20% (δ-selinene synthase) and less than 5% (γ-humulene synthase) of the total mixture [120]. Mechanistic schemes to rationalize multiple product formation by each enzyme were formulated (Schemes 12 and 13; these schemes also indicate relative product abundances) [120], yet it is important to emphasize that each turnover cycle results in a single product. For δ-selinene synthase, greater than 90% of the products are derived by ionization and cyclization of FPP to yield the *E,E*-germacradienyl cation, while for γ-humulene synthase, all of the products can be rationalized as arising via ionization and isomerization of FPP to NPP followed by cyclization, most often to the *cis*-humulyl or *Z,E*-germacradienyl cations.

Alternate deprotonations of the *E,E*-germacradienyl cation by δ-selinene synthase (Scheme 12), prior to or following a 1,3-hydride shift, generate germacrene A, *E,E*-germacrene B, or germacrenes C and D, respectively, and each of these olefins can be released as products or further protonated leading to bicyclic skeletons. Protonations at C6 followed by C2-C7 closure lead to α-selinene (from germacrene A), selina-3,7(11)-diene (from *E,E*-germacrene B), siberene (from germacrene D), and the principal product δ-selinene (from germacrene C). Protonations at C3 promote C6-C2 closure leading to guaia-6,9-diene (from germacrene C) or, following a 1,2-hydride shift, to α-guaiene (from germacrene A). Protonation of germacrene D at C15 promotes C6-C1 closure to generate the cadinyl cation which may undergo alternate deprotonations to yield the diastereomers, α-amorphene and α-cadinene, or δ-amorphene and δ-cadinene, or, following C2-C7 closure from the *si* or *re* face of C2, α-ylangene and α-copaene, respectively. Generation of the humulyl cation leads to α-humulene or, after C2-C10 closure, β-caryophyllene. The only products of δ-selinene synthase that require preliminary isomerization to NPP are *Z,E*-germacrene B and β-bisabolene which result from deprotonation of the *Z,E*-germacradienyl cation and bisabolyl cation, respectively.

The primary product of γ-humulene synthase (Scheme 13) results from C11-C1 closure of ionized NPP, followed by a 1,3-hydride shift and deprotonation. C6-C1 closure of the humulyl C1 cation generates the himachalyl cation, with alternate deprotonations affording α-, β-, and γ-himachalene. Further cyclizations of the himachalyl cation result in tricyclic olefin formation; C3-C7 closure and a Wagner-Meerwein rearrangement lead to longifolene and longicyclene, whereas C2-C7 closure leads to α- and β-longipinene. Deprotonation of the *Z,E*-germacradienyl cation yields *Z,E*-germacrene B or, following a 1,3-hydride shift and alternate deprotonations, germacrene D or bicyclogermacrene. Each of these enzyme-bound olefinic intermediates is protonated at C6, followed by either C2-C7 closure leading to siberene (from germacrene D) or, after a 1,2-hydride shift and 1,2-methyl migration, to β-gurjunene (from bicyclogermacrene). Generation of α- and δ-amorphene via C15 protonation of germacrene D is the same as for δ-selinene synthase, although the intermediacy of this olefin is not obligate as C6-C1 closure of the secondary germacradienyl cation also affords the cadinyl cation. C2-C7 closure from the *si* or *re* faces of C2 of the cadinyl cation leads to the tricyclic diastereomers α- and β-ylangene or α- and β-copaene, respectively. C3-C7 closure of the cadinyl cation may be followed by a Wagner-Meerwein rearrangement leading to sativene, or the reaction may be terminated

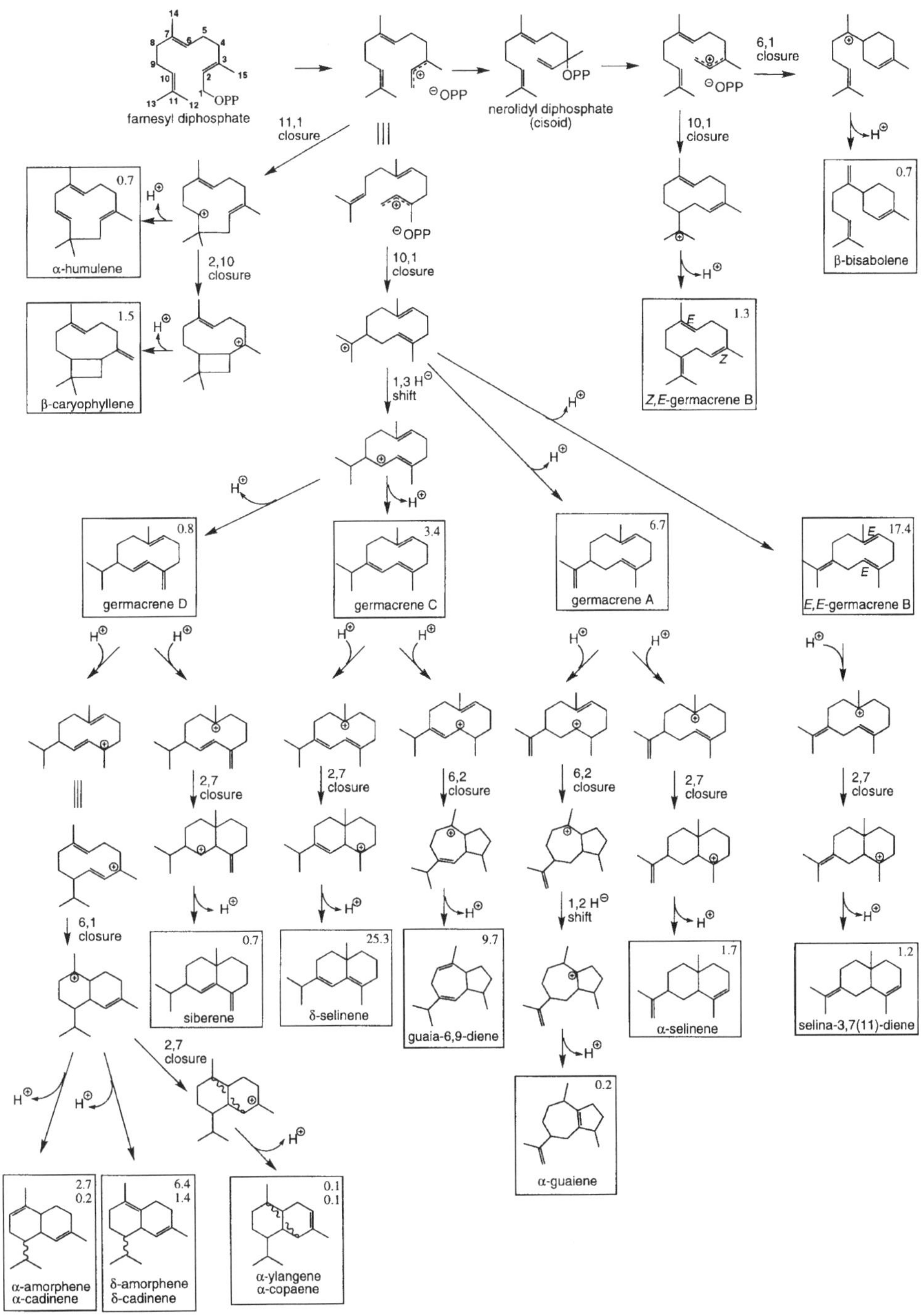
farnesyl diphosphate
OPP
nerolidyl diphosphate (cisoid)
6,1 closure
11,1 closure
10,1 closure
2,10 closure
1,3 H⊖ shift
α-humulene
β-caryophyllene
β-bisabolene
Z,E-germacrene B
germacrene D
germacrene C
germacrene A
E,E-germacrene B
2,7 closure
6,2 closure
1,2 H⊖ shift
siberene
δ-selinene
guaia-6,9-diene
α-selinene
selina-3,7(11)-diene
α-guaiene
α-amorphene
α-cadinene
δ-amorphene
δ-cadinene
α-ylangene
α-copaene

Scheme 12

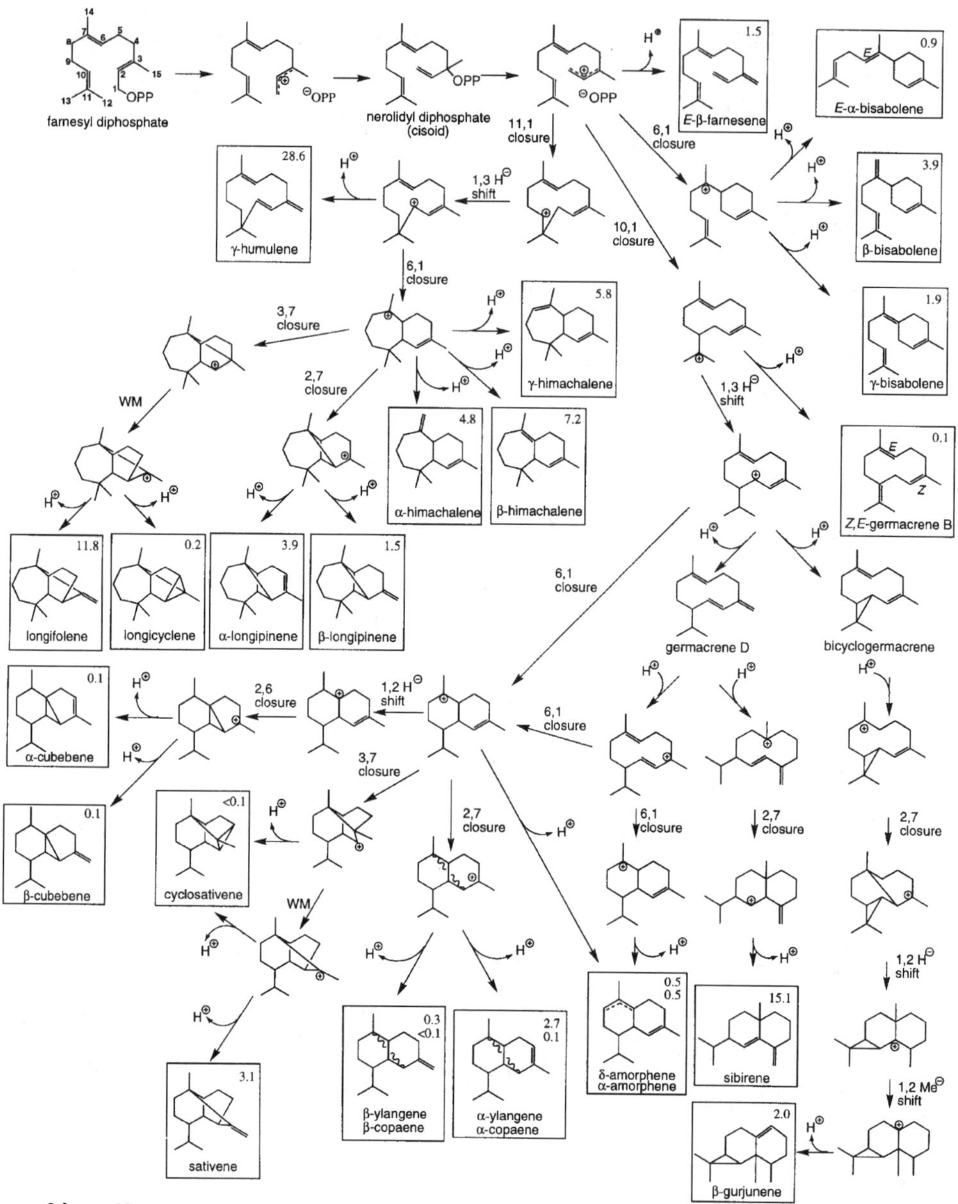

Scheme 13

by deprotonation-mediated cyclopropyl ring closure to generate the tetracyclic olefin cyclosativene. Alternately, the cadinyl cation may undergo a 1,2-hydride shift, followed by C2-C6 closure, and alternate deprotonations to yield α- and β-cubebene. The structurally simplest olefin products are generated by alternate deprotonations of the bisabolyl cation (formed by C6-C1 closure of NPP) that yield *E*-α-bisabolene, β-bisabolene and γ-bisabolene; direct deprotonation of the nerolidyl cation results in *E*-β-farnesene.

The tandem pair of arginine residues which serve in the isomerization of GPP to LPP catalyzed by monoterpene cyclases is not entirely conserved among the sesquiterpene synthases which employ a similar isomerization of FPP to NPP prior to cyclization. Formation of many of the products of γ-humulene synthase requires the isomerization step, and the lack of this element indicates an alternate mechanism for the isomerization. *E*-α-Bisabolene synthase from grand fir must also catalyze the preliminary isomerization of FPP to NPP, and this enzyme possesses an appropriately placed tandem arginine motif, as does δ-selinene synthase which cyclizes FPP directly with no intervening isomerization. Notably, many of the recombinant sesquiterpene cyclases thus far characterized (including γ-humulene synthase and δ-selinene synthase) are capable of converting GPP to a cyclic olefin, most commonly limonene [120–123]. This observation suggests that some sesquiterpene synthases may utilize the same mechanism for isomerization as the monoterpene synthases (involving the tandem arginines) while others have evolved a different, presently uncharacterized, mechanism for the isomerization step.

Deprotonation of the bisabolyl cation by δ-selinene synthase yields β-bisabolene, whereas for γ-humulene synthase alternate deprotonations of this cation lead to β-, γ-, and *E*-α-bisabolene. The latter enzyme may possess additional bases capable of catalyzing the alternate deprotonations or it may allow greater flexibility in the orientation of the intermediate at the active site. An inducible *E*-α-bisabolene synthase from grand fir generates the product in excess of 99% purity [123], suggesting an active site, unlike those of δ-selinene synthase and γ-humulene synthase, that enforces conformational rigidity on the substrate and reacting intermediates to direct high fidelity.

Besides the highly conserved DDxxD element found in all terpene cyclases and prenyltransferases, the notable presence of additional DDxxD motifs in δ-selinene synthase (residues 359–363 and 475–479) and in γ-humulene synthase (residues 487–491) may provide a rationale for multiple product formation in these cyclases resulting from diphosphate binding at these alternate sites. The binding of the diphosphate to (or between) different DDxxD elements would change the orientation of the olefinic chain within the active site, giving rise to alternate routes for product formation and providing a means for generating multiple products.

Consistent with the participation of arginine residues in diphosphate orientation and ionization by monoterpene and sesquiterpene synthases [91, 117] is the substitution of a highly conserved RxR motif by RxC in γ-humulene synthase (residues 306–308). Perhaps this substitution in γ-humulene synthase perturbs diphosphate binding in much the same way as was suggested for the central arginine mutation in the DRRYR motif of trichodiene synthase which resulted in multiproduct formation [117].

In accord with the high level of sequence similarity between δ-selinene synthase and γ-humulene synthase, all carbocation deprotonations result by abstraction from a limited number of common carbon atoms (Scheme 12 and 13). It is likely that a single active site base can mediate deprotonations from spatially adjacent carbon atoms of these carbocationic intermediates. In the case of

δ-selinene synthase and γ-humulene synthase, this would require three such bases, each positioned to abstract a proton from the C2, C4 and C15 atoms, the C6, C8 and C14 atoms, or the C9, C10, C11 and C13 atoms (numbering based upon the FPP skeleton).

Alternate protonations of neutral intermediates is another mechanism utilized by these cyclases for the generation of multiple products, and is most notable in the germacrene protonations catalyzed by δ-selinene synthase. The structural basis for protonation of a germacrene intermediate was recently proposed based on the Nt-EAS crystal structure [62], and consists of a protonation triad composed of two aspartates and a tyrosine (D444, D525, and Y520 in Nt-EAS) which facilitate the protonation at C6 of germacrene A leading to 5-*epi*-aristolochene (Scheme 11). This mechanism involves the removal of the hydroxyl proton from Y520 by the D444 carboxylate, thereby promoting a Y520-mediated proton abstraction from D525. This positions Y520 in such a manner as to facilitate protonation of germacrene A at C6, after which Y520 is reprotonated by D444. The relevance of this protonation triad for other sesquiterpene cyclases was noted in the case of germacrene C synthase from *Lycopersicon esculentum* [122]. This enzyme accumulates germacrene C, which is considered to be the result of the replacement of a triad aspartate with an asparagine (at a position corresponding to D444 of Nt-EAS). Conservation of this triad among other sesquiterpene cyclases, including Hm-VS, which like Nt-EAS requires a C6 protonation of germacrene A (Scheme 11), and the cotton δ-cadinene synthases, which may protonate germacrene D at C15, is consistent with the role of this motif. The conspicuous absence of the triad from enzymes which do not undergo intermediate protonations, such as β-farnesene synthase and *E*-α-bisabolene synthase, as well as in germacrene C synthase, further supports the role of this protonation triad in sesquiterpene biosynthesis. Thus, it is clear that δ-selinene synthase and γ-humulene synthase contain several structural features which promote multiple substrate binding orientations, and provide the means to catalyze alternate deprotonations of carbocationic intermediates and alternate protonations of neutral intermediates that lead to multiproduct formation.

5
Diterpene Synthases

Analogous to the electrophilic reactions catalyzed by the monoterpene and sesquiterpene cyclases, cyclization of GGPP by the diterpene cyclases can be initiated by ionization of the diphosphate ester followed by hydride shifts, methyl migrations, Wagner-Meerwein rearrangements and further cyclizations of the resulting carbocation, as well as internal deprotonations and reprotonations. However, diterpene cyclases may also cyclize GGPP by a protonation-initiated mechanism which is sometimes coupled to a second cyclization via ionization of the diphosphate. Diterpenes fall into three cyclic structural classes: bicyclic, polycyclic, and macrocyclic [17] and cDNAs encoding each of these types of diterpene cyclases have been isolated (Scheme 14), including representatives which initiate cyclization by ionization or protonation, or which couple the two reaction mechanisms.

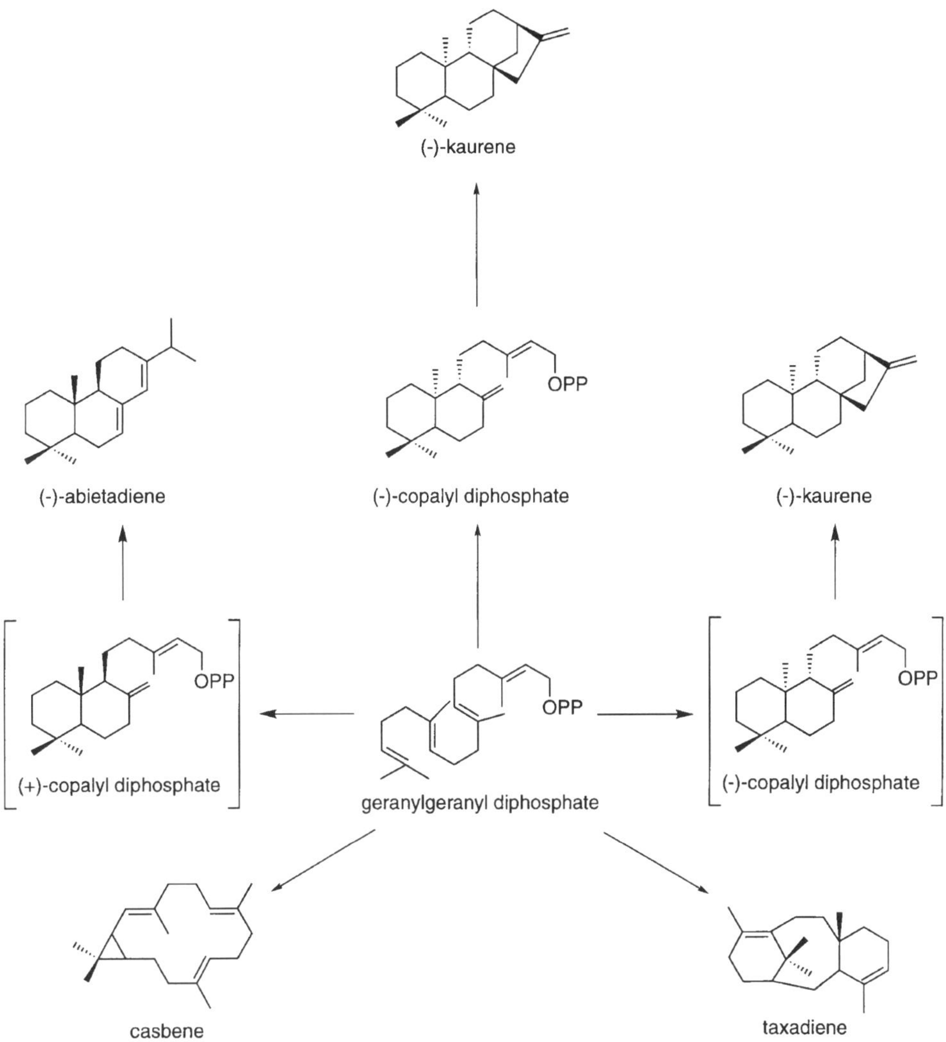

Scheme 14

Casbene synthase is a typical ionization-dependent type cyclase, the olefin product of which functions as an antimicrobial agent in castor bean (*Ricinus communis*) [124]. Purification of this cyclase [125] led to the isolation and sequencing of a full-length cDNA species [126] and a comparison with other terpene cyclases identified a typical *N*-terminal plastidial transit peptide, several conserved histidine, cysteine, and arginine residues, as well as the requisite DDxxD motif. Functional overexpression [127] and site-directed mutagenesis of several histidines, a cysteine, and residues of the DDxxD motif have been carried out with the casbene synthase preprotein [100]. Glutamyl replacements of the first and second aspartates of the DDxxD element showed a modest increase in

K_m for GGPP, and a decrease in k_{cat}, in the mutant synthases, while replacement of the third aspartate had no effect on kinetic parameters of this mutant enzyme. Glutamine replacements of three highly conserved histidines, as well as tryptophan substitutions for a conserved cysteine, had no effect on catalysis or apparent substrate binding in the corresponding mutant enzymes. Although histidines and cysteines have been implicated in terpene cyclase catalysis, the positions of these target residues (relative to the crystal structure of Nt-EAS) indicate that they are not likely at the active site, consistent with the observed results. The effect of these alterations on turnover number of the mutant enzymes may be masked, however, if product release is the rate-limiting step in catalysis, as for several sesquiterpene cyclases and a prenyltransferase (see above).

Taxadiene synthase from *Taxus brevifolia* catalyzes the committed step in the production of the potent anticancer drug paclitaxel (Taxol) [128]. This cyclase exhibits typical properties, including a preference for Mg^{2+}as the required cation, an alkaline pH optimum, no monovalent cation requirement, and one or more essential histidines, cysteines, and arginines as determined by inhibition studies [129]. Of these essential residues, substrate protected against inactivation only by cysteine-directed reagents. The cDNA for this cyclase was obtained using a similarity-based PCR cloning approach [130], and the deduced sequence encodes a typical plastidial transit peptide. Comparisons with other terpene cyclases identified highly conserved histidines, cysteines and arginines, as well as the expected DDxxD motif. A related DSYDD sequence is positioned about 100 amino acids amino terminal to the DDxxD motif; the possible role of this element is uncertain.

The mechanism of taxadiene synthase has been delineated (Scheme 15) [131], and involves initial ionization of the GGPP and C14-C1 closure to generate a 14-membered macrocyclic cation with charge residing at C15, from which ring A originates by C10-C15 closure leading to the enzyme-bound intermediate (1*S*)-verticillene formed transiently upon deprotonation. Internal reprotonation at C7 (C6 of GGPP) and C3-C8 closure (C2-C7 of GGPP) leads to formation of rings B and C, with a final deprotonation to release taxa-4(5),11(12)-diene as the

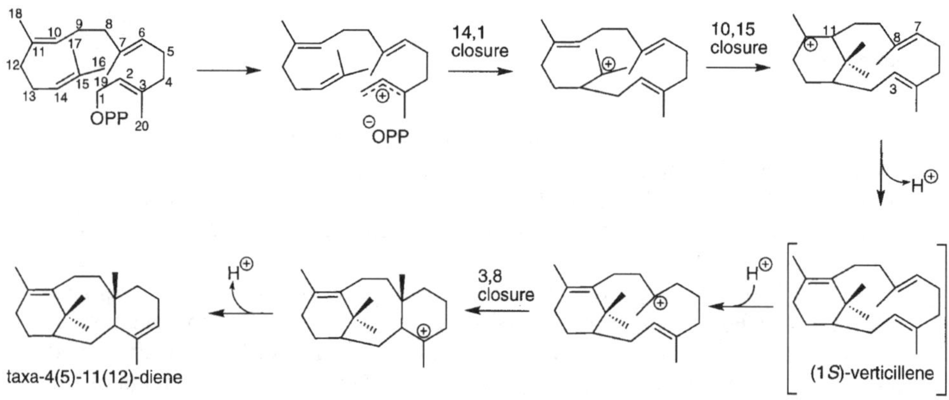

Scheme 15

product. The proton transfer from C11 to C7 (taxane numbering) was confirmed to be intramolecular based upon deuterium labeling studies [131].

The gibberellin family of compounds, derived from the olefin (–)-kaurene, have been identified in some fungi, and all plants where they function as phytohormones. In plants, (–)-kaurene is formed by two diterpene cyclases working in sequence, copalyl diphosphate synthase (*ent*-kaurene synthase A or (–)-CPP synthase) and *ent*-kaurene synthase (*ent*-kaurene synthase B or (–)-KS). (–)-CPP synthase catalyzes the protonation-initiated cyclization of GGPP to (–)-CPP, while (–)-KS cyclizes (–)-CPP into (–)-kaurene via an ionization-initiated mechanism (see Scheme 18).

Several highly conserved plant cDNAs [132–134], and one poorly conserved fungal cDNA [135], encoding (–)-CPP synthases have been isolated. None of these sequences contains the DDxxD motif, consistent with the fact that these synthases initiate cyclization by protonation of the terminal double bond of GGPP rather than by ionization which depends upon binding and reaction of the diphosphate ester. A related aspartate-rich motif, DxDDTA(V/M), is conserved in these cyclases, as well as in some triterpene cyclases [136], and is of significance because these triterpene cyclases also catalyze protonation-initiated cyclizations of squalene (Scheme 16) and 2,3-oxidosqualene (Scheme 17) to form polycyclic products. Mechanism-based inhibition of triterpene cyclases [137, 138] and site-directed mutagenesis of this aspartate-rich motif [139] identified a critical function for the aspartates, with a suggested role in the stabilization of carbocation intermediates of the reaction sequence [137, 140].

cDNAs encoding (–)-KS have been isolated from two plant sources [141, 142]; as expected, both have the requisite DDxxD element required for ionization of (–)-CPP, and they lack the DxDDTA(M/V) motif found in cyclases dependent upon the protonation-initiated reaction. Recently, a cDNA encoding a bifunctional (–)-KS was obtained from the fungus *Pheosphaeria* sp. L487 [143]. This single protein catalyzes the two-step conversion of GGPP to (–)-CPP and of (–)-CPP to (–)-kaurene; a transformation that requires two distinct enzymes in higher plants (See scheme 18B). No significant sequence similarity is apparent between this bifunctional (–)-kaurene synthase and any other terpene cyclase, with the exception of the only other fungal diterpene cyclase thus far isolated, (–)-CPP synthase [135]. However, both DxDDTA(V/M) and DDxxD motifs are conserved, consistent with the cyclizations by both mechanisms catalyzed by this bifunctional enzyme.

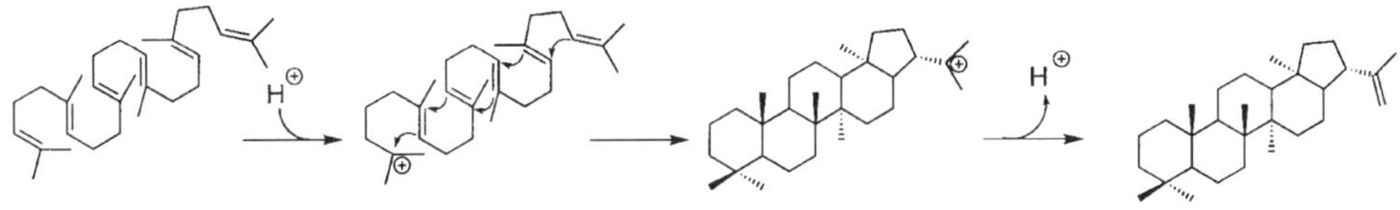

squalene

hopene

Scheme 16

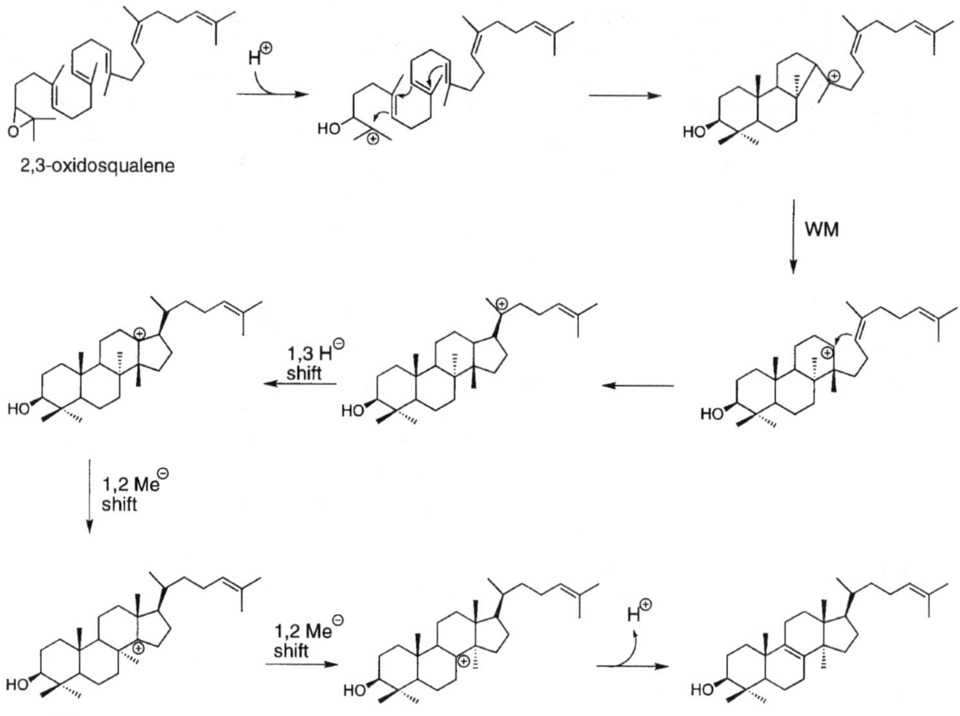

Scheme 17

5.1 Abietadiene Synthase

Abietic acid, a common component of conifer rosin, is formed from (–)-abieta-7(8),13(14)-diene ((–)-abietadiene) (Scheme 18 A) [144]. The cyclization of GGPP to abietadiene requires several discrete steps and enzyme-bound intermediates, yet is catalyzed by a single protein, abietadiene synthase. The mechanism of this enzyme involves protonation-induced cyclization of GGPP to enzyme-bound (+)-CPP, followed by typical ionization of (+)-CPP and cyclization to the transient olefin intermediate, pimaradiene, which undergoes reprotonation at the vinyl substituent, methyl migration and deprotonation to (–)-abietadiene [145]. Similarly, the route to (–)-kaurene involves protonation-initiated cyclization of GGPP to yield (–)-CPP, followed by ionization-initiated cyclization of (–)-CPP to a pimarenyl cation, methyl migration and deprotonation to (–)-kaurene (Scheme 18 B) [20].

Characterization of the constitutive and wound-inducible abietadiene synthase from *Pinus contorta* (lodgepole pine) and *Abies grandis* (grand fir), respectively, showed a pH optimum of 7.8, a preference for Mg^{2+} over Mn^{2+} and Fe^{2+} as the required metal ion, and no monovalent cation requirement [145]. Studies with alternate substrates demonstrated that (+)-CPP was efficiently cyclized whereas (–)-CPP was ineffectual [146]. The pimaradiene intermediate, when

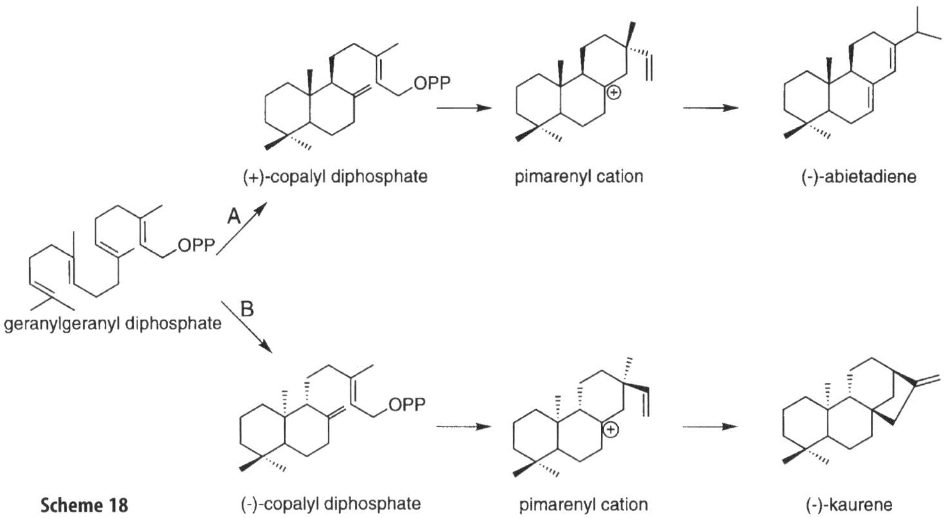

Scheme 18

tested as a substrate, was inactive, suggesting that substantial conformational change must occur at the active site during the reaction which prohibits entry of this exogenous transient intermediate. Active-site modifications of abietadiene synthase identified essential histidine and cysteine residues but only cysteine inactivation was prevented by the substrate, consistent with the results of similar studies on gymnosperm monoterpene cyclases [145].

The abietadiene synthase from fir was purified and internal amino-acid sequence information was employed to isolate the corresponding clone from a wound-induced stem cDNA library [147]. The cDNA encodes an 868 residue protein (99,536 Da) bearing a plastidial transit peptide and exhibiting the greatest sequence similarity to the only other known gymnosperm diterpene cyclase, taxadiene synthase from *Taxus brevifolia* [130]. Preliminary studies have identified a highly active truncation starting at V85 which approximates the preprotein-mature protein junction.

As with many other terpene cyclases, abietadiene synthase has an appropriately placed DDxxD motif (for diphosphate binding) as well as the DxDDTA (M/V) motif identified in the CPP synthases and in some triterpene synthases. Because both of the latter two enzyme types initiate cyclization via protonation, it has been suggested that the DXDDTA(M/V) motif of abietadiene synthase may function directly in the protonation-initiated cyclization of GGPP or indirectly in cation stabilization during the coupled cyclizations of GGPP and of (+)-CPP *en route* to (–)-abietadiene [147]. The proposed mechanism of abietadiene synthase [145] (Scheme 19) involves initial protonation of GGPP at C14 which promotes attack by the C10-C11 double bond on the cationic center at C15 and results in the formation of the A ring with cationic center at C11 (GGPP numbering). A similar attack of the C6-C7 double bond at C11 results in the formation of the B ring intermediate (with carbocation center at C7) which, following deprotonation, yields (+)-CPP. Subsequent ionization of (+)-CPP leads

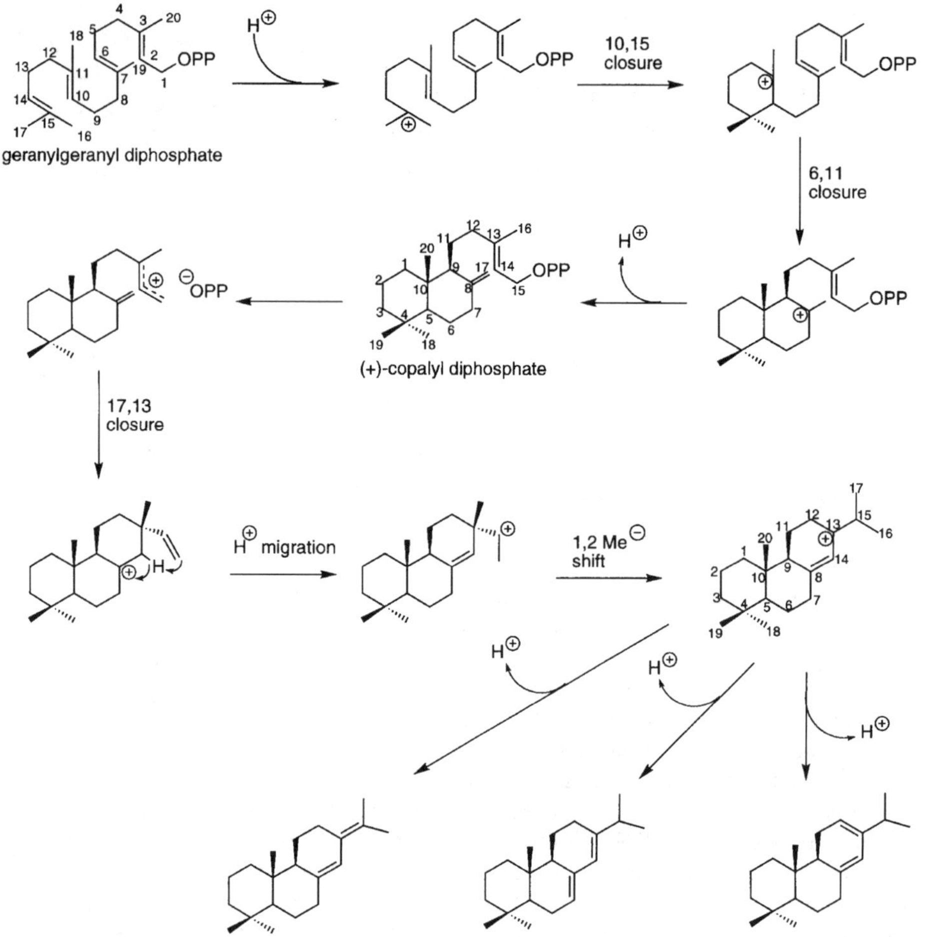

Scheme 19 (-)-abieta-8(14),13(15)-diene (-)-abieta-7(8),13(14)-diene (-)-abieta-8(14),12(13)-diene

to C17-C13 closure (note the (+)-CPP numbering) to generate ring C. This is followed by an intramolecular proton shift from C17 to C15 to generate the pimaradienyl intermediate, from which a 1,2-methyl migration and deprotonation provides (–)-abieta-7(8),13(14)-diene. Recent evidence suggests that (–)-abieta-8(14),13(15)-diene and (–)-abieta-8(14),12(13)-diene may also be generated by this cyclase via alternate deprotonations of the terminal abietane carbocation (Scheme 19). Studies on the closure to the C ring with concomitant rearrangement of the resulting pimaradienyl cation intermediate have utilized [19-^{2}H]-GGPP and *E*-[17-^{2}H]-CPP (deuterated at equivalent positions) to unambiguously demonstrate, by NMR analysis of the product, the intramolecular transfer of the deuterium from C19 of GGPP and from C17 of CPP to the isopropyl *pro-S* methyl (C16) of abietadiene (Scheme 19) [146]. It should be recalled that taxadiene synthase employs a similar intramolecular proton shift, via the transient (1*S*)-verticillene intermediate, in the formation of the terminal olefin (compare Scheme 15 with Scheme 19) [131].

6
Tertiary Structure of 5-*epi*-Aristolochene Synthase and Prediction of Structure/Function Relationships

Because of the striking similarities in properties, mechanisms and primary structures between the plant monoterpene, sesquiterpene, and diterpene synthases [21], the recent crystallization and X-ray diffraction analysis of 5-*epi*-aristolochene synthase (Nt-EAS) [62] is of central importance in providing the first three-dimensional model with which fundamental and common structural themes may be proposed. This model provides a convenient starting point for discussing general structure/function relationships of all terpene cyclases.

The Nt-EAS structure consists of 22 alpha helices with connecting loops; only the *N*-terminal region (residues 1–36) is unstructured. The structure can be divided into two domains, an *N*-terminal domain consisting of 8 short helices and connecting loops (residues 1–220) and a catalytic C-terminal domain consisting of 14 helices (residues 231–548). Searches of the sequence databases using the *N*-terminal domain do not find significant similarity with any other type of protein; however, the tertiary structure of this domain, consisting of a twisted α-barrel, shares similarity with two glycosyl hydrolases [62] and a domain of squalene-hopene cyclase and a protein-farnesyl transferase [148]. The significance of this resemblance is not known. In Nt-EAS, this domain makes contact with the active-site domain via a single hydrogen bond and has no apparent catalytic function. Whether this domain performs a structural role, is involved in protein-protein interactions, or provides a regulatory function is uncertain; however, its sheer size and level of conservation suggest that it is important.

The catalytic domain of Nt-EAS comprises a two-layered barrel of alpha helices with a deep hydrophobic active site cavity into which binds the olefinic moiety of the farnesyl diphosphate substrate. At the opening to the active site reside two Mg^{2+}-coordination sites defined by D301 and D305 of the DDxxD motif and two arginines, R441 and R264 (Nt-EAS numbering). This region of concentrated positive charge, localized at the opening of the active site, likely provides electrostatic interaction to facilitate binding of the diphosphate moiety. Upon substrate binding, closure of the active site by inward movement of the J/K loop (residues 521–534) is proposed to initiate the movement of the A/C loop (residues 255–265) and thereby promote ionization of FPP via interaction with R264. Hydrogen bonding of R266 to S21 of the *N*-terminal unstructured segment promotes extensive ordering of this region and may be critical in helping to close the active site upon substrate binding. Simultaneous with substrate binding is the coordination of a third Mg^{2+} ion with E402 and D444 to provide further positive charge in this region to promote ionization and to further direct the diphosphate moiety away from the newly formed carbocation, thereby preventing premature quenching of the reaction. Upon C10-C1 ring closure in the farnesyl substrate, the pi bonds of Y527 are positioned to stabilize the newly formed C11 carbocation, thus permitting an intramolecular proton transfer from C12 to C6. This transfer is assisted by a D444-Y520-D525 deprotonation/reprotonation triad in which D525 abstracts a C12 proton and D444 removes the hydroxyl proton from Y520, which allows Y520 to accept the D525 proton and insert it at C6. C2-C7 closure, followed by a 1,2-hydride shift, a 1,2-methyl migration, and a final deprotonation, involving a catalytic tryptophan (W273), complete the reaction.

Alignment of Nt-EAS with the gymnosperm and angiosperm (–)-limonene synthases, δ-selinene and γ-humulene synthases, and (–)-abietadiene synthase are shown in Fig. 2, in which important comparative features are highlighted. Based on the Nt-EAS model and the placement of identical amino acid residues, the conservation of the highly charged active site opening, including the DDxxD element and the two arginines positioned roughly thirty-five residues *N*-terminal to the aspartate rich motif, suggests that these residues play the same role in other synthases as they do in Nt-EAS, namely assistance in binding the diphosphate moiety of the substrate to promote ionization.

Positioned directly over the Nt-EAS active site opening is the unstructured amino terminal segment, which in the monoterpene cyclases contains the tandem arginines involved in the isomerization step required by this class of enzyme. The gymnosperm and angiosperm limonene synthases differ in the type (and position) of active site bases relative to Nt-EAS; the gymnosperm type contains at least one arginine (R531 and/or R492 in the *Abies grandis* enzyme) and the angiosperm type contains at least one histidine (H579 in the *Mentha spicata* enzyme).

As noted, the multiproduct sesquiterpene cyclases each have additional DDxxD motifs and, based upon the alignments with Nt-EAS (Fig. 2), one of the additional motifs, located at residues 475–479 of δ-selinene synthase and 487–491 of γ-humulene synthase, aligns with a region involved in Mg^{2+} binding within the active site cavity of Nt-EAS, suggesting that this motif may provide productive diphosphate binding. Binding of FPP at the common, highly conserved DDxxD element or the additional DDxxD motif will lead to at least two alternate orientations of the olefinic substrate chain, and thus contribute to multiple product formation by these grand fir cyclases.

Abietadiene synthase has at least one essential, substrate-protectable cysteine, and C606, C658, C785 and C862 are likely candidates based on the alignment with Nt-EAS. However, the DIDDTAM motif, predicted to play a role in protonation-initiated cyclization, is located outside the putative active site based on the alignment (Fig. 2). This suggests that the DIDDTAM motif is not involved in catalysis or, more likely, that the active site of abietadiene synthase is significantly different from Nt-EAS.

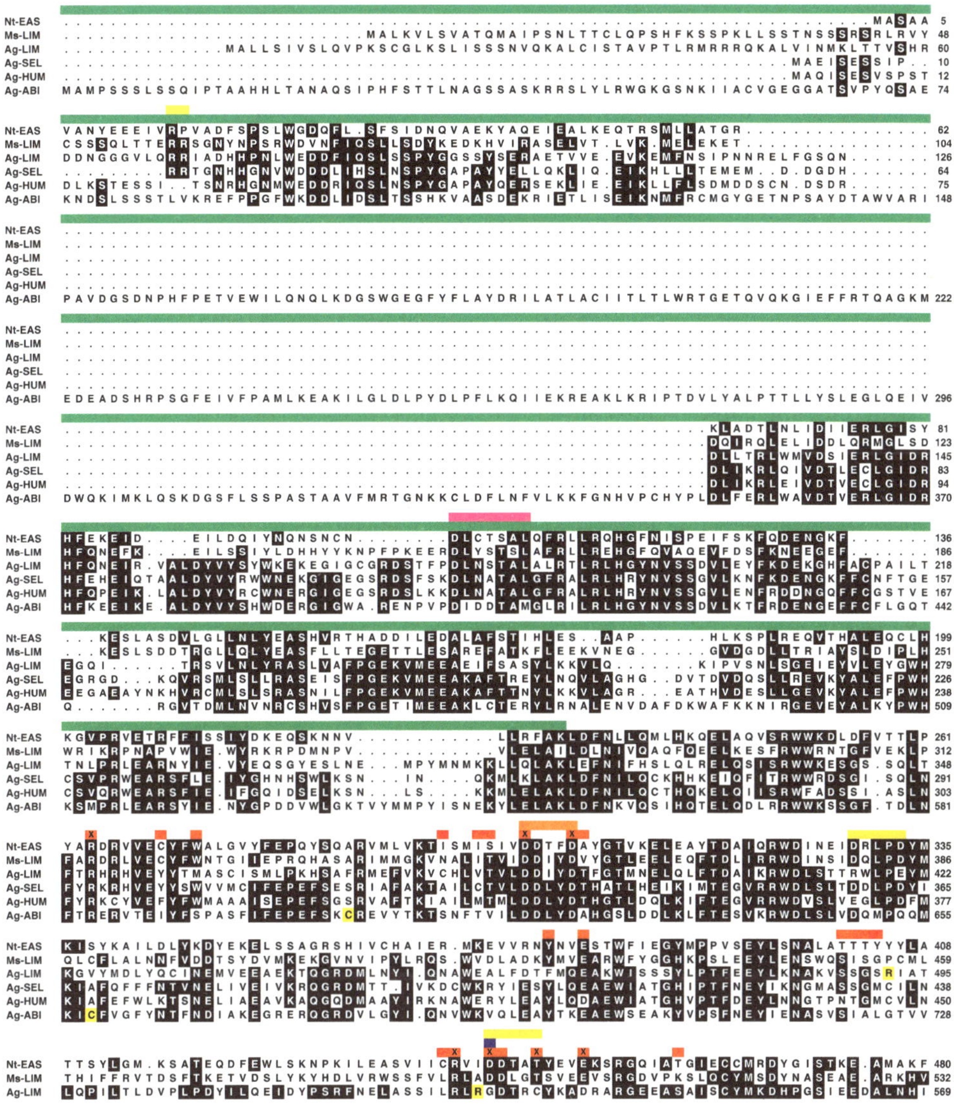

Fig. 2. Comparison of the deduced amino-acid sequences of 5-*epi*-aristolochene synthase from *Nicotiana tabacum* (Nt-EAS), (–)-(4*S*)-limonene synthase from *Mentha spicata* (Ms-LIM), (–)-(4*S*)-limonene synthase from *Abies grandis* (Ag-LIM), δ-selinene synthase from *Abies grandis* (Ag-SEL), γ-humulene synthase from *Abies grandis* (Ag-HUM), and (–)-abietadiene synthase from *Abies grandis* (Ag-ABI). Manual adjustments were made to the sequence alignment generated from the GCG Pileup program. Features are indicated as follows: *N*-terminal domain (green bar), DxDDTAM of Ag-ABI (pink bar), residues which line Nt-EAS active site cavity (red bars), Nt-EAS residues involved in Mg^{2+} coordination (indicated with an X), completely conserved DDxxD motif (orange bar), additional DDxxD motifs of Ag-SEL and Ag-HUM and tandem arginine element (yellow bar), and Nt-EAS residues of the protonation triad (blue bar). Individual residues highlighted in yellow are predicted active site residues of the indicated cyclase

7
Perspectives

Over thirty plant terpene synthases have been cloned to date and provide valuable primary sequence data from which conserved structural features may serve as guides for probing functional relationships using site-directed mutagenesis. These cyclases have been classified into six phylogenetic categories based on percent identity, and are grouped according to the designations *tpsa* through *tpsg* [21]. Heterologous overexpression of cloned cDNAs provides the opportunity to examine structure-function relationships in greater detail via manipulation of primary structure (i.e. random or site-directed mutagenesis, truncations, deletions, domain-swapping, etc.). Most critically, high level expression and purification of terpene cyclases from the various classes will lead to solution of crystal structures and allow detailed insight into terpene cyclase chemistry.

Acknowledgments. The authors would like to thank K. Walker and M. Wise for helpful discussions and K.W., M.W., J. Crock and R. Peters for critical reading of the manuscript. R.C. would like to acknowledge the members of his laboratory, past and present, for their contributions to this work.

8
References

1. Buckingham J (ed) (1998) Dictionary of natural products on CD-ROM, vol 6.1. Chapman & Hall, London
2. Nes WR, McKean ML (1977) Biochemistry of steroids and other isopentanoids. University Park Press, Baltimore
3. Rohmer M, Knani M, Simonin P, Sutter B, Sahm H (1993) Biochem J 295:517
4. Lichtenthaler HK, Schwender J, Disch A, Rohmer M (1997) FEBS Lett 400:271
5. Schwender J, Seemann M, Lichtenthaler HK, Rohmer M (1996) Biochem J 316:73
6. Adam K-P, Thiel R, Zapp J, Becker H (1998) Arch Biochem Biophys 354:181
7. Lichtenthaler HK, Rohmer M, Schwender J (1997) J Physiol Plant 101:643
8. Eisenreich W, Sagner S, Zenk MH, Bacher A (1997) Tetrahedron Letts 38:3889
9. Eisenreich W, Menhard B, Hylands PJ, Zenk MH, Bacher A (1996) Proc Natl Acad Sci USA 93:6431
10. Kleinig H (1989) Annu Rev Plant Phys Plant Mol Biol 40:39
11. Poulter CD, Rilling HC (1981) Prenyl transferases and isomerase. In: Porter JW, Spurgeon SL (eds) Biosynthesis of isoprenoid compounds. John Wiley and Sons, New York, Vol 1, p 161
12. Davisson VJ, Neal TR, Poulter CD (1985) J Am Chem Soc 107:5277
13. Saito A, Rilling HC (1981) Arch Biochem Biophys 208:508
14. Croteau R, Cane DE (1985) Monoterpene and sesquiterpene cyclases. In: Law JH, Rilling HC (eds) Methods in enzymology – steroids and isoprenoids. Academic Press, New York, p 383
15. Croteau R (1987) Chem Rev 87:929
16. Cane DE (1990) Chem Rev 90:1089
17. West CA (1981) Biosynthesis of diterpenes. In: Porter JW, Spurgeon SL (eds) Biosynthesis of isoprenoid compounds. Wiley, New York, Vol 1, p 375
18. Wise ML, Croteau R (1999) Monoterpene biosynthesis. In: Cane DE (ed) Comprehensive natural products chemistry:isoprenoids including carotenoids and steroids. Elsevier Science, Oxford, Vol 2, p 97

19. Cane DE (1999) Sesquiterpene biosynthesis:cyclization mechanisms. In: Cane DE (ed) Comprehensive natural products chemistry:isoprenoids including carotenoids and steroids. Elsevier Science, Oxford, Vol 2, p 155
20. MacMillan J, Beale MH (1999) Diterpene biosynthesis. In: Cane DE (ed) Comprehensive natural products chemistry:isoprenoids including carotenoids and steroids. Elsevier Science, Oxford, Vol 2, p 217
21. Bohlmann J, Meyer-Gauen G, Croteau R (1998) Proc Natl Acad Sci USA 95:4126
22. Bohlmann J, Steele CL, Croteau R (1997) J Biol Chem 272:21784
23. Savage TJ, Hatch MW, Croteau R (1994) J Biol Chem 269:4012
24. Lewinsohn E, Gijzen M, Croteau R (1992) Arch Biochem Biophys 293:167
25. Rajaonarivony JIM, Gershenzon J, Croteau R (1992) Arch Biochem Biophys 296:49
26. Croteau R, Karp F (1979) Arch Biochem Biophys 198:512
27. Poulose AJ, Croteau R (1978) Arch Biochem Biophys 191:400
28. Alonso WR, Croteau R (1991) Arch Biochem Biophys 286:511
29. Croteau R, Felton M, Ronald RC (1980) Arch Biochem Biophys 200:534
30. Hallahan TW, Croteau R (1988) Arch Biochem Biophys 264:618
31. Gambliel H, Croteau R (1982) J Biol Chem 257:2335
32. Bohlmann J, Phillips M, Ramachandiran V, Katoh S, Croteau R (1999) Arch Biochem Biophys 368:232
33. Colby SM, Alonso WR, Katahira EJ, McGarvey DJ, Croteau R (1993) J Biol Chem 268:23016
34. Wise ML, Savage TJ, Katahira EJ, Croteau R (1998) J Biol Chem 273:14891
35. Yuba A, Yazaki K, Tabata M, Honda G, Croteau R (1996) Arch Biochem Biophys 332:280
36. Dudareva N, Cseke L, Blanc VM, Pichersky E (1996) Plant Cell 8:1137
37. Gambliel H, Croteau R (1984) J Biol Chem 259:740
38. Wagschal K, Savage TJ, Croteau R (1991) Tetrahedron 47:5933
39. Wagschal KC, Pyun H-J, Coates RM, Croteau R (1994) Arch Biochem Biophys 308:477
40. Croteau R, Wheeler CJ, Cane DE, Ebert R, Ha H-J (1987) Biochemistry 26:5383
41. Croteau R, Karp F (1977) Arch Biochem Biophys 179:257
42. Croteau R, Alonso WR, Koepp AE, Johnson MA (1994) Arch Biochem Biophys 309:184
43. Croteau R, Felton NM, Wheeler CJ (1985) J Biol Chem 260:5956
44. Croteau R, Satterwhite DM, Cane DE, Chang CC (1988) J Biol Chem 263:10063
45. LaFever RE, Croteau R (1993) Arch Biochem Biophys 301:361
46. Croteau R (1992) Biosynthesis of thujane monoterpenes. In: Hopp R, Mori K (eds) Recent developments in flavor and fragrance chemistry. VCH, Weinheim Germany, p 263
47. Croteau R (1992) Monoterpene biosynthesis:cyclization of geranyl pyrophosphate to (+)-sabinene. In: Teranishi R, Takeoka GR, Guntert M (eds) Flavor precursors:thermal and enzymatic conversions. American Chemical Society, Washington DC, ACS Symp Series 490, p 8
48. Savage TJ, Croteau R (1993) Arch Biochem Biophys 305:581
49. Hallahan TW, Croteau R (1989) Arch Biochem Biophys 269:313
50. Satterwhite DM, Wheeler CJ, Croteau R (1985) J Biol Chem 260:13901
51. Croteau R, Gershenzon J, Wheeler CJ, Satterwhite DM (1990) Arch Biochem Biophys 277:374
52. Wheeler CJ, Croteau R (1986) Arch Biochem Biophys 246:733
53. Wheeler CJ, Croteau R (1987) Proc Natl Acad Sci USA 84:4856
54. Croteau R (1986) Arch Biochem Biophys 251:777
55. Poulter CD, Argyle JC, Mash EA (1978) J Biol Chem 253:7227
56. Cane DE, Pawlak JL, Horak RM, Hohn TM (1990) Biochemistry 29:5476
57. Stork G, White WN (1956) J Am Chem Soc 78:4609
58. Cramer F, Rittersdorf W (1967) Tetrahedron 23:3015
59. Vial MV, Rojas C, Portilla G, Chayet L, Perez LM, Cori O (1981) Tetrahedron 37:2351
60. Chayet L, Rojas MC, Cori O, Bunton CA, McKenzie DC (1984) Bioorg Chem 12:329
61. Tarshis LC, Yan M, Poulter CD, Sacchettini JC (1994) Biochemistry 33:10871
62. Starks CM, Back K, Chappell J, Noel JP (1997) Science 277:1815
63. Lesburg CA, Zhai G, Cane DE, Christianson DW (1997) Science 277:1820

64. Suelter CH (1970) Science 168:789
65. O'Brien MC, McKay DB (1995) J Biol Chem 270:2247
66. Wilbanks SM, McKay DB (1995) J Biol Chem 270:2251
67. Croteau R, Alonso WR, Koepp AE, Shim J-H, Cane DE (1993) Arch Biochem Biophys 307:397
68. McGeady P, Croteau R (1995) Arch Biochem Biophys 317:149
69. Rajaonarivony JIM, Gershenzon J, Miyazaki J, Croteau R (1992) Arch Biochem Biophys 299:77
70. Savage TJ, Ichii H, Hume SD, Little DB, Croteau R (1995) Arch Biochem Biophys 320:257
71. Wheeler CJ, Croteau R (1987) J Biol Chem 262:8213
72. Wheeler CJ, Croteau R (1988) Arch Biochem Biophys 260:250
73. Cori O, Chayet L, De la Fuenta M, Fernandez LA, Hashagen U, Perez L, Portilla G, Rojas MC, Sanchez G, Vial MV (1980) Mol Biol Biochem Biophys 32:97
74. Rojas MC, Chayet L, Portilla G, Cori O (1983) Arch Biochem Biophys 222:389
75. Croteau R, Wheeler CJ, Aksela R, Oehlschlager AC (1986) J Biol Chem 261:7257
76. McGeady P, Pyun H-J, Coates RM, Croteau R (1992) Arch Biochem Biophys 299:63
77. Croteau R, Satterwhite DM, Cane DE, Chang CC (1986) J Biol Chem 261:13438
78. Croteau R, Satterwhite DM, Wheeler CJ, Felton NM (1989) J Biol Chem 264:2075
79. Kjonaas R, Croteau R (1983) Arch Biochem Biophys 220:79
80. Haley RC, Miller JA, Wood HCS (1969) J Chem Soc (C) 264
81. Hiraga Y, Shi W, Ito DI, Ohta S, Suga T (1993) J Chem Soc Chem Commun 1370
82. Pyun H-J, Coates RM, Wagschal KC, McGeady P, Croteau R (1993) J Org Chem 58:3998
83. Suga T, Hiraga Y, Ahara M, Izumi S (1992) J Chem Soc Chem Commun 1556
84. Leopold MF, Epstein WW, Grant DM (1988) J Am Chem Soc 110:616
85. Alonso WR, Rajaonarivony JIM, Gershenzon J, Croteau R (1992) J Biol Chem 267:7582
86. Gershenzon J, McCaskill D, Rajaonarivony JIM, Mihaliak C, Karp F, Croteau R (1992) Anal Biochem 200:130
87. Turner G, Gershenzon J, Nielson EE, Froehlich JE, Croteau R (1999) Plant Physiol 120:879
88. Wan J, Blakeley SD, Dennis DT, Ko K (1996) J Biol Chem 271:31227
89. Keegstra K, Olsen JJ, Theg SM (1989) Annu Rev Plant Physiol Plant Mol Biol 40:471
90. von Heijne G, Steppuhn J, Herrmann RG (1989) Eur J Biochem 180:535
91. Williams DC, McGarvey DJ, Katahira EJ, Croteau R (1998) Biochemistry 37:12213
92. Chen A, Kroon PA, Poulter CD (1994) Protein Sci 3:600
93. Cori O, Chayet L, Perez LM, Rojas MC, Portilla G, Holuigue L, Fernandez LA (1981) Arch Biol Med Exp 14:129
94. Tarshis LC, Proteau PJ, Kellogg BA, Sacchettini JC, Poulter CD (1996) Proc Natl Acad Sci USA 93:15018
95. Song L, Poulter CD (1994) Proc Natl Acad Sci USA 91:3044
96. Marrero OF, Poulter CD, Edwards PA (1992) J Biol Chem 267:21873
97. Cane DE, Xue Q, Fitzsimons BC (1996) Biochemistry 35:12369
98. Cane DE, Zue Q, Van Epp JE (1996) J Am Chem Soc 118:8499
99. Cane DE, Xue Q (1996) J Am Chem Soc 118:1563
100. Huang K-X, Huang Q-L, Scott AI (1998) Arch Biochem Biophys 352:144
101. Cane DE, Chiu H-T, Liang P-H, Anderson KS (1997) Biochemistry 36:8332
102. Mathis JR, Back K, Starks C, Noel J, Poulter CD, Chappell J (1997) Biochemistry 30:8340
103. Cane DE, Pargellis C (1987) Arch Biochem Biophys 254:421
104. Hohn TM, Plattner RD (1989) Arch Biochem Biophys 272:137
105. Hohn TM, VanMiddlesworth F (1986) Arch Biochem Biophys 251:756
106. Croteau R, Munck SL, Akoh CC, Fisk HJ, Satterwhite DM (1987) Arch Biochem Biophys 256:56
107. Dehal SS, Croteau R (1988) Arch Biochem Biophys 261:346
108. Munck SL, Croteau R (1990) Arch Biochem Biophys 282:58
109. Vögeli U, Freeman JW, Chappell J (1990) Plant Physiol 93:182
110. Cane DE (1981) Biosynthesis of sesquiterpenes. In: Porter JW, Spurgeon SL (eds) Biosynthesis of isoprenoid compounds. Wiley, New York, Vol 1, p 283

111. Cane DE, Ha H, Pargellis C, Waldmeier F, Swanson S, Murphy PPN (1985) Bioorg Chem 13:246
112. Cane DE, McIlwaine DB, Harrison PHM (1989) J Am Chem Soc 111:1152
113. Cane DE, Tandon M (1995) J Am Chem Soc 117:5602
114. Cane DE, Oliver JS, Harrison PHM, Abell C, Hubbard BR, Kane CT, Lattman R (1990) J Am Chem Soc 112:4513
115. Harrison PHM, Oliver JS, Cane DE (1988) J Am Chem Soc 110:5922
116. Cane DE, Prabhakaran PC, Salaski EJ, Harrison PHM, Noguchi H, Rawlings BJ (1989) J Am Chem Soc 111:8914
117. Cane DE, Shim JH, Xue Q, Fitzsimmons BC, Hohn TM (1995) Biochemistry 34:2480
118. Back K, Chappell J (1996) Proc Natl Acad Sci USA 93:6841
119. Steele CL, Lewinsohn E, Croteau R (1995) Proc Natl Acad Sci USA 92:4164
120. Steele CL, Bohlmann J, Crock J, Croteau R (1998) J Biol Chem 273:2078
121. Crock J, Wildung M, Croteau R (1997) Proc Natl Acad Sci USA 94:12833
122. Colby SM, Crock J, Dowdle-Rizzo B, Lemaux PG, Croteau R (1998) Proc Natl Acad Sci USA 95:2216
123. Bohlmann J, Crock J, Jetter R, Croteau R (1998) Proc Natl Acad Sci USA 95:6756
124. Sitton D, West CA (1975) Phytochemistry 14:1921
125. Moesta P, West CA (1985) Arch Biochem Biophys 238:325
126. Mau CJD, West CA (1994) Proc Natl Acad Sci USA 91:8497
127. Hill AM, Cane DE, Mau CJD, West CA (1996) Arch Biochem Biophys 336:283
128. Koepp AE, Hezari M, Zajicek J, Stofer Vogel B, LaFever RE, Lewis NG, Croteau R (1995) J Biol Chem 270:8686
129. Hezari M, Lewis NG, Croteau R (1995) Arch Biochem Biophys 322:437
130. Wildung MR, Croteau R (1996) J Biol Chem 271:9201
131. Lin X, Hezari M, Koepp AE, Floss HG, Croteau R (1996) Biochemistry 35:2968
132. Sun T-P, Kamiya Y (1994) Plant Cell 6:1509
133. Benson RJ, Johal GS, Crane VC, Tossberg JT, Schnable PS, Meeley RB, Briggs SP (1995) Plant Cell 7:75
134. Ait-Ali T, Swain SM, Reid JB, Sun TP, Kamiya Y (1997) Plant J 11:443
135. Tudzynski B, Kawaide H, Kamiya Y (1998) Curr Genet 34:234
136. Ochs D, Kaletta C, Entian K-D, Beck-Sickinger A, Poralla K (1992) J Bacteriol 174:298
137. Abe I, Prestwich GD (1994) J Biol Chem 269:802
138. Corey EJ, Cheng H, Baker CH, Matsuda SPT, Li D, Song X (1997) J Am Chem Soc 119:1289
139. Feil C, Sussmuth R, Jung G, Poralla K (1996) Eur J Bioch 242:51
140. Abe I, Prestwich GD (1999) Squalene epoxidase and oxidosqualene:lanosterol cyclase-key enzymes in cholesterol biosynthesis. In: Cane DE (ed) Comprehensive natural products chemistry:isoprenoids including carotenoids and steroids. Elsevier Science, Oxford, Vol 2, p 267
141. Yamaguchi S, Saito T, Abe H, Yamane H, Murofushi N, Kamiya Y (1996) Plant J 10:203
142. Yamaguchi S, Sun T-P, Kawaide H, Kamiya Y (1998) Plant Physiol 116:1271
143. Kawaide H, Imai R, Sassa T, Kamiya Y (1997) J Biol Chem 272:21706
144. Funk C, Croteau R (1994) Arch Biochem Biophys 308:258
145. LeFever RE, Stofer Vogel B, Croteau R (1994) Arch Biochem Biophys 313:139
146. Ravn MM, Coates RM, Jetter R, Croteau R (1998) Chem Commun 21
147. Stofer Vogel B, Wildung M, Vogel G, Croteau R (1996) J Biol Chem 271:23262
148. Wendt KU, Schulz GE (1998) Structure 6:127

Biosynthesis of Prenylated Alkaloids Derived from Tryptophan

Robert M. Williams · Emily M. Stocking · Juan F. Sanz-Cervera

Department of Chemistry, Colorado State University, Fort Collins, Colorado 80523, USA
E-mail: rmw@chem.colostate.edu

The biosynthesis of prenylated indole alkaloids and related natural substances derived from tryptophan is reviewed. The families of compounds covered in this review include the brevianamides, austamides, paraherquamides, marcfortine, roquefortine, aszonalenin, echinulin, verruculogen, the fumitremorgins, α-cyclopiazonic acid, and the ergot alkaloids. Although other families of naturally occurring prenylated indole alkaloids exist, such as the iridoids, this review is intended to examine the biosynthesis of the groups selected based on their structural and biogenetic similarities. In addition, the biosynthesis of the families selected for this chapter have not, to the best of our knowledge, been previously reviewed.

Keywords: Prenylated indole alkaloids, Biosynthesis, Paraherquamides, Brevianamides, Roquefortine, Echinulin, Verruculogen, Cyclopiazonic acid, Ergot alkaloids

1 Introduction

Tryptophan constitutes a key primary metabolite from which countless secondary metabolic indole alkaloids and related nitrogenous substances are biosynthesized. Nature has created a particularly fascinating array of structurally diverse and interesting natural alkaloids through sequestering and adorning the basic tryptophan core with isoprenic building blocks. The rich nucleophilic chemistry of the indole ring, wherein all positions of this heterocyclic nucleus are susceptible to electrophilic attack, has been extensively exploited by Nature to craft a bewildering array of structurally intriguing natural substances through the use of the basic isoprene building blocks coupled with some seemingly simple oxidative "finishing" work on the prenylated substrates. This article will review the biosynthesis of several members of this large family of natural products. In many instances, there exists little or no experimental data upon which the retro-biosynthetic analyses have been based. As is the case for the biosynthesis of secondary metabolites in general, where very few biosynthetic pathways have been elucidated in detail, many of the studies presented below can only sketch a fairly vague picture of what is likely to be going on inside the cell producing these substances. Future work will certainly clarify and realign many of the pathways outlined below. It is hoped that this overview will form the basis upon which researchers interested in this field, may gain insight and inspiration from others attempting to unravel Nature's closely guarded chemical synthesis secrets. Many of the structures covered in this review have also inspired significant synthetic work which will not be reviewed or cited here.

At the end of the article a brief overview of the biosynthesis of tryptophan itself, as well as the currently known pathways to produce the C_5 isoprenoids, is described. The purpose of this is to give a working pictorial reference source for the primary metabolites as this bears so frequently on the labeling and incorporation studies used to determine the secondary metabolic pathways.

2 Isoprenylated Bicyclo [2.2.2] Alkaloids

A number of structurally interesting natural alkaloids have been isolated from various fungi that contain the unique bicyclo [2.2.2] ring system constituted mainly from tryptophan, proline, and substituted proline derivatives where the olefinic unit of the isoprene moiety has been formally oxidatively cyclized across the α-carbon atoms of a cyclic dipeptide (a piperazinedione or diketopiperazine). The family of alkaloids that shares this unusual structural core includes the paraherquamides, brevianamides, marcfortines, asperparalines (aspergillimides), sclerotamides, and several other natural metabolites assigned identifying numbers. The biosynthesis of these substances will be described in structural sub-groups sharing the most common structural and biosynthetic features.

2.1 The Brevianamides and Austamides

The brevianamides constitute a small but structurally interesting family of indole alkaloids constituted from tryptophan, proline and one isoprene unit (Fig. 1). Brevianamide A was originally isolated from cultures of *Penicillium brevicompactum* by Birch and Wright in 1969 [1]. Brevianamide A has also been found in cultures of *Penicillium viridicatum* [2] and *Penicillium ochraceum* [3]. The structure and absolute stereochemistry of brevianamide A was secured through X-ray crystallography by Coetzer in 1974 on the semi-synthetic derivative, 5-bromobrevianamide A [4]. Due to biosynthetic commonalties with the brevianamides to be discussed below, the austamides will also be included in this family. Austamide was first isolated by Steyn from cultures of *Aspergillus ustus* in 1971 [5]; subsequently, 12,13-dihydroaustamide and "deoxyaustamide" were isolated from the same fungi by Steyn in 1973 [6]. The absolute configuration of austamide was subsequently secured through X-ray crystallography by Coetzer and Steyn in 1973 on the semi-synthetic derivative, 5-bromo-12*S*-tetrahydroaustamide [7].

Following the original isolation, Birch and associates reported the isolation and structural elucidation of brevianamides B–F from *Penicillium brevicompactum* [8, 9]. Brevianamide A is the major fluorescent metabolite produced by this fungus; a minor metabolite named brevianamide B, that also exhibited the characteristic ψ-indoxyl chromophore, was assigned a structure that was epimeric to brevianamide A at the spiro-ψ-indoxyl quaternary center. The structural assignment for brevianamide B was based primarily on the semi-synthetic conversion of natural (+)-brevianamide A into brevianamide B by borohydride reduction and acid dehydration to deoxybrevianamide A; subsequent air oxidation provided (–)-brevianamide B [9] (Scheme 1).

Birch and associates did not record optical rotation data for natural or semi-synthetic brevianamide B and this oversight later proved to constitute the basis of an intriguing stereochemical puzzle; the complete stereochemical elucidation of brevianamide B will be discussed below.

Brevianamides C and D, also isolated in minor amounts from these *Penicillium* sp., were postulated by Birch and Russell [9] to be artifacts of isolation since visible light irradiation of brevianamide A produces brevianamides C and D in virtually quantitative yield. When cultures of *Penicillium brevicompactum* were grown in the dark, only brevianamides A and B were detected, thus verifying the hypothesis that brevianamides C and D are photochemical artifacts [9] and are not true metabolites of enzymic origin. A mechanism for the photochemical formation of brevianamides C and D from brevianamide A is provided in Scheme 2.

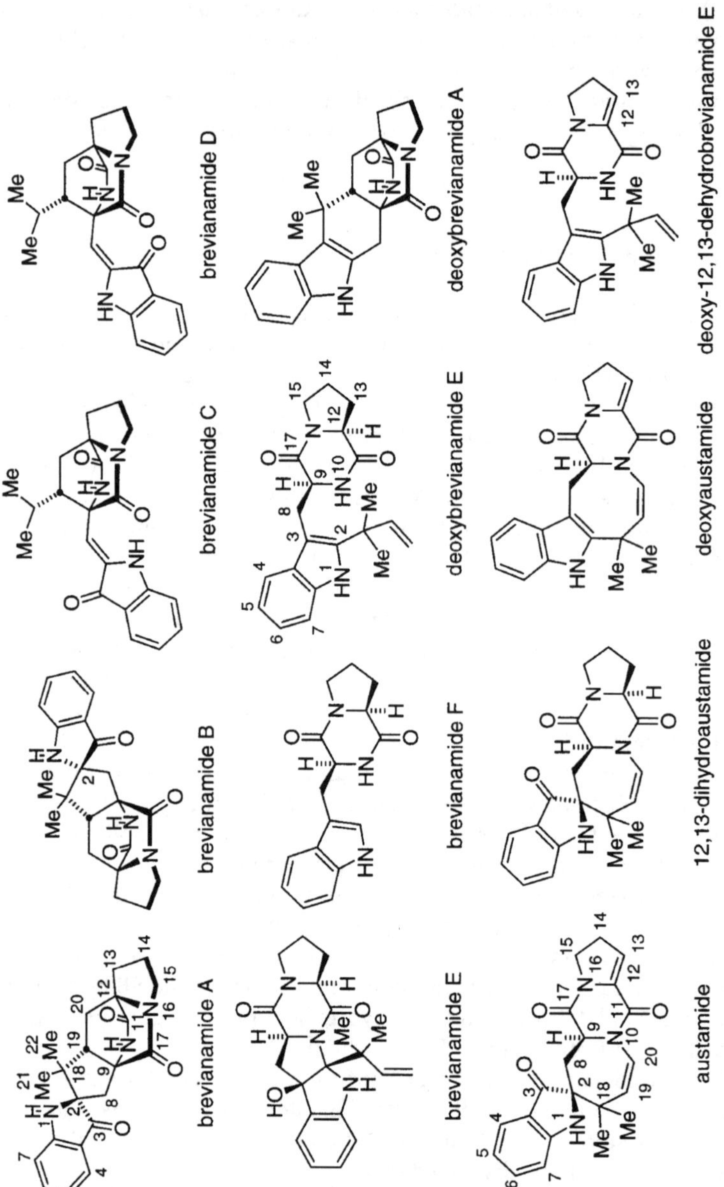

Fig. 1. Structures of the brevianamides and austamides

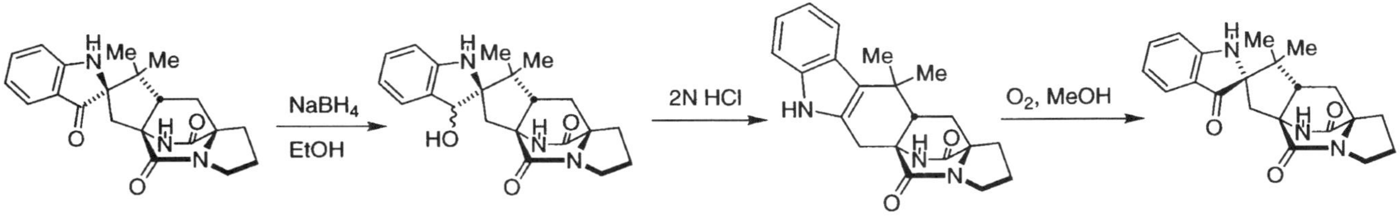

(+)-brevianamide A

deoxybrevianamide A

(-)-brevianamide B

Scheme 1. Conversion of (+)-brevianamide A into (−)-brevianamide B [9]

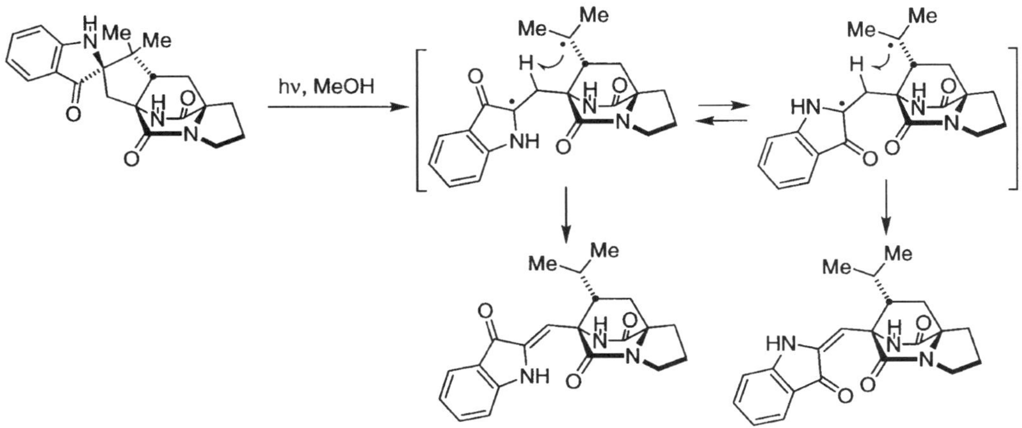

brevianamide C

brevianamide D

Scheme 2. Photochemical formation of brevianamides C and D [9]

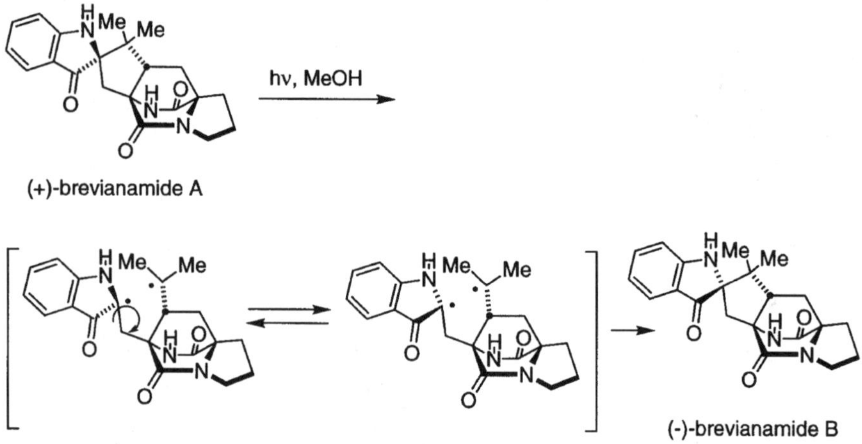

Scheme 3. Proposed mechanism for the photochemical formation of brevianamide B from A [9]

Interestingly, Birch and Russell reported that during the early stages of photolysis of brevianamide A into brevianamides C and D with white light, traces of brevianamide B could be detected in the mixture. A mechanism for this transformation was postulated as shown in Scheme 3. However, the absolute stereochemistry of the brevianamide B produced under these conditions was not determined. As will become clear below, the *absolute* stereochemistry of the brevianamide B produced from brevianamide A, either via the photochemical route, or via the chemical redox route through deoxybrevianamide A as described above, must be as depicted in Schemes 1 and 3 giving (–)-brevianamide B.

The structure and absolute configuration of deoxybrevianamide E was confirmed by a total synthesis reported by Ritchie and Saxton [10, 11] and the absolute stereochemistry of brevianamide E was similarly secured by a synthesis reported by Kametani et al. [12].

The first significant biosynthetic experiment conducted in this family was reported by Birch and Wright in 1970 [8] wherein DL-[3-^{14}C]-tryptophan, [2-^{14}C]-mevalonic acid lactone, Na[2-^{14}C]acetate, and L-[U-^{14}C]-proline were shown to be incorporated into brevianamide A in significant amounts as shown in Table 1. It was also shown that [$^{14}CH_3$]-methionine was not incorporated into

Table 1. Biosynthetic feeding experiments with radioactive precursors to *Penicillium brevicompactum* [8]

Substrate	Activity of brevianamide A (cts/100 s)	Incorporation (% activity fed)
DL-[3-^{14}C]-tryptophan	1430	6×10^{-1}
[2-^{14}C]-mevalonic lactone	28	3×10^{-3}
Na[2-^{14}C]acetate	425	25×10^{-2}
[$^{14}CH_3$]-methionine	0	0
L-[U-^{14}C]-proline	4880	9×10^{-2}

brevianamide A. The acetate and mevalonate incorporations also firmly establish the origin of the isoprene unit as being derived from the classical mevalonate pathway.

Subsequently, Birch and associates reported the biosynthetic incorporation of doubly radiolabeled brevianamide F (*cyclo*-L-tryptophyl-L-proline) into brevianamide A in *Penicillium brevicompactum* [13]. Thus, feeding experiments were conducted with DL-[*methylene*-^{14}C]-tryptophan, L-[5-^{3}H]-proline, DL-[2-^{14}C]-mevalonic acid lactone, and *cyclo*-L-[*methylene*-^{14}C]-tryptophyl-L-proline as shown by the resulting data in Table 2. The specific molar incorporations were determined by feeding precursors of certain specific activities and recrystallizing the brevianamide A produced under these conditions to constant activity without the addition of cold, unlabeled brevianamide A.

Table 2. Biosynthetic feeding experiments with radioactive precursors to *Penicillium brevicompactum* [13]

Precursor	Activity fed (μCi)	% Total incorporation	% Specific incorporation
DL-[*methylene*-^{14}C]-tryptophan	100 100	5.1	 17.6
L-[5-^{3}H]-proline	100	0.40	
DL-[2-^{14}C]-mevalonic acid lactone	75	0.16	
cyclo-L-[*methylene*-^{14}C]-tryptophyl-L-proline	12.9 12.5	1.8	 14.0

To demonstrate that brevianamide F was incorporated as an intact unit without any hydrolytic degradation and recombination of the constituent amino acids, *cyclo*-L-[*methylene*-^{14}C]-tryptophyl-L-[5-^{3}H]-proline was prepared (^{3}H:^{14}C ratio = 3.67:1) and fed in two batches of ~100 mg to *Penicillium brevicompactum*. The brevianamide A isolated from these experiments had a ^{3}H:^{14}C ratio = 3.82:1 and 3.74:1 and total incorporations of 3.2% and 3.6%, respectively. The high specific incorporation of 14.0% confirms that brevianamide F is a biosynthetic precursor (Scheme 4).

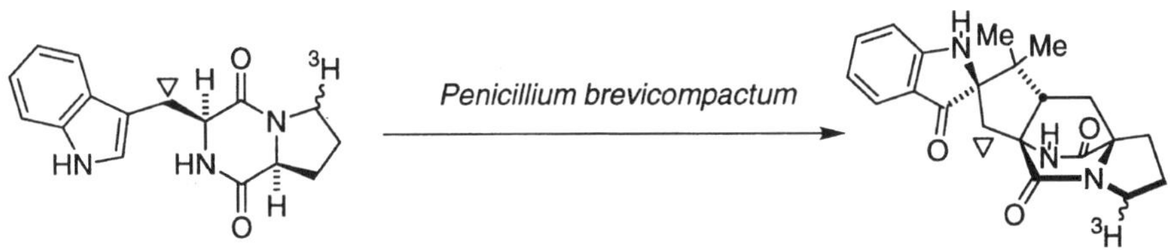

cyclo-L-[*methylene*-^{14}C]-tryptophyl-L-[5-^{3}H]-proline
(brevianamide F, ▽ = ^{14}C)

brevianamide A

Scheme 4. Incorporation of brevianamide F into brevianamide A [13]

Based on the co-occurrence of brevianamide E with austamide in cultures of *Aspergillus ustus* as reported by Steyn [6], Birch and associates postulated the logical biosynthetic sequence involving: the coupling of L-proline and L-tryptophan → brevianamide F → deoxybrevianamide E → brevianamide A.

The introduction of the isoprene unit has been a subject of much speculation. Birch et al. speculated that: "The reversed unit could be attached directly through a cation or might be biogenetically transferred in the brevianamides from the dioxopiperazine nitrogen atom, either in an indole or an indoxyl precursor" [13]. This possibility, shown in Scheme 5, amounts to a formal S_N2' transfer of a "normal" prenylated precursor to the "reverse" prenylated metabolite. Birch et al. further speculated that a second possibility is ring formation leading to austamide by the simultaneous addition of carbon and nitrogen in a carbene-type reaction, which was suggested by Birch et al. for the introduction of an isoprene unit into a non-phenolic benzene ring. The fascinating reverse prenylation reaction has also been discussed in the context of the biosynthesis of roquefortine, echinulin, and the paraherquamides and will be readdressed below.

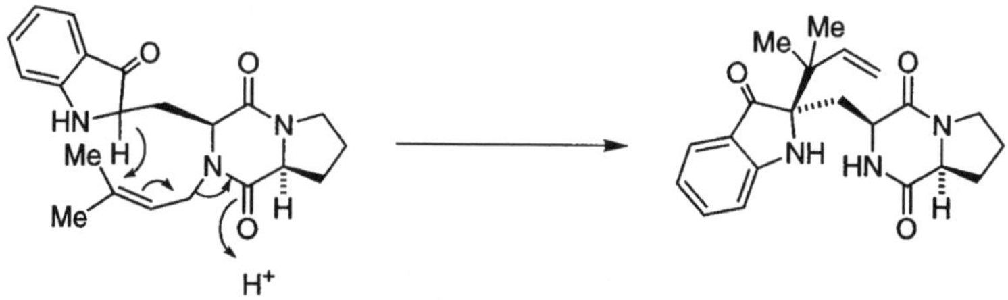

Scheme 5. Reverse prenylation proposal of Birch et al. [13]

The most intriguing feature of the final assembly of brevianamides A and B is the mechanism of formation of the novel bicyclo [2.2.2] ring system common to this family of metabolites. Porter and Sammes [14] were the first to suggest the involvement of an intramolecular Diels-Alder reaction for the formation of this ring system and drew the structure shown in Fig. 2 to illustrate this provocative idea.

Experimental support [14] for this notion was further presented by the Diels-Alder cycloadditions of dimethylacetylene dicarboxylate and norbornadiene with the pyrazine derivatives shown in Scheme 6.

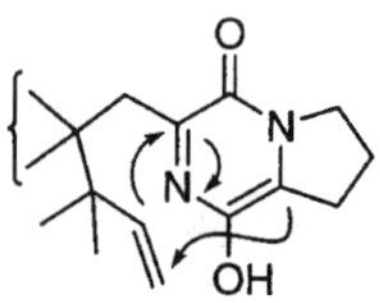

Fig. 2. Porter and Sammes Diels-Alder proposal [14]

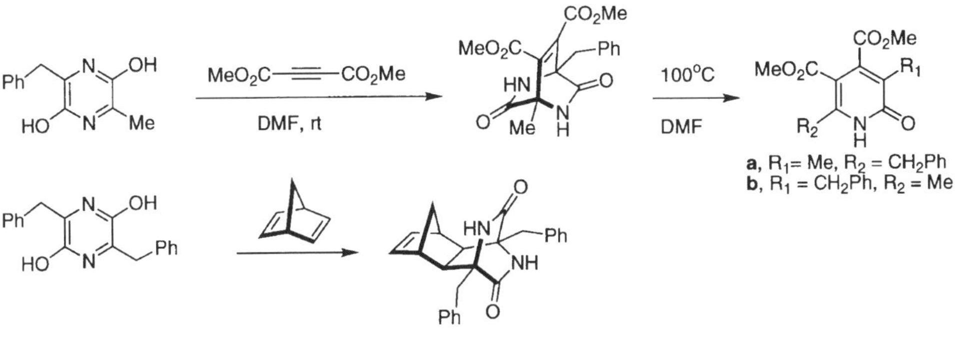

Scheme 6. Biomimetic Diels-Alder cycloadditions [14]

Shortly after the publication of the Diels-Alder hypothesis by Porter and Sammes, Birch [15] speculated on the possible modes of formation of the bicyclo [2.2.2] ring system and stated: "An alternative suggestion (Porter and Sammes 1970) is the occurrence of a Diels-Alder type reaction on a pyrazine, which is perhaps unlikely in view of the unactivated nature of the double bond." Birch offered two other mechanisms, one involving the intermediacy of an epidithiapiperazinedione as shown in Fig. 3, (although he also admits that no sulfur compounds could be detected in the mold products) and also alludes to a radical process; no details are given on the latter concept.

No other publications on the biogenesis of the brevianamides appeared in the literature until the late 1980s when a total synthesis of (–)-brevianamide B was reported by Williams et al. [16, 17]. Due to the very low levels of brevianamide B produced by *Penicillium brevicompactum* relative to brevianamide A, the authentication of the synthetic (–)-brevianamide B was secured through the semi-synthetic conversion of natural (+)-brevianamide A into (–)-brevianamide B as originally described by Birch and Russell [9]. The semi-synthetic and totally synthetic brevianamide B samples both exhibited the same specific optical rotation ($[\alpha] = -124°$) and CD spectra. These workers found that *natural* brevianamide B obtained directly from cultures of *Penicillium brevicompactum* had a specific optical rotation $[\alpha] = +124°$, and a CD spectrum with Cotton effects of *exactly opposite sign* to that of the synthetic and semi-synthetic materials. The fact that the optical rotation of natural (+)-brevianamide B was of equal magnitude and opposite sign to that of semi-synthetic and totally synthetic (–)-brevianamide B suggests that the natural metabolite is

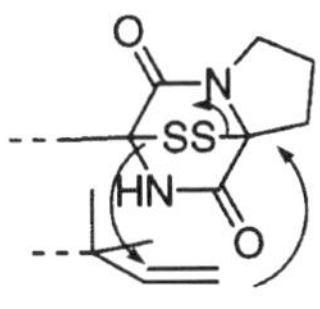

Fig. 3. Epidithiapiperazinedione proposal of Birch [15]

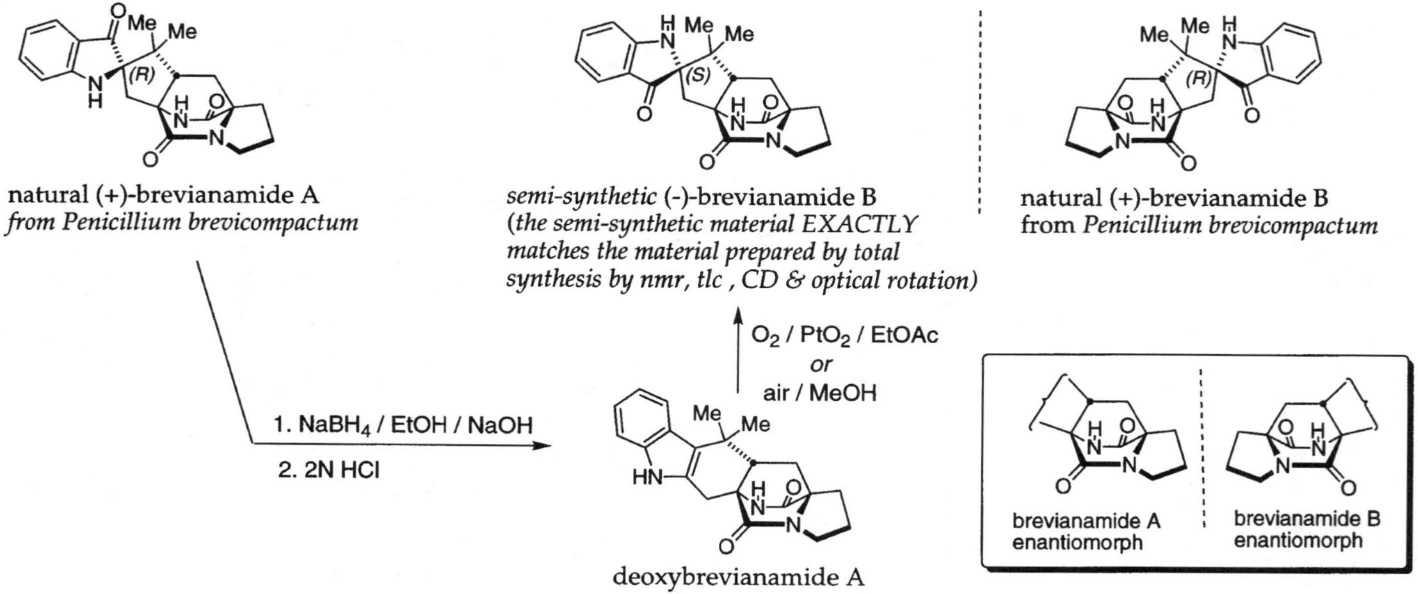

Scheme 7. Stereochemical relationships of the brevianamides as determined by Williams et al. [16, 17]

optically pure; the significance of this will be discussed in more detail below. This rather unexpected finding mandates that the absolute configuration of brevianamide A and brevianamide B are *enantiomorphic* with respect to the bicyclo [2.2.2] moiety but that each metabolite has the (R)-absolute configuration at the sprio-ψ-indoxyl stereogenic center; these relationships are summarized in Scheme 7.

Based on this unusual finding, Williams et al. proposed a biogenetic pathway that could accommodate the formation of the two enantiomorphic bicyclo [2.2.2] ring systems in the respective natural metabolites as shown in Scheme 8 [18]. Reverse prenylation of brevianamide F as outlined by Birch to deoxybrevianamide E followed by two-electron oxidation of the tryptophyl moiety and enolization of the proline residue would provide an *achiral* azadiene. Cycloaddition can, in principle, occur from either face of the azadiene system, thus yielding the *racemic* (or partially racemic) hexacyclic indole cycloaddition products. Since both natural brevianamides A and B have the (R)-absolute stereochemistry at the indoxyl spiro-center, the involvement of an (R)-selective indole oxidase which only recognizes the binding orientation of the 2,3-disubstituted indole was invoked to effect the "resolution" of the hypothetical racemic Diels-Alder adduct into two optically pure, diastereomeric hydroxyindolenines. Subsequent pinacol-type ring contraction in a stereospecific manner would thus furnish the respective optically pure (+)-brevianamides.

To test this hypothesis, Williams et al. synthesized the hypothetical Diels-Alder cycloadduct in *racemic* form bearing a ^{13}C label at the tryptophyl methylene carbon [19]. Biosynthetic feeding of this potential precursor suspended in DMSO to cultures of *Penicillium brevicompactum* yielded both brevianamides A and B but, within the limits of experimental error, there was no evidence for the incorporation of the labeled precursor into either metabolite by NMR and/or mass spectral analysis (see Scheme 9). In addition, culture filtrates and mycelia from *Penicillium brevicompactum* were extracted and examined for the presence of the hypothetical Diels-Alder cycloaddition product, yet evidence for the existence of this substance was not obtained.

Although these results do not rigorously exclude the possibility of the hexacyclic indole as a biosynthetic precursor to the brevianamides (poor solubility, cell penetration, uptake, k_Ds, etc. may limit the access of the labeled, synthetic precursor to the biosynthetic machinery in the cells) enough doubt had been cast on the pathway depicted in Scheme 8 such that alternatives were pursued.

The synthesis of [8-methylene-$^3H_2$]deoxybrevianamide E was carried out by Williams et al. [19] to probe the intermediacy of deoxybrevianamide E in the biosynthesis of brevianamides A, B, and E. It was found that this substance, when fed to cultures of *Penicillium brevicompactum*, was very efficiently incorporated into brevianamide E, brevianamide A, and brevianamide B. The data is reproduced in Table 3.

When radioactive brevianamide E was re-fed to the cultures, there was no significant incorporation into either brevianamide A or B, clearly indicating that brevianamide E is a dead-end, shunt metabolite. Based on the available data, these workers proposed the biosynthetic pathway illustrated in Scheme 10 that can accommodate the formation and stereochemistry of all of the relevant

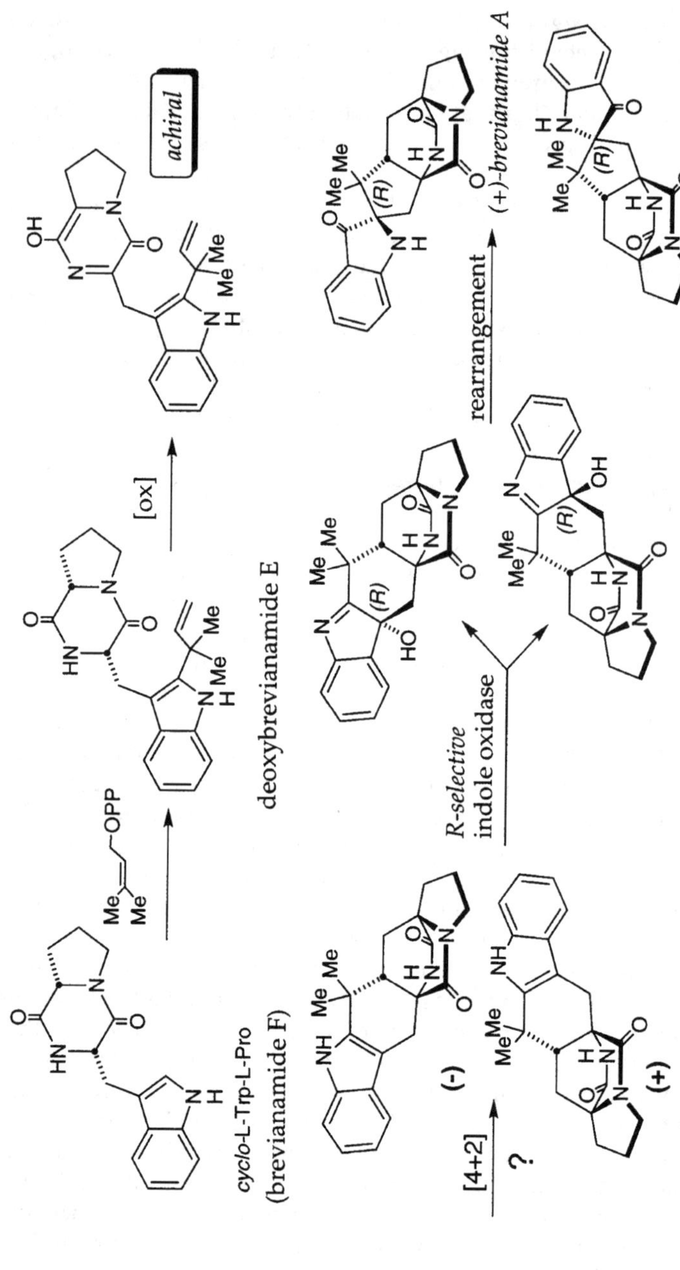

Scheme 8. Early proposal for brevianamide biosynthesis by Williams et al. [18]

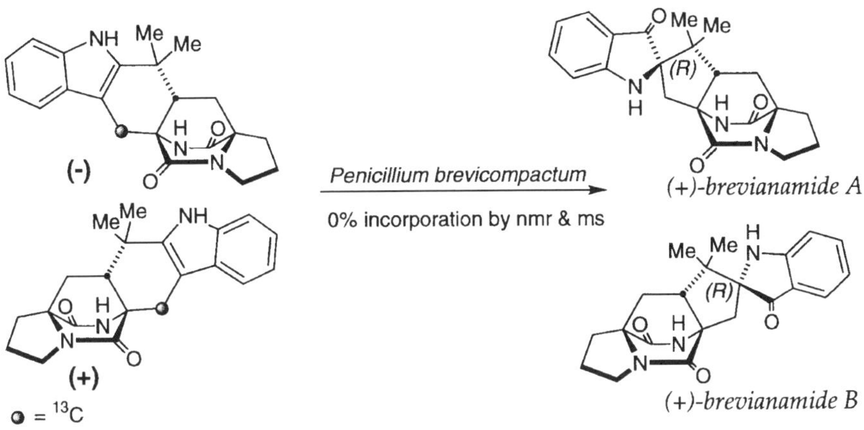

Scheme 9. Biosynthetic feeding of the hexacyclic indole [19]

Table 3. Incorporation of [8-*methylene*-$^3H_2$]deoxybrevianamide E into brevianamides A, B, and E [19]

Metabolite	% Specific incorporation	% Total incorporation
brevianamide A	12.1	7.8
brevianamide B	1.4	0.9
brevianamide E	38.5	24.9

metabolites. Thus, reverse prenylation of brevianamide F (*cyclo*-L-Trp-L-Pro) to deoxybrevianamide E followed by an (R)-selective oxidation at the 3-position of the indole gives an *(R)*-hydroxyindolenine that suffers one of two fates: (1) *irreversible* nucleophilic addition of the tryptophyl amide nitrogen forming brevianamide E or (2) a stereospecific pinacol-type rearrangement to an as yet, unidentified ψ-indoxyl (Scheme 10, box). This hypothetical substance was envisioned to constitute the penultimate intermediate that must undergo three final reactions to the brevianamides: (1) two-electron oxidation of the tryptophyl α-carbon, (2) enolization of the prolyl methine, and (3) intramolecular Diels-Alder cycloaddition from two of the four possible diastereomeric transition states to yield brevianamides A and B.

Based on the existence of numerous hydroxamate-containing secondary metabolites such as mycelianamide, it can be further proposed that the tryptophyl oxidation proceeds through a hydroxamic acid intermediate (shown in Scheme 10). This functionality, an excellent chelator of (a non-heme) Fe(III), for instance, could further function to assist the elimination of water to form the imino portion of the azadiene. Since the prolyl methine (pKa > 30) must also be removed at the active site of an enzyme, a metal-coordinated amide might also be essential to lower the activation barrier for the deprotonation step. Finally, the intramolecular Diels-Alder cycloaddition might enjoy a catalytic effect of azadiene-metal coordination.

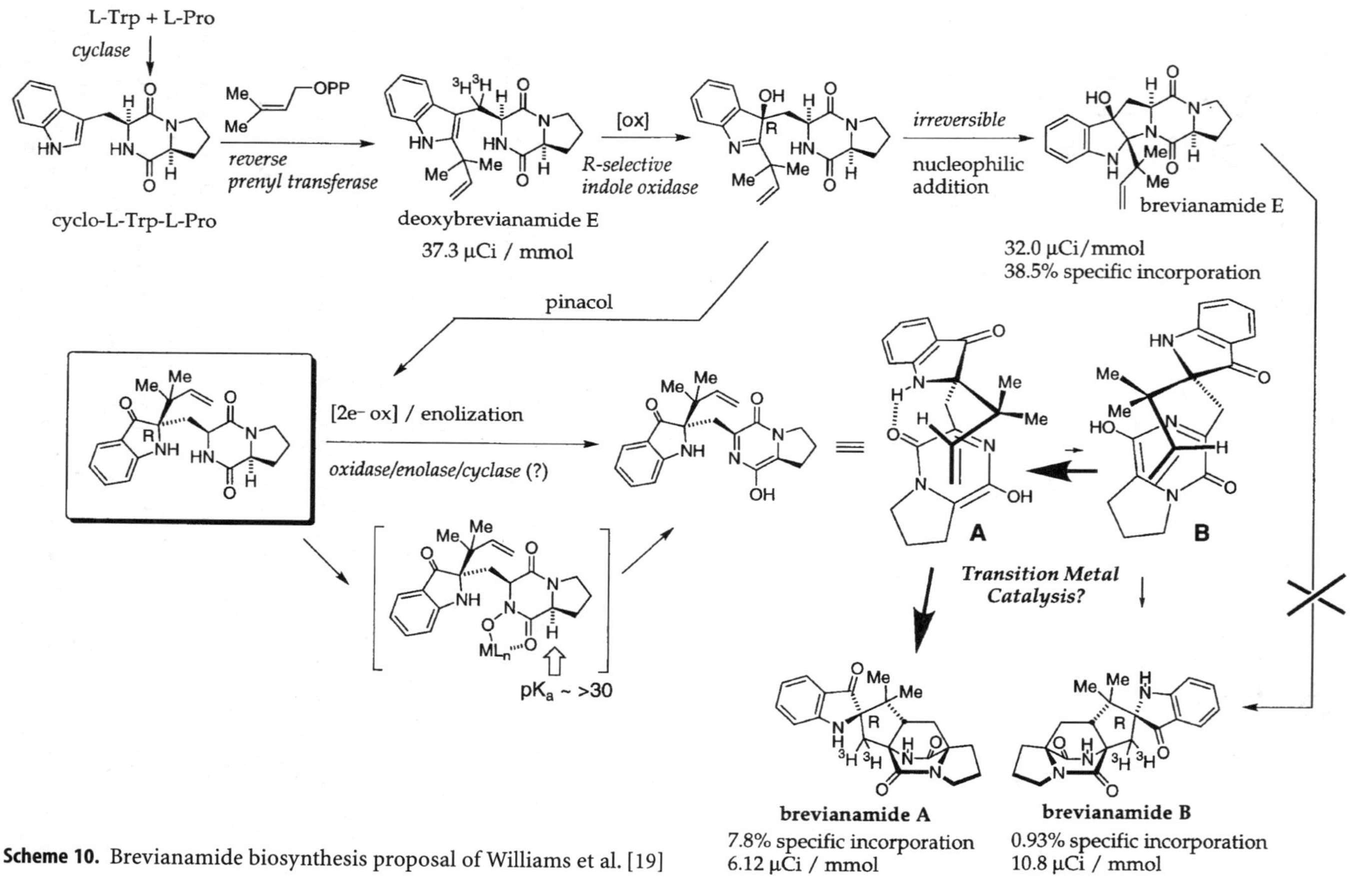

Scheme 10. Brevianamide biosynthesis proposal of Williams et al. [19]

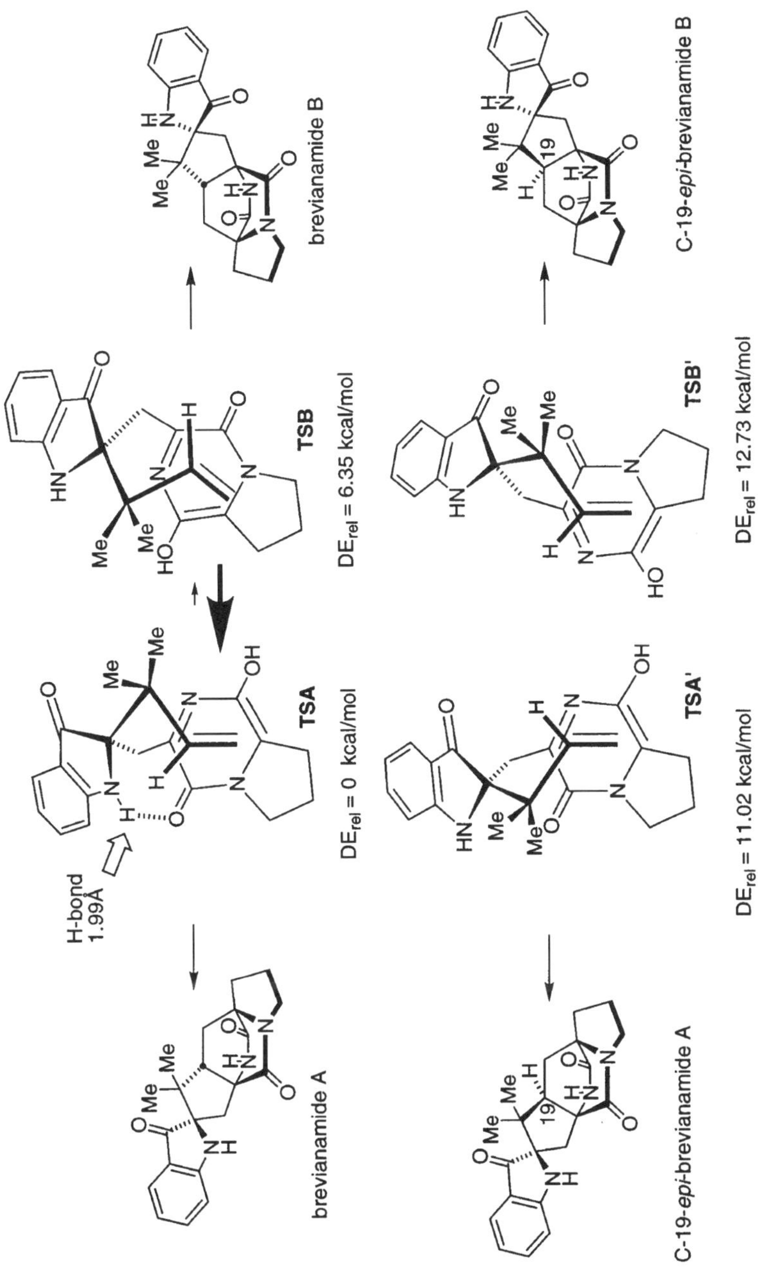

Scheme 11. *Ab initio* calculations on the putative Diels-Alder cycloadditions by Domingo et al. [20]

Table 4. Calculated potential energy barriers and relative energies (kcal/mol) for TSA, TSB, TSA' and TSB'

	3–21G/3–21G		6–31G*/3–21G	
	ΔE_a	ΔE_{rel}	ΔE_a	ΔE_{rel}
TSA	31.81	0.00	38.68	0.00
TSB	39.42	7.61	45.03	6.35
TSA′	44.17	12.36	49.71	11.02
TSB′	47.07	15.26	51.41	12.73

In order to gain some insight into the putative biosynthetic Diels-Alder cycloaddition, Domingo et al. carried out *ab initio* calculations on the four possible transition structures from the azadiene as summarized in Scheme 11 [20]. The calculations were performed with analytical gradients at *ab initio* 3–21 G and 6–31 G* basis sets within Hartree-Fock procedures. This Diels-Alder reaction appears to be controlled by frontier molecular orbitals $HOMO_{(dieneophile)}/LUMO_{(azadiene)}$. All four diastereomeric transition structures were examined and it was found that the $lowest_{(rel)}$ energy transition structure is TSA which leads to brevianamide A (see Table 4). The next highest energy transition structure is TSB, which lies approximately 6–7 kcal/mol above TSA; this transition structure would produce brevianamide B. An intramolecular H-bond was found in TSA between the indoxyl N-H and the tryptophyl carbonyl group which can account for a large fraction of the relative stabilization energy of this transition structure. After TSB, the next highest energy transition structures are TSA′ and TSB′, which would lead to C-19-*epi*-brevianamide A and C-19-*epi*-brevianamide B, respectively. Williams et al. have previously described the total synthesis of C-19-*epi*-brevianamide A [17, 21] and, using this authentic specimen, have been unable to detect this substance as a metabolite of *Penicillium brevicompactum* [22]. This theoretical study therefore predicts that the relative distribution of brevianamides A and B from *Penicillium brevicompactum* is a direct manifestation of the relative energies of the four diastereomeric transition structures: brevianamide A is the major metabolite, brevianamide B is a minor metabolite and neither of the C-19-*epi*- structures are found.

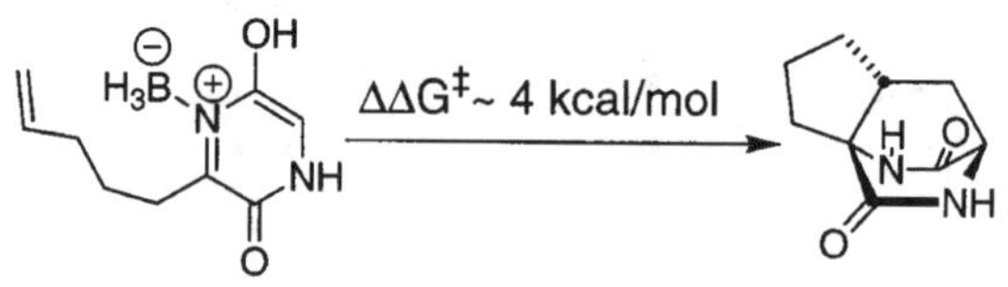

Scheme 12. Lewis acid activation of the proposed biosynthetic intramolecular [4+2] cycloaddition [20]

Domingo et al. also examined the effect of a Lewis acid (BH_3) on the simpler model system shown in Scheme 12. It was found that Lewis acid complexation lowered the potential energy barrier by 3–5 kcal/mol relative to the parent system. The speculation presented above regarding the complexation of the aza-diene system to a non-heme metal, such as Fe(III), appears to be supported by this theoretical insight.

Further work needs to be done to confirm the pathway outlined in Scheme 10 and in particular to elucidate the possible involvement of protein organization and/or catalysis of the biosynthetic reactions culminating in the formation of brevianamides A and B.

Very little has appeared in the literature concerning the biosynthesis of austamide. The occurrence of deoxybrevianamide E and deoxyaustamide in cultures extracts of *Apergillus ustus* suggests the following biosynthetic pathway (Scheme 13). Reverse prenylation of *cyclo*-L-Trp-L-Pro (brevianamide F) furnishes deoxybrevianamide E. Oxidative ring closure of the tryptophyl amide nitrogen across the isoprene-derived olefinic moiety and desaturation of the proline ring would yield the eight-membered ring metabolite deoxyaustamide. Finally, oxidation at the 3-position of the 2,3-disubstituted indole followed by a pinacol-type rearrangement would provide the spiro ψ-indoxyl austamide.

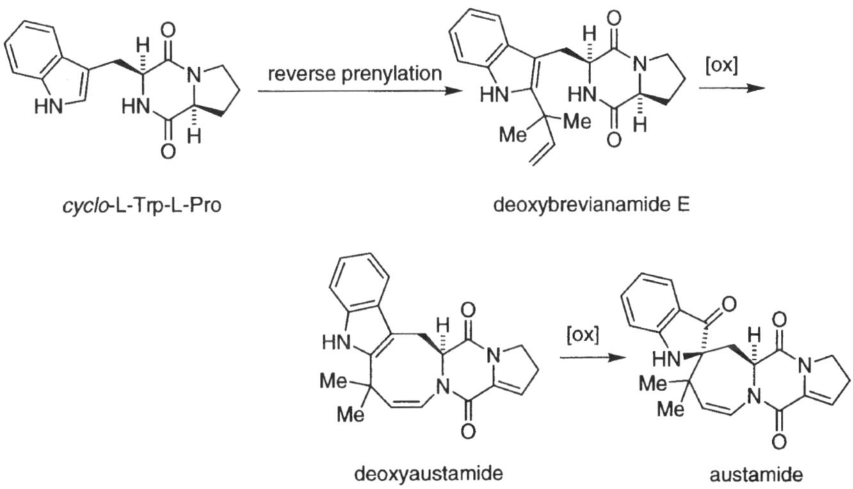

Scheme 13. Possible biosynthetic pathway to austamide

Harrison has suggested that fumitremorgin B and austamide may arise biosynthetically from an isoprene-derived imine such as that shown in Scheme 14 [23]. Cyclization of the imine either via path A or path B result in the formation of the fumitremorgin B and deoxyaustamide ring systems, respectively.

A simple experimental system designed to test this hypothesis was conducted as shown in Scheme 15 [23]. Condensation of L-tryptophan methyl ester and tryptamine with 3-methylbut-2-enal gave the corresponding imines which were recal-

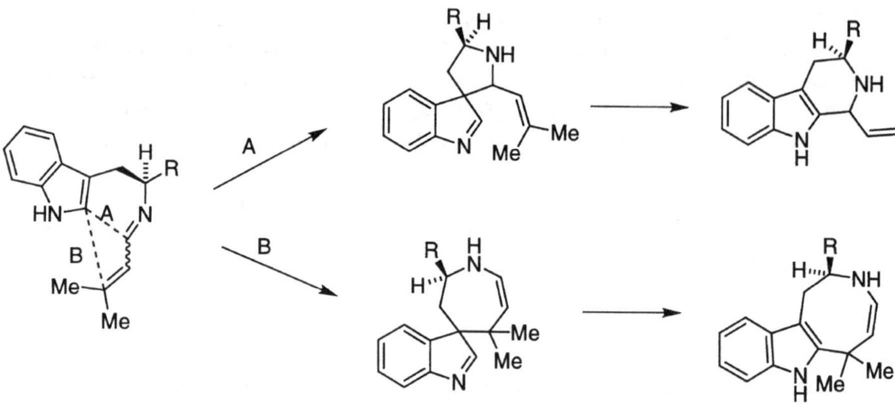

Scheme 14. Possible biosynthetic modes of ring closure to austamide and fumitremorgin [23]

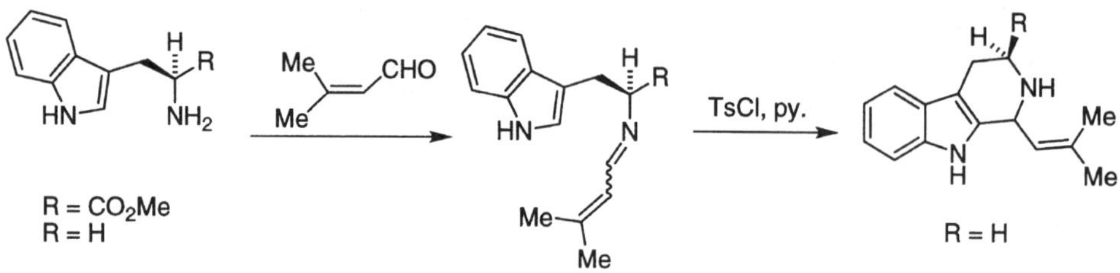

Scheme 15

citrant to cyclization under acidic or neutral conditions. The tryptamine-derived imine was induced to undergo the Pictet-Spengler cyclization in the presence of *p*-toluenesulfonyl chloride in pyridine to the tetrahydro-*β*-carboline derivative (Scheme 15). In neither case were products derived by pathway B (Scheme 14) detected and no subsequent work on the biosynthesis of austamide has been published.

It might also be parenthetically noted that the absolute stereochemistry of austamide was determined by X-ray analysis on 5-bromo-12*S*-tetrahydroaustamide [7] (obtained from austamide by hydrogenation followed by bromination) and that the absolute configuration at the spiroindoxyl stereogenic center (C-2) was determined to be *S*. This assignment was based on the hydrolytic liberation of *S*-proline from 12,13-dihydroaustamide; the remaining stereogenic centers at C-9 and C-2 (the spiro center) were assigned relative to that for C-12. This is in marked contrast to brevianamides A and B which both possess the R-absolute configuration at the corresponding spiroindoxyl stereogenic center. Assuming that this assignment is correct, it is interesting that *Aspergillus ustus* must therefore have a *pro-S*-selective indole oxidase and that *Penicillium brevicompactum* utilizes a *pro-R*-selective indole oxidase.

2.2 The Paraherquamides, Marcfortines, and Related Alkaloids

The paraherquamides, are a structurally complex family of alkaloids comprised of two isoprene units, tryptophan, and various proline derivatives. Many members of this family display potent anthelmintic and antinematodal activities and have been under intensive investigation for use in veterinary medicine to treat various intestinal parasites [24–28]. The parent and most potent member, paraherquamide A, was isolated from cultures of *Penicillium paraherquei* as first described by Yamazaki and Okuyama in 1981 [29]. The simplest member, paraherquamide B, plus five other structurally related paraherquamides C–G were isolated from *Penicillium charlesii* (*fellutanum*) (ATCC 20841) in 1990 at Merck & Co. [30] and concomitantly at SmithKline Beecham [31a]. Subsequently, three additional related compounds, VM55596, VM55597, and VM55599 were discovered by the same group at SmithKline Beecham from *Penicillium* strain IMI 332995 [31b]. Very recently, a Pfizer group [32] reported the isolation of anthelmintic metabolites VM54159, SB203105, SB200437 along with the non-tryptophan-derived metabolites possessing the common bicyclo [2.2.2] nucleus aspergillimide (identical to asperparaline A [33]) and 16-keto-aspergillimide from *Aspergillus* strain IMI 337664. This report and a recent paper by Whyte and Gloer [34] describing the isolation of sclerotamide from *Aspergillus sclerotiorum* constitute the first examples of paraherquamide derivatives isolated outside of the *Penicillia*.

The structurally related alkaloid marcfortine, first described by Polonsky et al. in 1980 [35] is rendered from two isoprene units, tryptophan and pipecolic acid. The structures of the members of this family of mold metabolites are presented in Fig. 4. The paraherquamides differ with respect to substitution and oxygenation in the proline ring and the prenylated oxindole ring; paraherquamide B is the simplest member of the paraherquamide family, being comprised of the amino acids proline, tryptophan, and two isoprene units.

Finally, Zeeck et al. isolated the interesting substances aspergamide A and B from *Aspergillus ochraceus* [36]. These metabolites, including VM55599, are the only natural members of the paraherquamide family that do not have a spiro-ring system but instead a 2,3-disubstituted indole-derived system.

The biosynthesis of marcfortine has been investigated at the Pharmacia and Upjohn company [37]. Industrial interest in the biosynthesis of marcfortine A, which does not display the potency that paraherquamide A possesses, is presumably due to the report from the Pharmacia and Upjohn group demonstrating that marcfortine A can be semi-synthetically converted into paraherquamide A [38]. Kuo et al. found that marcfortine is derived from L-tryptophan (oxindole moiety), L-methionine (via SAM methylation at the γ-N; C29), L-lysine (pipecolic acid residue), and acetate (isoprene units) (Scheme 16) [37].

These results clearly demonstrate that the isoprene moieties are of mevalonate origin. The pipecolic acid moiety, being derived from L-lysine, can arise via two biochemical pathways as shown in Scheme 17. In order to discriminate between these pathways (A and B), Kuo et al. fed individually [α-^{15}N]- or [ε-^{15}N]-L-lysine to a marcfortine A-producing *Penicillium* strain UC7780 and found that

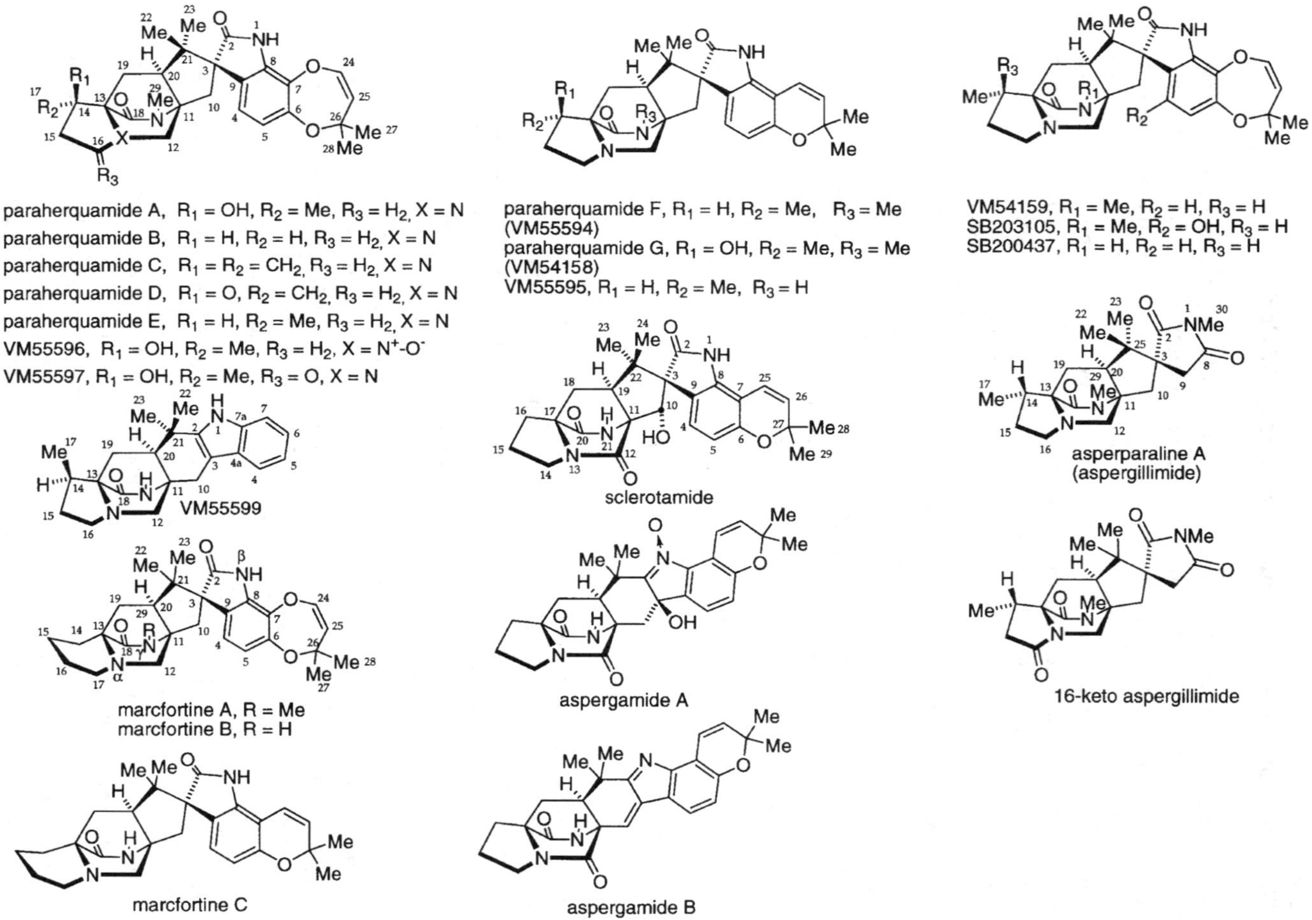

Fig. 4. Structures of the paraherquamides and related alkaloids

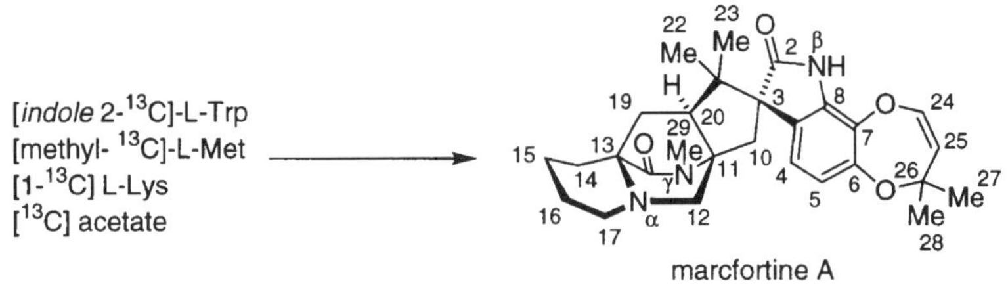

Scheme 16. Biosynthetic constitution of marcfortine A [37]

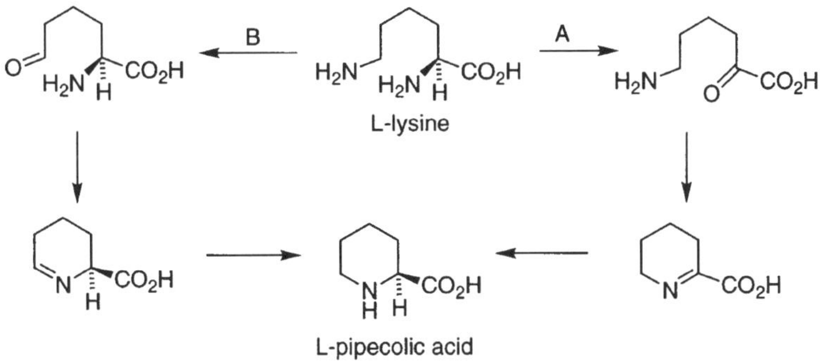

Scheme 17. Possible biosynthetic pathways from L-lysine to pipecolic acid [37b]

the α-amino group (pathway A) is lost in the biosynthetic production of the pipecolic acid moiety [37b]. Specifically, it was found that [α-^{15}N]-L-lysine gave incorporation at the α-, β-, and γ-nitrogen atoms with a rate of 1.8%, 13%, and 9%, respectively. Incorporation of [ε-^{15}N]-L-lysine at the α-, β-, and γ-nitrogen atoms gave incorporation rates of 54%, 2.7%, and 3.9%, respectively. The enrichments were measured by FAB-MS, ^{1}H NMR and HMBC experiments. The authors state that this is the first example of the HMBC technique being successfully employed to quantitate the enrichment of an isotope. The low levels of ^{15}N-enrichment at the other positions (β- and γ-) was attributed to lysine catabolism and reconstitution.

The origin of the β-methylproline moiety present in paraherquamide A and several congeners has recently been investigated. Examination of the absolute stereochemistry of paraherquamide A, which possesses the (S)-absolute stereochemistry at C-14, led Williams et al. to speculate that the methylated proline may be derived from L-isoleucine as opposed to proline and S-adenosylmethionine (SAM) and this possibility was experimentally tested in *Penicillium fellutanum* (ATCC 20841) as shown in Fig. 5 [39]. The position of ^{13}C incorporation in paraherquamide A was determined using ^{13}C NMR and the percentage of the labeled amino acid incorporated was also determined using ^{13}C NMR.

These workers found that 1-[^{13}C]-L-tryptophan was incorporated as expected (2.5%) with the label at C-12. The methyl-[^{13}C]-methionine was not incorporated in the β-methyl-proline ring but rather only at C-29, the *N*-methyl position of the monoketopiperazine ring (0.6%). Feeding of 1-[^{13}C]-L-isoleucine to *Peni-*

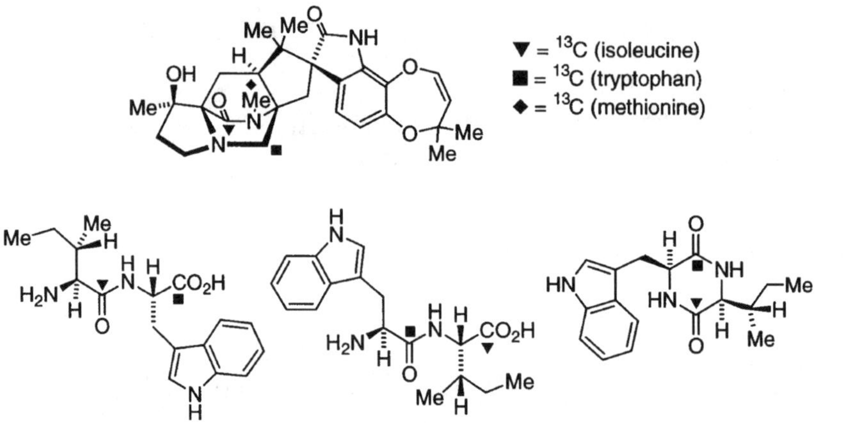

[1-^{13}C]-L-Ile-[1-^{13}C] -L-Trp   [1-^{13}C] -L-Trp-[1-^{13}C]-L-Ile   *cyclo*-[2-^{13}C] -L-Trp-[2-^{13}C]-L-Ile

Fig. 5. Biosynthetic derivation of the β-methyl proline moiety of paraherquamide A [39]

cillium fellutanum (ATCC 20841), followed by harvesting the cells and isolation of paraherquamide A revealed that the labeled L-isoleucine was incorporated into the monoketopiperazine ring system in high isotopic yield (3.3 – 3.7 %) with the label at C-18.

Since L-isoleucine forms the β-methylproline ring of paraherquamide A, *cyclo*-L-Trp-L-β-methylproline or *cyclo*-L-Trp-L-Ile are plausible precursors. There are numerous possible sequences of events that might occur in the formation of the final β-methylproline ring system. Formation of the dipeptides NH_2-L-Ile-L-Trp-COOH or NH_2-L-Trp-L-Ile-COOH and dehydration to *cyclo*-L-Trp-L-Ile followed by oxidation of the terminal carbon of L-Ile and cyclization to form the β-methylproline moiety would result in *cyclo*-L-Trp-L-β-methylproline. Another possibility involves oxidation of the L-Ile followed by cyclization and reduction to afford β-methylproline followed by coupling to L-Trp to give *cyclo*-L-Trp-L-β-methylproline. Many other possibilities exist that would involve formation of the β-methylproline ring at a later stage.

Williams et al. investigated the simplest of these possibilities by synthesizing and feeding doubly labeled NH_2-1-[^{13}C]-L-Ile-1-[^{13}C]-L-Trp-COOH; NH_2-1-[^{13}C]-L-Trp-1-[^{13}C]-L-Ile-COOH and 2,5-[$^{13}C_2$]-*cyclo*-L-Trp-L-Ile to *P. fellutanum* [39]. After growing and harvesting the cells, 1.2 – 1.8 % incorporation at C-18 and 0.4 – 0.9 % incorporation at C-12 was evidenced by ^{13}C NMR. The ^{13}C NMR spectra of the paraherquamide A so produced did not provide compelling evidence for site-specific incorporation of both labels from the intact dipeptides and there was no significant incorporation of 2,5-[$^{13}C_2$]-*cyclo*-L-Trp-L-Ile. This low level of incorporation is more consistent with dipeptide hydrolysis, re-incorporation of the individual amino acids presumably coupled with additional metabolic degradation, and reconstitution of ^{13}C-enriched building blocks. Based on these observations, it seems likely that β-methylproline is formed prior to conjugation to tryptophan although other possibilities exist.

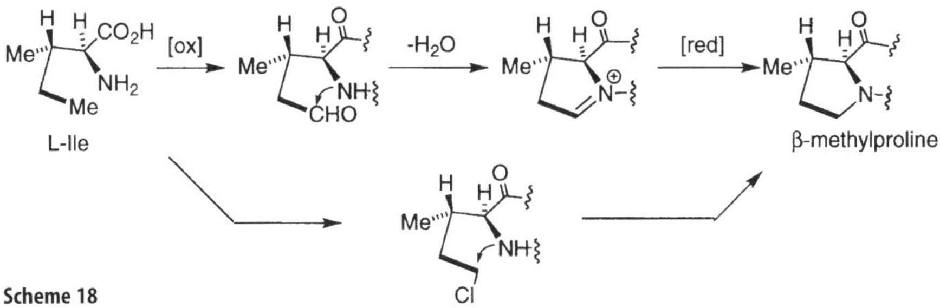

Scheme 18

Oxidative cyclization of the nitrogen atom onto the C-5-methyl group of isoleucine appears to be a unique biosynthetic transformation. Two reasonable pathways, depicted in Scheme 18 would involve 4-electron oxidation of the distal side-chain methyl group to an aldehyde followed by cyclization and loss of water to produce an iminium species; subsequent reduction (or in the case of VM55597, oxidation) furnishes the β-methylproline derivative. Alternatively, direct functionalization of the C-5 methyl group, for example, by chlorination and closure can also be envisioned to lead to β-methylproline; a related example of this type of reaction has been observed by Arigoni and Looser [40] in the biosynthesis of victorin C.

It is also interesting to note that β-methylproline residues can be found in several peptide antibiotics such as bottromycin, scytonemin A, roseotoxin B in addition to several members of the paraherquamide family. Arigoni and Kellenberger [41] have recently shown that the methyl group in the β-methylproline in bottromycin, a metabolite of *Streptomyces bottropenis*, is derived from *S*-adenosylmethionine via methylation of proline (Scheme 19) and have proposed a radical mechanism for this reaction.

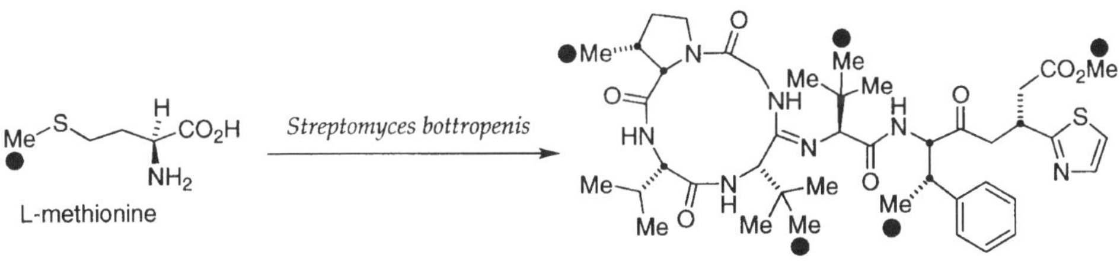

Scheme 19

The identification of L-isoleucine as the biosynthetic building block of paraherquamide A has posed an interesting stereochemical paradox with respect to the origin and stereochemistry of the β-methylproline in the natural metabolite VM55599 which is produced by a paraherquamide-producing fungi, *Penicillium* strain IMI332995. The relative stereochemistry of VM55599 was determined by ^{1}H NMR nOe experiments by a Smith-Kline Beecham group [31b] but the absolute stereochemistry has not been determined. Significantly, the methyl group in the *β*-methylproline ring of VM55599 is disposed *syn-* to the bridging isoprene moiety whereas, in paraherquamides A, E, F, G, VM55596, VM55597, VM55595, VM54159, SB203105, SB200437, and the aspergillimides, the methyl group of the *β*-methylproline ring is disposed *anti-* to the bridging isoprene moiety of the bicyclo [2.2.2] ring system.

Using the assumption that the absolute stereochemistry at C-20 of VM55599 is S- (the same as that found in the paraherquamides), the stereochemistry at C-14 was assigned as R-, which is the *opposite* to that found in paraherquamides. The side-chain stereochemistry of L-isoleucine is preserved in the biosynthesis of paraherquamide A with hydroxylation at C-14 proceeding with net *retention* of configuration. If L-isoleucine is also the precursor to the *β*-methylproline ring of VM55599, it must follow that the bicyclo [2.2.2] ring system of this compound must be *enantiomorphic* to that of the paraherquamides (see Fig. 6). Alternatively, VM55599 may be derived from L-*allo*-isoleucine; this would result in the (R)-stereochemistry at C-14 and would accommodate the same bicyclo [2.2.2] ring system absolute stereochemistry as paraherquamide A. Finally, since the methyl group of the *β*-methylproline ring of VM55599 is *syn-* to the isoprene unit comprising the bicyclo [2.2.2] ring system, this implies that the putative biosynthetic Diels-Alder cyclization to form this system occurs from the more hindered face of the azadiene system. The stereochemical paradox posed by VM55599 raises numerous interesting questions concerning the biogenesis of these substances: is VM55599 a biosynthetic precursor to various paraherquamide family members or is VM55599 a minor shunt metabolite with the opposite absolute stereochemistry of the bicyclo [2.2.2] ring system? The mechanism of formation of the bicyclo [2.2.2] ring system in both series continues to pose an interesting stereochemical and enzymological puzzle.

In the course of isotopic labeling studies aimed at examining the origin of the isoprene units in the paraherquamide structure, Williams et al. discovered an unexpected stereochemical distribution of the geminal methyl groups derived from DMAPP [42]. These workers carried out feeding experiments in *Penicillium*

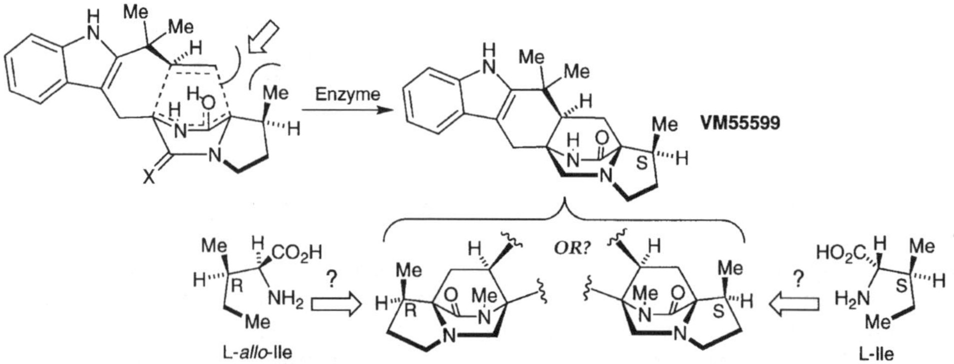

Fig. 6. The VM55599 stereochemical paradox [39]

fellutanum with [U-$^{13}C_6$]-glucose and [$^{13}C_2$]-acetate which aid in distinguishing between the classical mevalonate and deoxyxylulose pathways. Specific incorporation of intact C_2 units was observed in agreement with the mevalonic acid pathway. However, in the first C_5 fragment (C19 to C23) the observed couplings mean that C19 is coupled to C20, while C21 is coupled to C22 or C23 but not to both simultaneously. For the second C_5 unit (C24 to C28) the coupling constants show that C24 and C25 are coupled, while C26 is coupled to C27. In this case, although C-28 shows ^{13}C enhancement, C28 shows no coupling. It is significant that in the C_5 fragment formed by carbons C24 to C28, carbon C28 shows no coupling with C26, while C27 does (Fig. 7). This means that the methyl groups in DMAPP are not equivalent in the biosynthesis of the geminal methyl groups at C-21 of this metabolite. In contrast, in the other C_5 fragment, formed by carbons C19 to C23, both methyl groups show coupling with C20, although not simultaneously. The stereochemical integrity of DMAPP in forming the dioxepin ring system is thus left intact.

This unexpected result was interpreted to mean that the *reverse* prenyl transferase presents the olefinic π-system of DMAPP in a manner in which both faces of the π-system are susceptible to attack by the 2-position of the indole moiety. The simplest explanation is to invoke binding of the DMAPP in an "upside down" orientation relative to "normal" prenyl transferases which permits a facially non-selective S_N' attack on the π-system as shown in Scheme 20. It was speculated that in this situation the pyrophosphate group is likely anchored in the enzyme active site with the hydrophobic isopropenyl moiety being presented in a conformationally flexible (**A** ⇆ **B**) disposition with respect to the tryptophan-derived substrate (Scheme 20). This is in contrast to the normal mode of prenyl transfer where the nucleophilic displacement at the pyrophosphate-bearing methylene carbon occurs with inversion of stereochemistry at carbon with the hydrophobic tail of DMAPP buried in the enzyme active site.

In contrast, the methyl groups in the other C_5 unit (dioxepin moiety) are clearly differentiated; therefore, it is quite likely that this C_5 group (carbons 24 to 28) is introduced in the molecule via direct alkylation with DMAPP by a normal prenyl transferase followed by a stereospecific net oxidative addition to the olefinic system.

A plausible mechanism of several possibilities for the formation of this ring system, is via face-selective epoxidation followed by ring-opening and dehydration. Since this isoprene unit is introduced without loss of stereochemical integrity, it was deemed unlikely that the *Penicillia* sp. has a mechanism for scrambling the DMAPP via a dimethyl vinyl carbinol-type intermediate which would necessarily provide stereochemically scrambled isoprene equivalents to the cell's cytosolic pool of DMAPP.

These experiments clearly indicate that the C_5 units in paraherquamide A are introduced in stereofacially distinct manners. Since it has been established that prenylation of the indole moiety in the biosynthesis of the structurally related brevianamides occurs in an analogous fashion to that postulated for paraherquamide A [19], the prenyl transferase that installs this C_5 unit must display DMAPP to the 2-position of the indole in a π-facially indiscriminate manner.

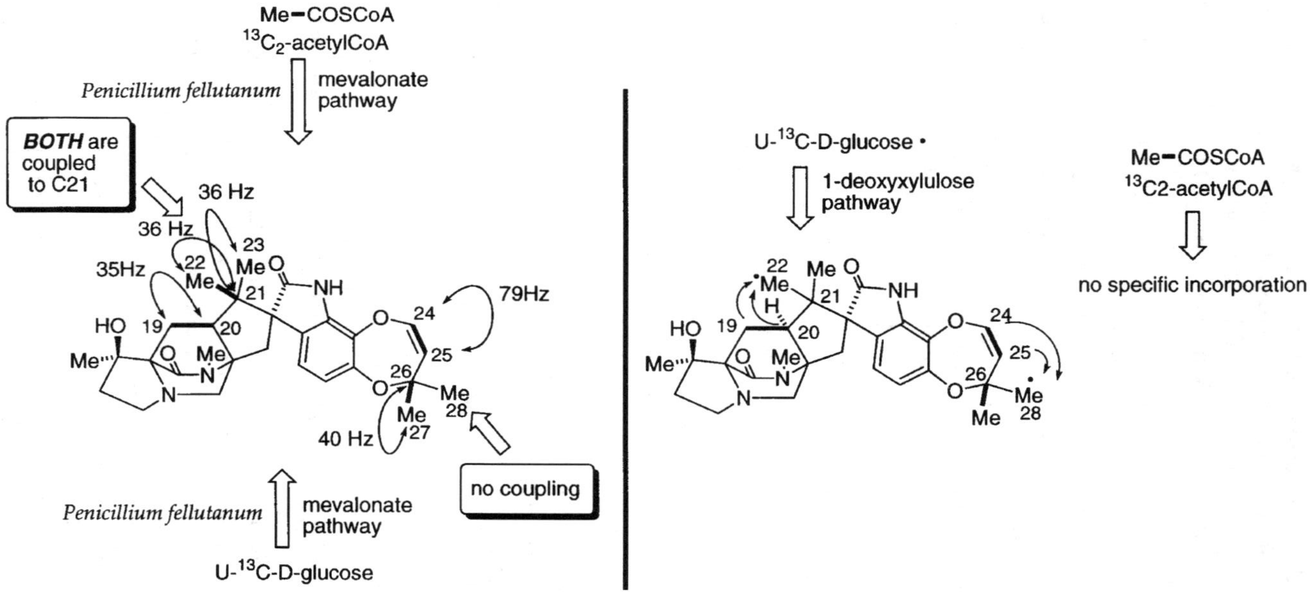

Fig. 7. ^{13}C labeling in paraherquamide A corresponding to a feeding experiment with [U-$^{13}C_6$]-D-glucose and [$^{13}C_2$]-acetate via the mevalonate and 1-deoxy-D-xylulose pathways. *Thick lines* represent intact acetate units and *arrows* represent couplings expected and observed in the ^{13}C-NMR spectrum

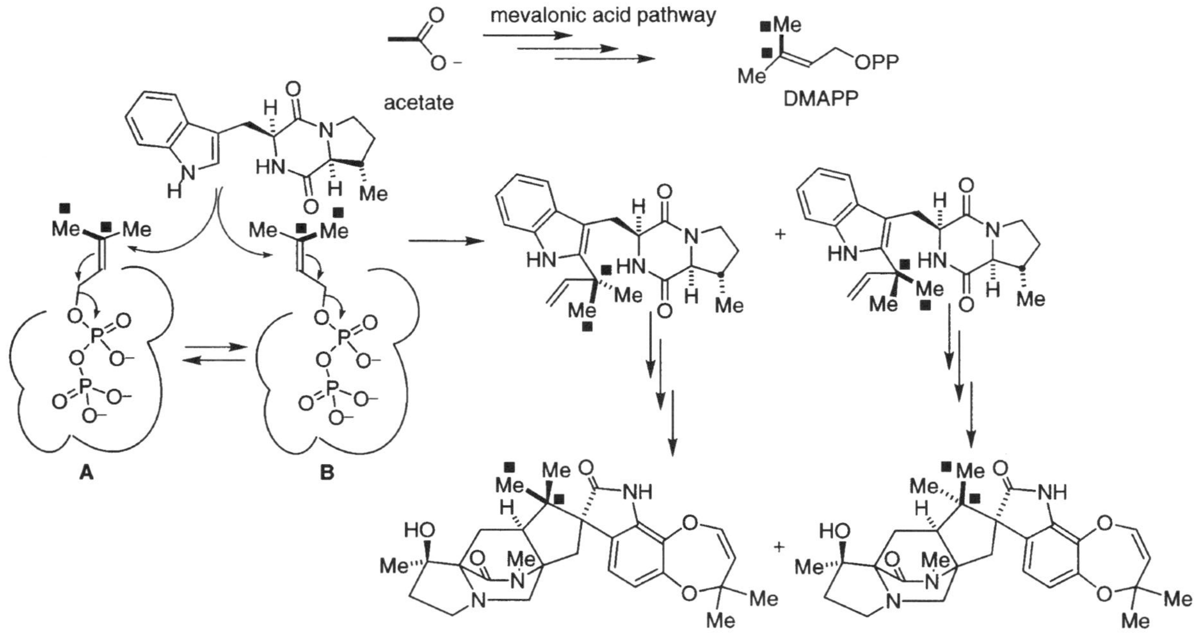

paraherquamide A

Scheme 20. A possible biosynthetic sequence that may explain why C22 and C23 are rendered equivalent in the biosynthesis of paraherquamide A. *Thick bonds with the black square* represent one intact C2 unit from acetate, incorporated in C3/C5 of individual DMAPP molecules, and in C21, C22, and C23 of paraherquamide A

This work demonstrates the first case where *both* a non face-selective *and* a face-selective addition to the tri-substituted olefinic portion of DMAPP has occurred within the same molecule. A related observation with respect to the loss of stereochemistry in the reverse prenylated indole alkaloids echinulin and roquefortine have been reported and are discussed elsewhere in this article.

3
Roquefortine

Roquefortine (Fig. 8) is a neurotoxic mold metabolite produced by *Penicillium roqueforti* that was first isolated by Polonsky et al. 1976 [43]. Roquefortine is also produced by *Penicillium crustosum* and *Penicillium oxalicum*. This substance is derived from the oxidative cyclization of tryptophan, histidine, and dimethylallyl pyrophosphate.

Barrow et al. [44] described the incorporation of [2-^{14}C]-labeled mevalonic acid, [*methylene*-^{14}C]-labeled tryptophan and [2-^{14}C]-labeled histidine into roquefortine in 0.08%, 0.15%, and 1.12%, respectively, utilizing cultures of *Penicillium roqueforti* (Scheme 21). Although the incorporation of mevalonate is somewhat low, later work by Gorst-Allman et al. [45] demonstrated the incorporation of [1,2-$^{13}C_2$]-labeled acetate, thus confirming the mevalonate origin of the isoprene unit.

The authors speculate about the incorporation of the "reverse" prenyl unit in this and related molecules such as echinulin, lanosulin, oxaline, austamide, and the brevianamides and infer the rearrangement of the *N*-prenyl to the "reverse" 2-prenyl structure as shown in Scheme 22. These workers investigated the involvement of the 2-position of the indole by deuterium labeling using an auxotrophic mutant strain of *Penicillium roqueforti* that was apparently deficient in anthranilate synthetase.

Utilizing multiply deuterated tryptophan it was found that, of the original >95% deuterium label at the 2-position in the starting tryptophan, <5% deuterium was observed in the isolated roquefortine at C6 (Scheme 23).

Based on this result, these workers speculate the involvement of the "reverse" prenylated diketopiperazine **A**, that subsequently undergoes a [1,2]-migration of the prenyl unit to the 3-position giving structure **B** (Scheme 24) which subsequently suffers cyclization and oxidation to roquefortine.

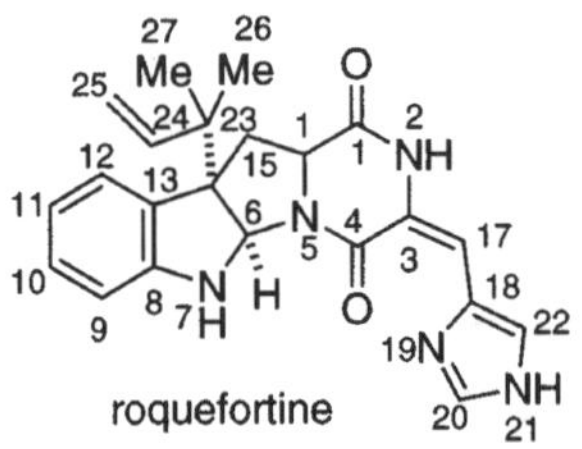

Fig. 8

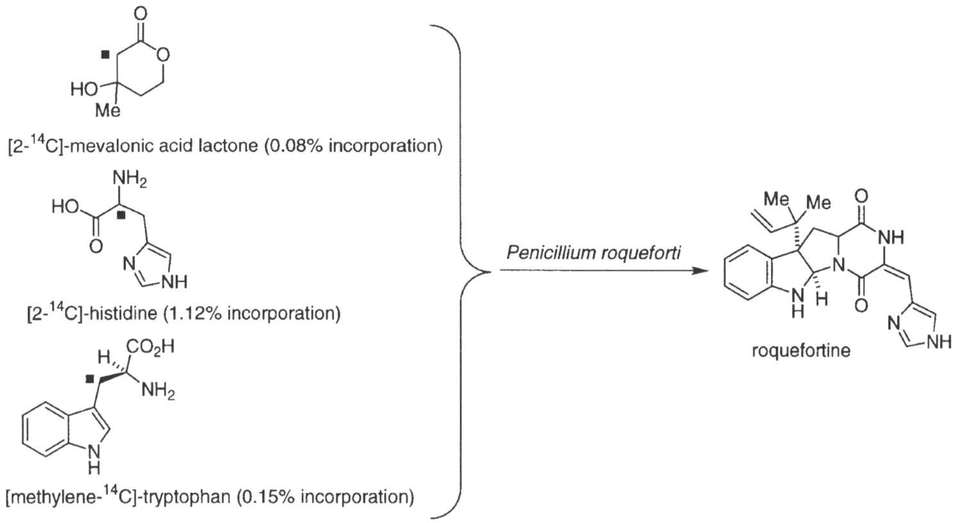

Scheme 21. Primary metabolite incorporations into roquefortine [44]

Scheme 22. Possible mechanism for introduction of the "reverse" prenyl substituent [44]

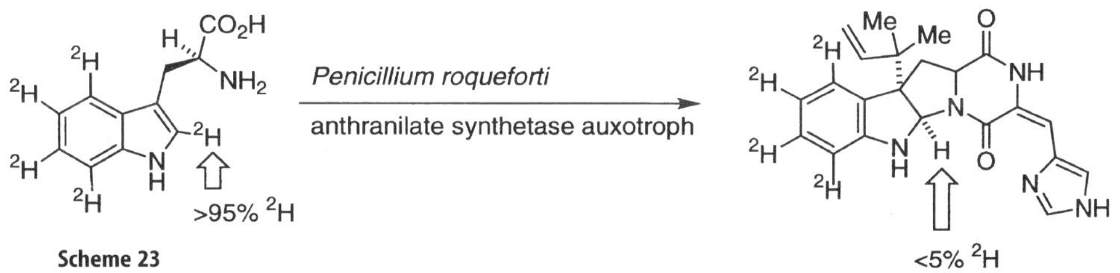

Scheme 23

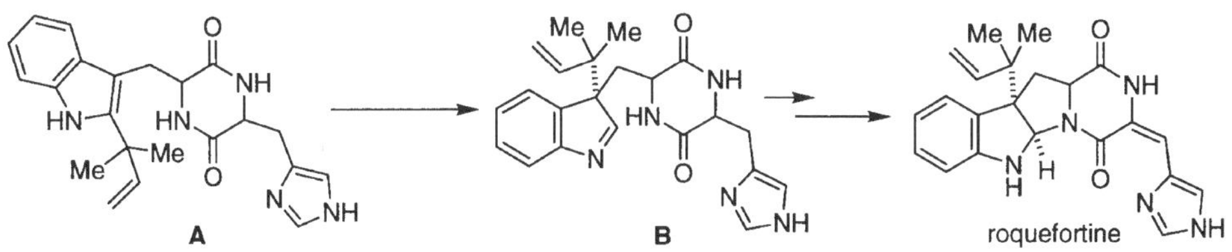

Scheme 24

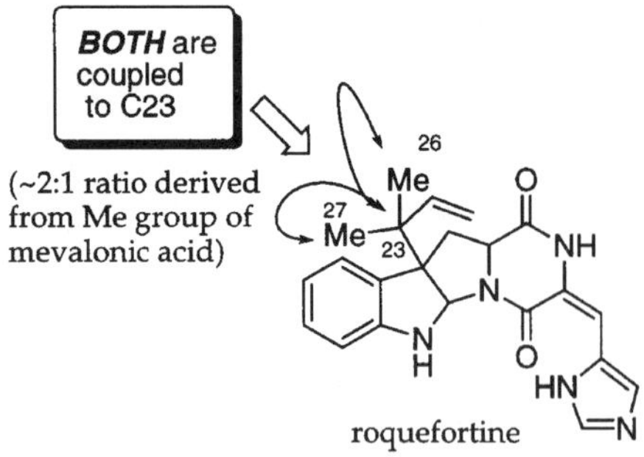

Fig. 9 roquefortine

In a particularly provocative study, Gorst-Allman et al. fed [1,2-^{13}C]-labeled acetate and [2,3-^{13}C]-labeled mevalonic acid lactone to cultures of *Penicillium roqueforti* and *Penicillium crustosum* [45] The ^{13}C NMR spectrum of the roquefortine showed that both C(26) and C(27) were coupled to C(23) (Fig. 9). This interesting result mandates that, during the incorporation of dimethylallyl pyrophosphate, the methyl groups become quasi-equivalent.

To rationalize this data, these workers speculate the intermediacy of the *N*-prenylated indole derivative **A** (Scheme 25); an aza-Claisen-type rearrangement would furnish **B** and a stereorandom [1,2]-migration of the isoprene moiety would result in the loss of the stereochemical integrity of the label in C-26 and C-27. An alternative pathway involving direct alkylation at C-3 of the

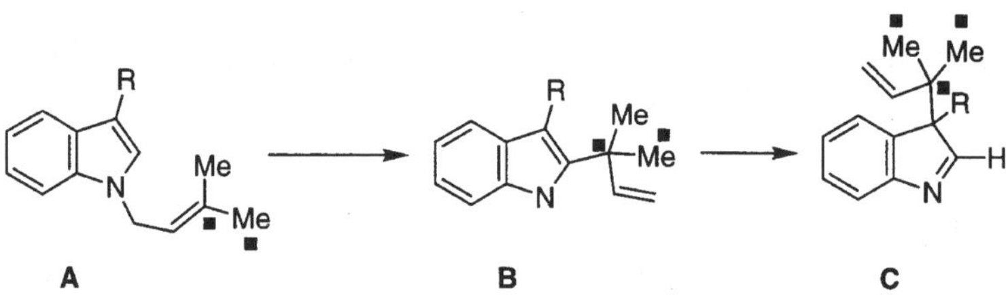

Scheme 25

indole by dimethylallyl pyrophosphate in a non-face-selective manner was also proffered. It was also observed that with [1,2-$^{13}C_2$]-labeled acetate as the precursor, the satellite resonances at δ 22.9 are approximately twice the intensity of the signal at δ 22.5; it was further pointed out that this ratio was reversed when mevalonic acid lactone was used as the precursor. The authors interpreted this data to mean that, the carbon atom resonating at δ 22.9 arises from C-3′ of mevalonic acid lactone and the resonance at δ 22.5 corresponds to C-2 of mevalonic acid lactone.

This data can also be interpreted to accommodate a mechanism where there is a partially non-face-selective transfer of the isoprene moiety from putative intermediate **B** or, alternatively, that the introduction of the isoprene unit occurs directly at C(3) of the indole ring via a partially non-face-selective S_N2' attack on dimethylallyl pyrophosphate as proposed by Bhat et al. [46, 47]. A similar explanation has also been proposed by Williams et al. regarding the biosynthesis of paraherquamide A [42].

Bhat et al. [46, 47] re-investigated the fate of the C(2) hydrogen of the indole moiety of tryptophan in the biosynthesis of roquefortine in cultures of *Penicillium roqueforti* and the related metabolite aszonalenin in cultures of *Aspergillus zonatus* by using deuterium NMR and mass spectral analysis. These workers point out that in the study reported by Barrow et al. [44], where ^{1}H NMR was used to assay for deuterium by difference, that this method is likely to be insensitive to the presence of biosynthetically significant traces of deuterium at the C(6) position of roquefortine. Thus, feeding of L-[2,4,5,6,7-$^{2}H_5$] tryptophan to cultures of *Penicillium roqueforti* and *Aspergillus zonatus* provided roquefortine and aszonalenin, respectively (Scheme 26). In both cases, significant incorporation of deuterium at C(6) was observed by deuterium NMR and mass spectral analysis further revealed the presence of the d_5 species (8% for roquefortine and 3% for aszonalenin). This study stands in stark contradiction to the results and attendant interpretations of Barrow et al. [44] described above. In particular, the retention of the deuterium at C(6) rules out the involvement of a 2,3-disubstituted indole intermediate such as **A** (Scheme 24).

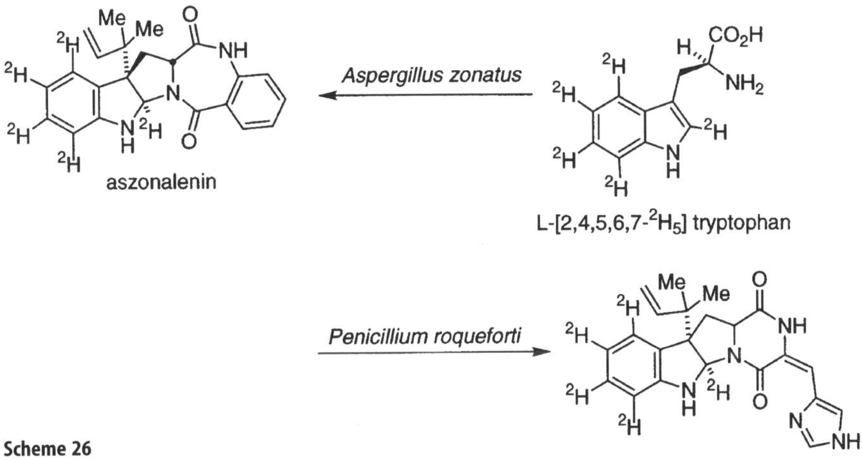

Scheme 26

In addition, Bhat et al. [47] point to the unlikely intermediacy of the enzyme-bound 2-substituted intermediate **A** (Fig. 10) which undergoes a thio-Claisen rearrangement followed by reductive removal of sulfur first proposed by Bycroft and Landon [48].

These workers therefore concluded that the simplest and most direct mechanism for the introduction of the dimethylallyl moiety, a simple S_N2' alkylation at C(3) of the indole with dimethylallyl pyrophosphate, appears the most appealing based on all the available data. These authors cite the observations that both diastereotopic methyls of the 1,1-dimethylallyl group in roquefortine [45] and echinulin are derived in unequal amounts from the methyl group of mevalonic

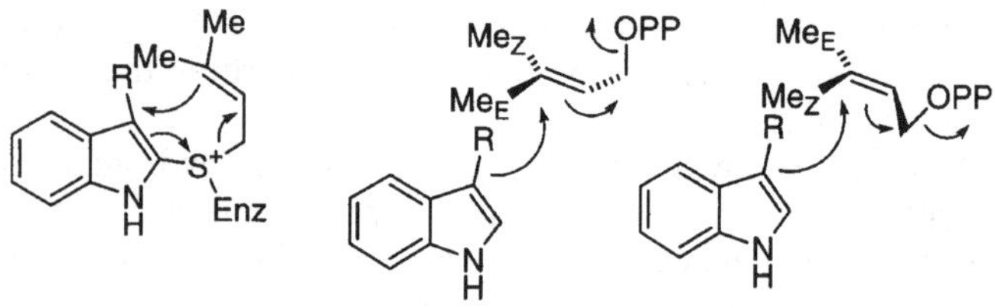

Fig. 10 A B C

acid. These observations are rationalized to occur via poor discrimination of the enzyme for the two orientations (**B** and **C**, Fig. 10) of dimethylallyl pyrophosphate.

The stereochemistry of the dehydrogenation of histidine into the dehydrohistidine moieties of roquefortine and oxaline, a related mold metabolite isolated from *Penicillium oxalicum*, was studied by Vleggaar and Wessels [49]. Feeding (2*S*, 3*S*)-[3-^{3}H] histidine (**a**) and (2*S*, 3*R*)-[3-^{3}H] histidine (**b**) to the respective *Penicillium* sp. in the presence of (2*S*)-[ring-2′-^{14}C] histidine as an internal standard to give comparable ^{3}H : ^{14}C ratios was carried out (Scheme 27). These workers observed good incorporations (1–4%) of the labeled histidines into the natural products.

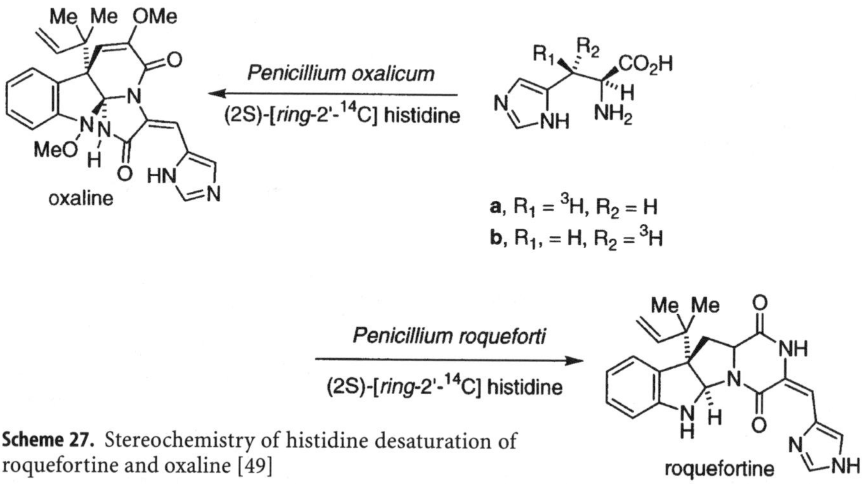

Scheme 27. Stereochemistry of histidine desaturation of roquefortine and oxaline [49]

The incorporation data shown in Table 5 clearly demonstrates that in both cases the *pro-S*-hydrogen is the one lost. Since the stereochemistry of the dehydroamino acid unit in both roquefortine and oxaline are of the E-configuration, a net *syn*-elimination of the 3-*pro-S* hydrogen and H(2) must occur during the biosynthesis of the dehydrohistidine unit (Scheme 28).

Steyn and Vleggaar [50] have also studied the biosynthetic relationship between roquefortine and oxaline. The co-occurrence of roquefortine and oxaline in cultures of *Penicillium oxalicum* led these workers to speculate that roque-

Table 5. Incorporation of [3-^{3}H]-histidine into roquefortine and oxaline [49]

Configuration	^{3}H:^{14}C ratio	^{3}H:^{14}C ratio (roquefortine)	^{3}H:^{14}C ratio (oxaline)
2*S*, 3*S* (**a**)	6:50	0:30 (4.6%)[a]	0:30 (4.6%)
2*S*, 3*R* (**b**)	6:50	6:12 (94.2%)	6:24 (96.0%)

[a] Figures in brackets are % ^{3}H retention.

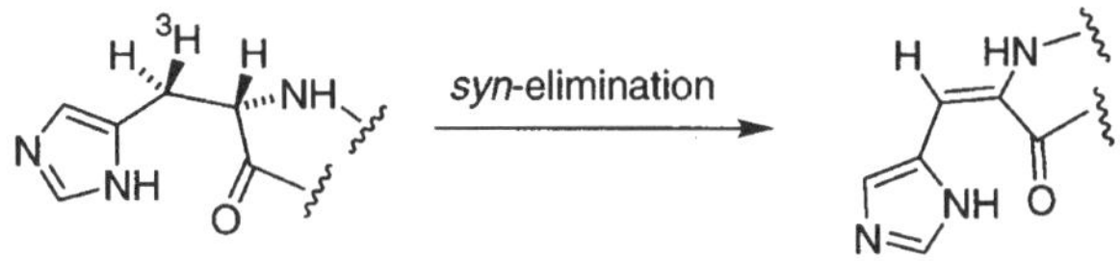

Scheme 28

fortine is a biosynthetic precursor to oxaline and proposed the biosynthetic transformations shown in Scheme 29. Feeding of (2*RS*)-[*indole*-2-^{13}C, 2-^{15}N] tryptophan admixed with (2*RS*)-[*benzene ring*-U-^{14}C] tryptophan (to provide an independent measure of dilution) to cultures of *Penicillium roqueforti* provided the labeled roquefortine (6.1% incorporation; dilution of 8.7). The origin of N(14) of oxaline was ascertained by feeding of (2*RS*)-[*indole*-2-^{13}C, 2-^{15}N] tryptophan admixed with (2*RS*)-[*benzene ring*-U-^{14}C] tryptophan to cultures of *Penicillium oxalicum* (the ^{14}C precursor was efficiently incorporated (3.0%) with low dilution (20.5) into oxaline).

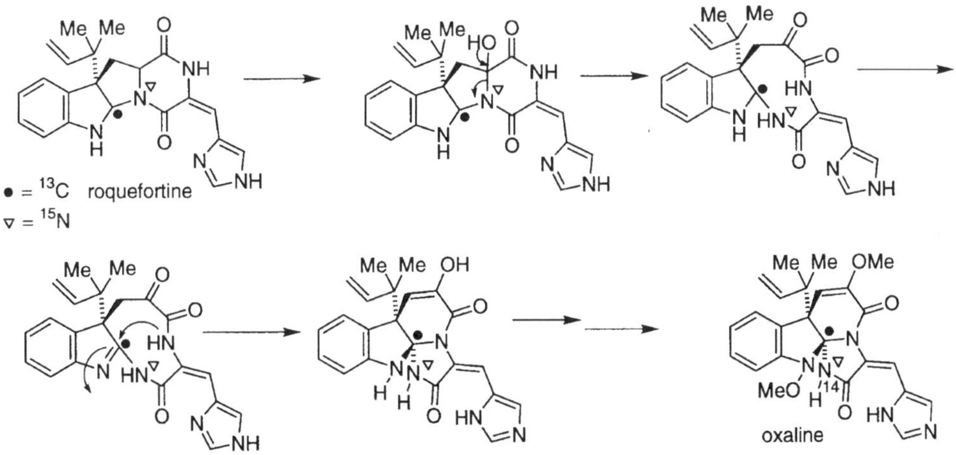

Scheme 29. Proposed biosynthetic relationship between roquefortine and oxaline [50]

These workers observed that in these feeding experiments a significant amount of the ^{15}N-label from the (2*SR*)-tryptophan is lost under the conditions of the feeding experiment. This was attributed to the conversion of the (2*R*)-tryptophan into indolylpyruvic acid and then to (2*S*)-tryptophan.

The intermediacy of roquefortine in the biosynthesis of oxaline was established next using [*benzene ring*-U-^{14}C, 6-^{13}C, 5-^{15}N] roquefortine obtained from cultures of *Penicillium crustosum* supplemented with (2*RS*)-[*indole*-2-^{13}C, 2-^{15}N] tryptophan (incorporation 4.3 %, dilution 10.1). The labeled roquefortine was very efficiently incorporated into oxaline (24.3 %) but the relatively high dilution of 29.8 precluded the observation of the one-bond (^{13}C-^{15}N)-coupling for the C(2) resonance in the ^{13}C NMR spectrum.

Mantle et al. investigated the biosynthesis of the complex alkaloids, the penitrems, and roquefortine in *Penicillium crustosum* using [2-^{14}C]-labeled mevalonic acid lactone [51]. The important finding in this study relevant to the biosynthesis of roquefortine, was the lack of scrambling of the label in the isoprene units of penitrem A (Scheme 30, Fig. 11). This rules out the possibility that the stereochemical integrity of the label is scrambled at the level of dimethylallyl pyrophosphate biosynthesis in the cellular pool. This finding indirectly gives additional credence to the S_N2' alkylation mechanism with poor facial discrimination for introduction of the reversed prenyl unit as proposed by Bhat et al. (see Fig. 10 B, C) [47].

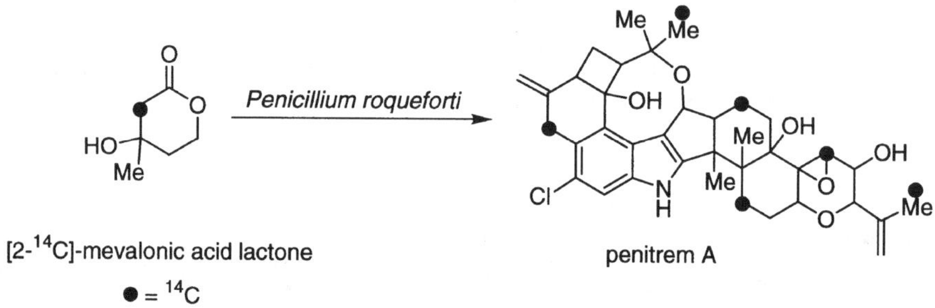

Scheme 30

4 Echinulin

Echinulin is a triisoprenylated cyclic dipeptide isolated from *Aspergillus amstelodami* and *Aspergillus echinulatus* (Fig. 12) [52]. Birch et al. established that the biosynthetic precursors to echinulin are L-tryptophan, L-alanine, and mevalonic acid [53] by feeding radioactive precursors to *Aspergillus amstelodami*. In these experiments it was found that [1-^{14}C] sodium acetate, DL-[1-^{14}C]-alanine, DL-[1-^{14}C]-tryptophan, [2-^{14}C]-mevalonic acid lactone, and [1-^{14}C]-glycine were incorporated into echinulin with incorporation rates of 4.25 %, 0.12 %, 1.36 %, 0.74 %, and 0.84 %, respectively (Scheme 31).

In addition, it was further established by Slater et al. that *cyclo*-L-alanyl-L-tryptohan are precursors to echinulin in vivo (Scheme 32) [54]. In this instance, the radioactive precursor *cyclo*-L-alanyl-L-[*methylene*-^{14}C]-tryptophan was fed to cultures of *Aspergillus amstelodami* and gave echinulin with high levels of incorporation (9–16 %).

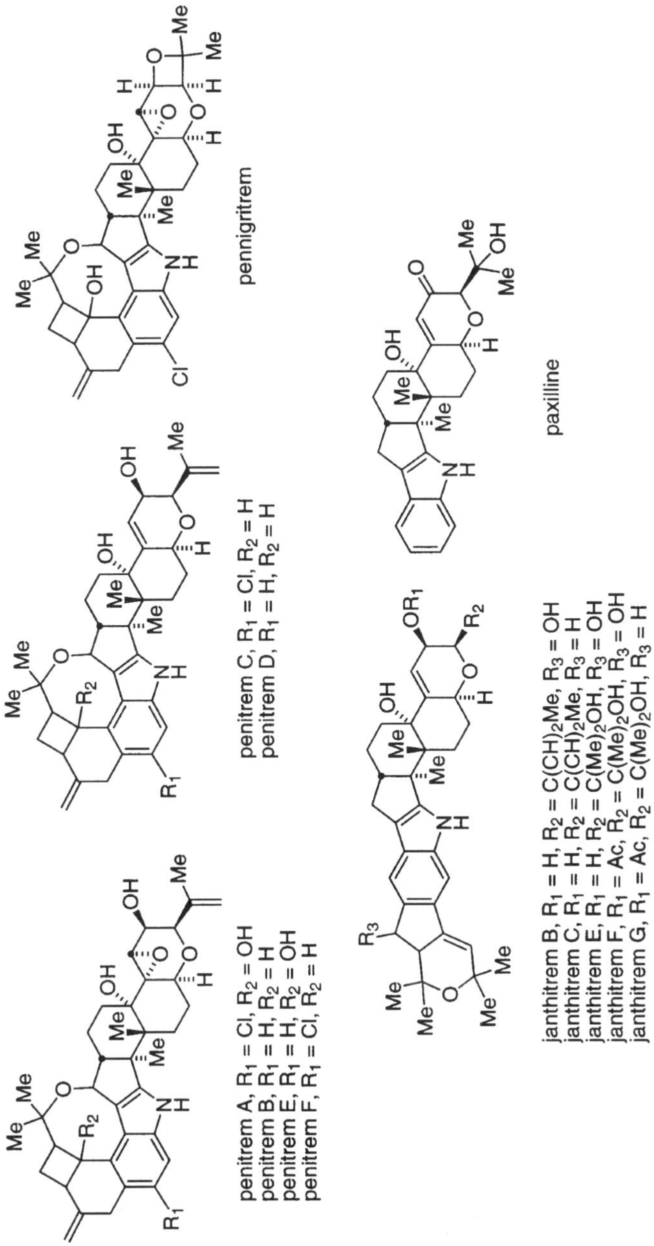

Fig. 11. Structures of the penitrems, pennigritrem, janthitrems, and paxilline [51b]

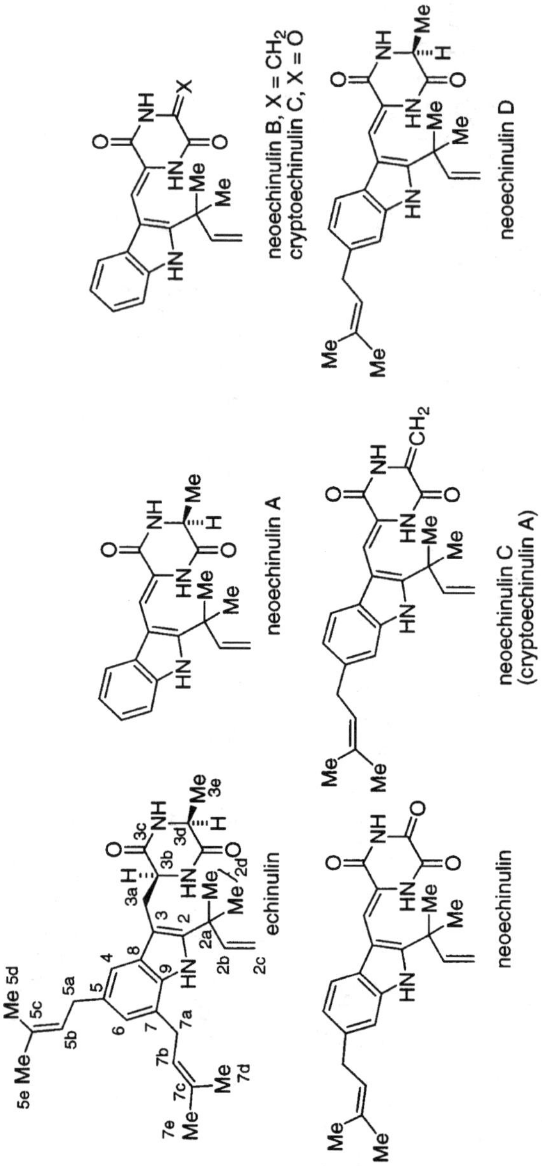

Fig. 12. Structures of the echinulins

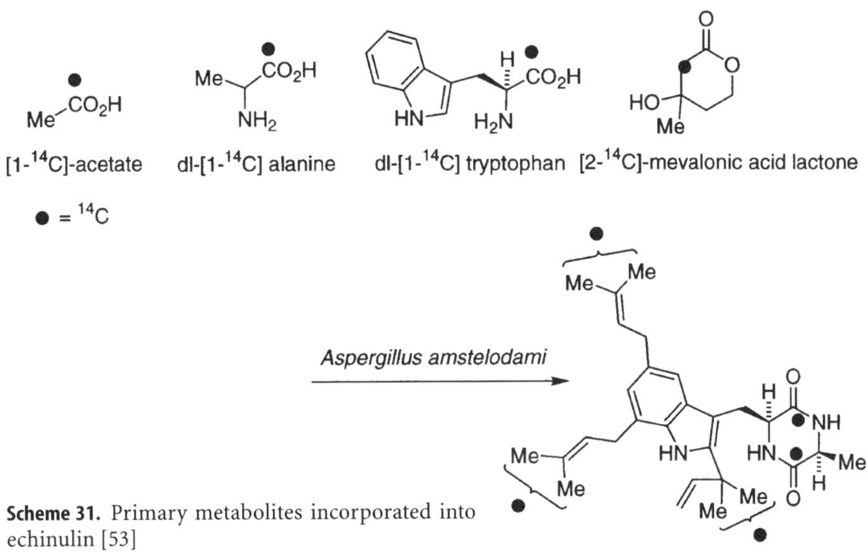

Scheme 31. Primary metabolites incorporated into echinulin [53]

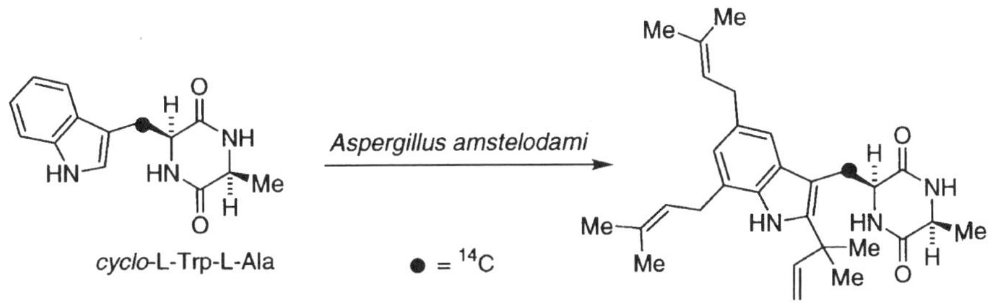

Scheme 32. Incorporation of *cyclo*-L-Trp-L-Ala into echinulin [54]

echinulin

A similar experiment was later reported by Marchelli et al. [55], wherein doubly labeled *cyclo*-L-[U-^{14}C]alanyl-L-[5,7-$^{3}H_2$]tryptophan and *cyclo*-L-[U-^{14}C] alanyl-D-[5,7-$^{3}H_2$]tryptophan were synthesized and fed to cultures of *Aspergillus amstelodami*. A series of echinulins were isolated and showed incorporations as follows: neoechinulin A (88%), neoechinulin B (43%), neoechinulin C (43%), neoechinulin D (34%), and neoechinulin (43%) respectively, relative to the radioactivity recovered in echinulin. The L,D-cyclic dipeptide was very inefficiently incorporated into the natural metabolites thus further confirming *cyclo*-L-alanyl-L-tryptophyl as a key biosynthetic precursor to this family of natural substances.

In 1973, Allen demonstrated that the monoprenylated cyclic dipeptide *cyclo*-L-alanyl-2-(1,1,-dimethylallyl)-L-tryptophan was incorporated into echinulin in cultures of *Aspergillus amstelodami* [56]. In this study, the reverse prenylated cyclic dipeptide was doubly labeled with ^{3}H and ^{14}C by treating a partially purified prenyl transferase obtained from *Aspergillus amstelodami* with *cyclo*-L-Ala-L-[3-^{14}C]-Trp and [1-^{3}H]-3-methyl-2-butenyl 1-pyrophosphate (Scheme 33)

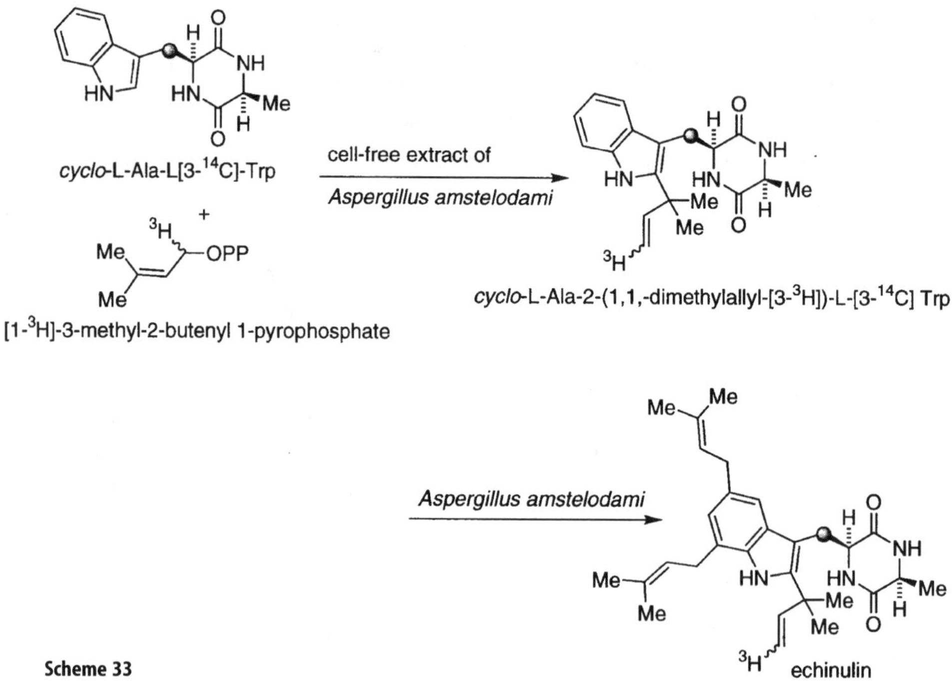

Scheme 33

[56a]. Feeding and incorporation of the doubly labeled substrate gave echinulin that, when recrystallized to constant specific activity, had the same $^{3}H/^{14}C$ ratio as the precursor, thus confirming this substance as a metabolic intermediate. This study demonstrates the important initial sequence of reactions that occur in the biosynthetic pathway with respect to isoprenylation. The first isoprene residue is incorporated in a "reverse" manner to the intact dioxopiperazine, *cyclo*-L-Ala-L-Trp; the subsequent "normal" isoprene transfer reactions occur by electrophilic aromatic substitution on the mono-isoprenylated product culminating in echinulin.

Shortly after the studies reported by Allen, Casnati et al. reported the fate of the aromatic protons in the indole ring during the subsequent isoprenylation steps [57]. This study was undertaken to investigate the possible role of arene oxides and the diagnostic NIH shift that was discovered in the early 1970s.

As shown in Scheme 34, [4,6-$^{3}H_2$]-tryptophan and [5,7-$^{3}H_2$]-tryptophan were synthesized and fed to cultures of *Aspergillus amstelodami*. The [5,7-$^{3}H_2$]-tryptophan was incorporated into echinulin and neoechinulin B with 2% and 103% retention of tritium activity, respectively. The [4,6-$^{3}H_2$]-tryptophan was incorporated into echinulin with 102% retention of tritium activity and into neoechinulin with 48% loss of tritium activity. These experiments are complementary and clearly demonstrate that the introduction of the isoprene units go via a direct electrophilic aromatic substitution reaction mechanism. It should also be noted that Fuganti et al. isolated cryptoechinulin from *Aspergillus amstelodami* during the course of their biosynthetic work on echinulin [58].

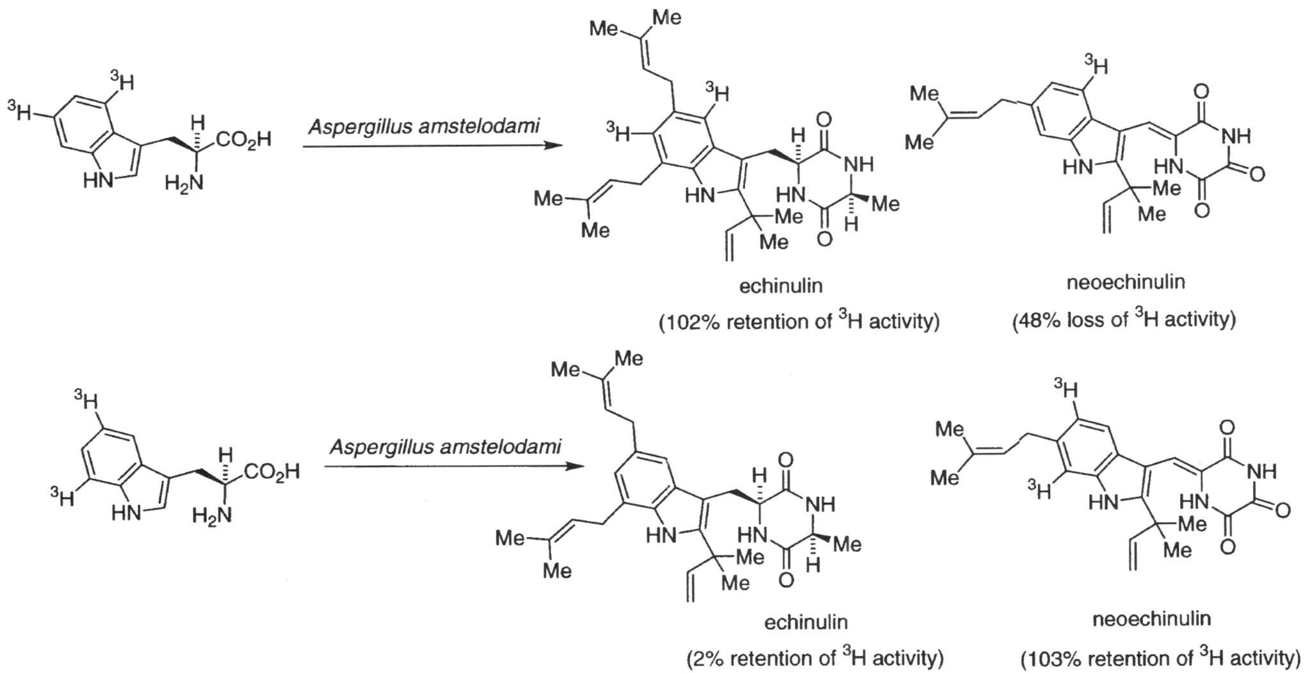

Scheme 34. Fate of aromatic protons during aromatic isoprenylation in echinulin biosynthesis [57]

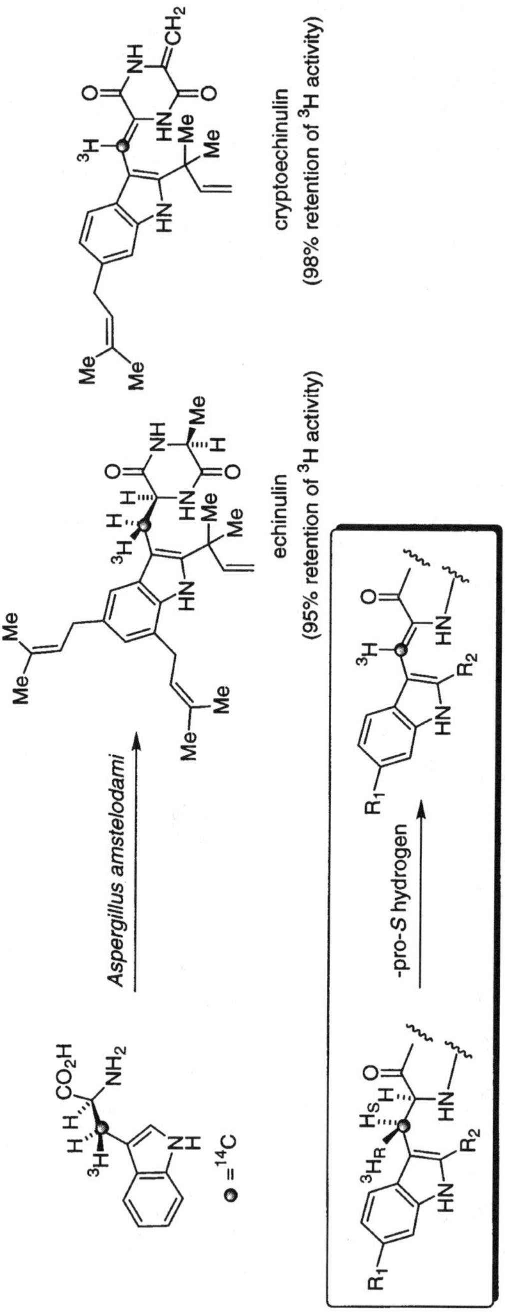

Scheme 35. Incorporation of stereospecifically labeled tryptophan into echinulin and cryptoechinulin [59]

The stereochemistry of the *α,β*-desaturation of cryptoechinulin was reported by Fuganti et al. and is summarized in Scheme 35 [59]. These workers prepared L-tryptophan stereospecifically labeled with tritium in the *β*-methylene position from labeled serine, indole, and fibre-entrapped tryptophan synthetase obtained from *E. coli*. When (3′*R*)[3′-^{3}H; 3′-^{14}C]-L-tryptophan was fed to *Aspergillus amstelodami,* incorporation into echinulin and cryptoechinulin took place with 95% and 98% retention of tritium activity, respectively. Feeding of (3′*S*)[3′-^{3}H; 3′-^{14}C]-L-tryptophan to *Aspergillus amstelodami* gave incorporation into echinulin and cryptoechinulin with 96% and 5% retention of tritium activity, respectively. Thus, in the desaturation reaction the *pro-S* hydrogen is stereospecifically removed.

Barrow et al. investigated the stereochemistry of the C-C bond-forming reaction in the aromatic isoprenylation in echinulin biosynthesis utilizing [(5*R*)-^{3}H] and [(5*S*)-^{3}H]-mevalonates [60]. In addition, feeding of [1,2-$^{13}C_2$]-acetate to *Aspergillus amstelodami,* showed that the (*E*)-methyl groups in the isoprene moieties are derived only from C-2 of mevalonic acid (Scheme 36). The (*E*)-methyl group was found to be enriched but not coupled to the adjacent olefinic center.

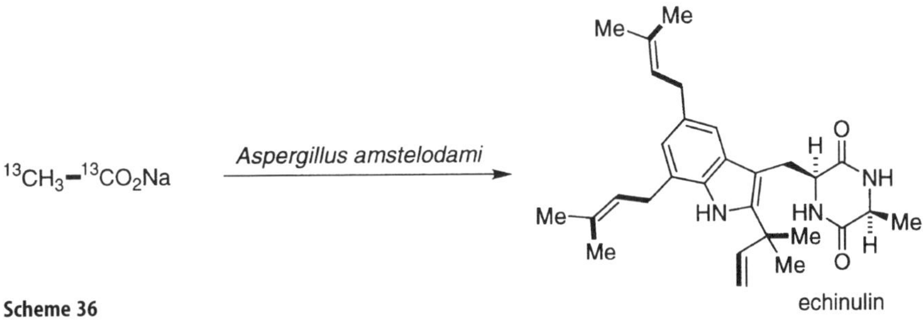

Scheme 36

Net inversion at the allylic pyrophosphate carbon of dimethylallyl pyrophosphate was demonstrated by chemical degradation of the echinulin produced from feeding the [(5*R*)-^{3}H] and [(5*S*)-^{3}H]-mevalonates as shown in Scheme 37 [60]. The isoprenyl units of echinulin were reduced with diimide and the reduction products were oxidized with fuming nitric acid giving isocaproic acid in ~5% overall yield. The isocaproic acid was converted into isopentylamine and oxidized with pea seedling diamine oxidase. Battersby has shown that this enzyme is completely stereospecific in the removal of the *pro-S* hydrogen from benzylamines, and the Barrow group confirmed this on independently synthesized stereospecifically labeled isopentylamines. The degradation product obtained from echinulin biosynthesized with added [(5*R*)-^{3}H]-mevalonate lost >90% of the tritium activity. Based on this result and those described above on aromatic isoprenylation, these workers conclude that the isoprenylation of the 5- and 7-positions of the indole ring occurs by direct electrophilic attack on dimethylallyl pyrophosphate and does not occur by rearrangement.

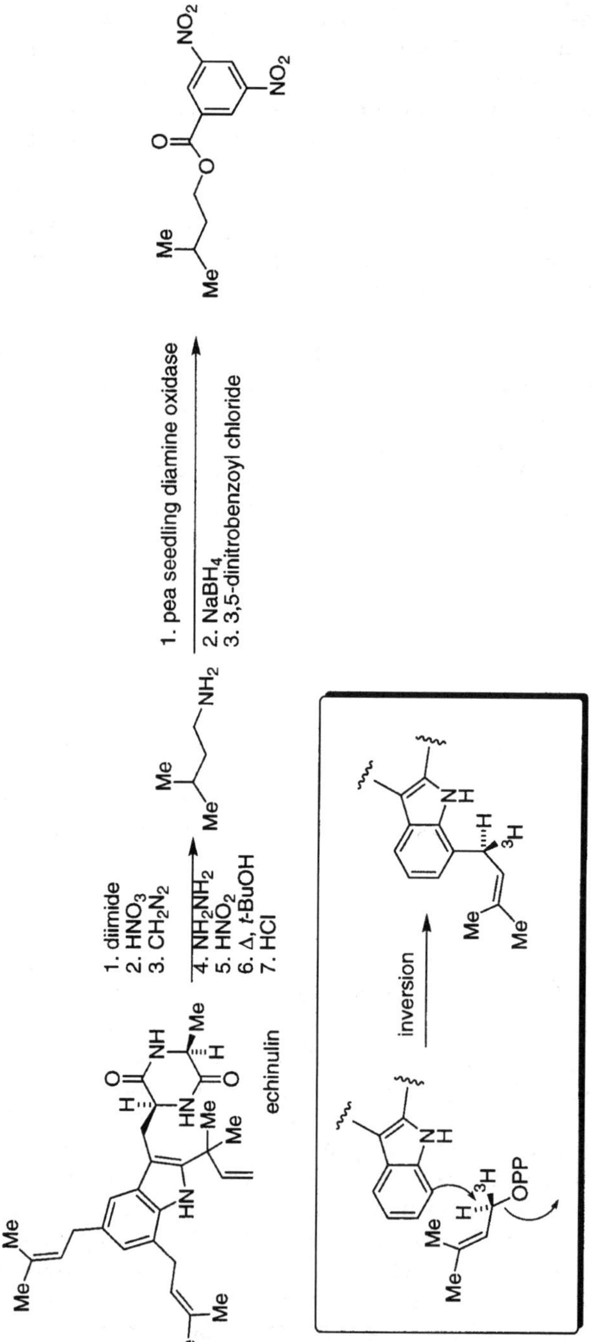

Scheme 37. Mechanism and stereochemistry of aromatic isoprenylation of echinulin [60]

Grundon et al. examined the possibility that the "reverse" prenyl unit was introduced via an indolic N-prenylated precursor followed by aza-Claisen rearrangement and 1,2-migration as shown in Scheme 38 [61]. These workers synthesized 1-([1-^{3}H]-3,3-dimethylallyl)-L-tryptophan and *cyclo*-L-alanyl-1-([1-^{3}H]-3,3-dimethylallyl)-L-tryptophan and fed these labeled substances to cultures of *Aspergillus amstelodami* and found that neither compound was incorporated in radiochemically significant amounts into echinulin.

A similar mechanism was also postulated by Barrow et al. for the incorporation of the "reversed" isoprene unit at the indolic 3-position of roquefortine (see Schemes 22 and 25) [44]; to date there is no experimental support published in the literature for the notion that the "reverse" prenyl unit at either the 2-position, as is found in echinulin, or the 3-position, as is found in roquefortine, is introduced indirectly at the indole nitrogen (or some other position) followed by migration. All evidence, both direct and indirect, points to the direct S_N' attack of the pertinent indole carbons on dimethylallylpyrophosphate.

Perhaps one of the most intriguing studies on the biosynthesis of echinulin was that reported by Harrison and Quinn concerning the introduction of the "reverse" isoprene unit [62]. This group noted that there are two interesting questions that can be posed concerning the 1,1-dimethylallyl group: (1) which of the two diastereotopic methyl groups (Me_R or Me_S) in the "reverse" isoprene unit (the 1,1-dimethylallyl group) of echinulin is derived from the methyl group of mevalonic acid, and (2) which of the two diastereotopic vinyl hydrogens (H_E or H_Z) is derived from the 5-*pro-R*-hydrogen of mevalonic acid? This study, reported in 1983, experimentally addressed the first question and they were able to demonstrate that the *pro-S* methyl group of the 1,1-dimethylallyl group of echinulin was derived primarily (89%) from the methyl group of mevalonic acid. The technique employed to answer this question (1) was secured via the administration of [Me-$^{2}H_3$]-mevalonic acid lactone to cultures of *Aspergillus amstelodami* and harvesting the echinulin isolated from this feeding experiment. The ^{2}H NMR spectrum of the echinulin produced showed two overlapping signals at δ 1.57 and δ 1.69 ppm of approximately equal intensity that were assigned to the labeled methyl of the 1,1-dimethylallyl group and the *Z*-methyl groups of the 3,3-dimethylallyl groups, respectively. The deuterated echinulin sample was subsequently degraded as shown in Scheme 39 to furnish labeled 2,2-dimethylbutan-1-ol. Utilizing the chiral shift reagent Eu(hfbc)$_3$, these workers were able to assign unequivocally the deuterated methyl group as being derived from the *pro-S*-methyl group.

Fuganti et al. studied the fate of stereospecifically labeled leucine incorporation into echinulin as shown in Scheme 40 [63]. Both the *pro-S* and *pro-R* ^{13}C-labeled leucines were synthesized from (2*RS*)[1-^{13}C]2-methyl-4-phenylbutyric acid. It was found that the label derived from the *pro-R* position, gave enhancements (in parentheses, the % enhancements are the results of two feeding experiments) to the signals assigned to C-2b (180%, 140%); C-5b, C-7b (190%, 160%); C-2d (80%, 40%); C-5e, C-7e (130%, 50%); C-5d, C-7d (70%, 50%). In the case of the feeding experiments via the *pro-S*-derived label, the enhancements were as follows: C-2b (30%, 40%); C-5b, C-7b (70%, 40%); C-2d (120%, 160%); C-5e, C-7e (50%, 60%); C-5d, C-7d (160%, 210%).

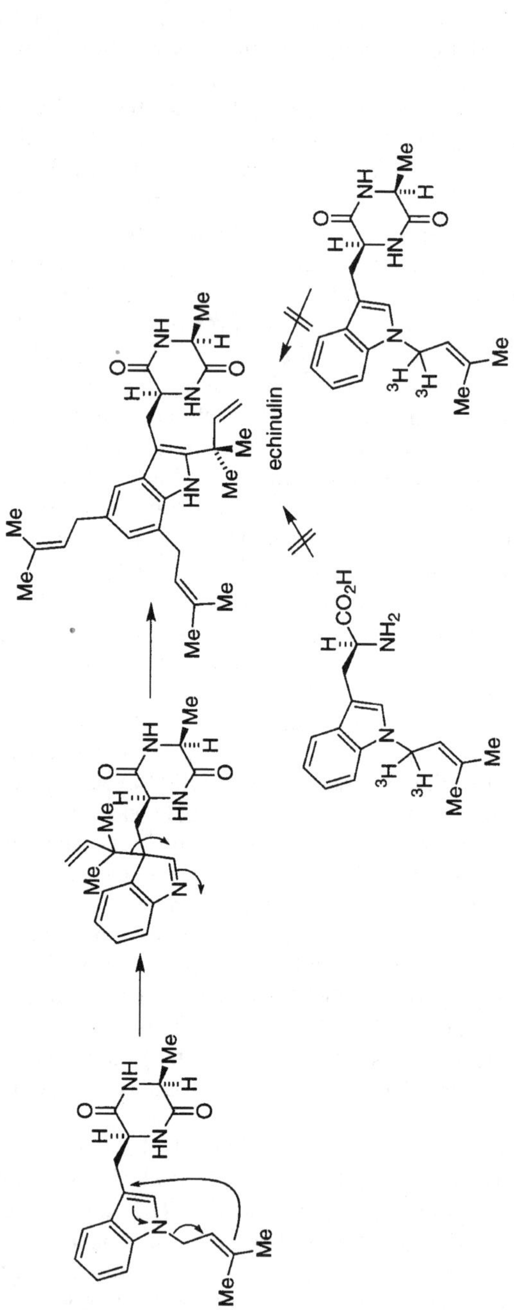

1-([1-^{3}H]-3,3-dimethylallyl)-L-tryptophan

cyclo-L-alanyl-1-([1-^{3}H]-3,3-dimethylallyl)-L-tryptophan

Scheme 38

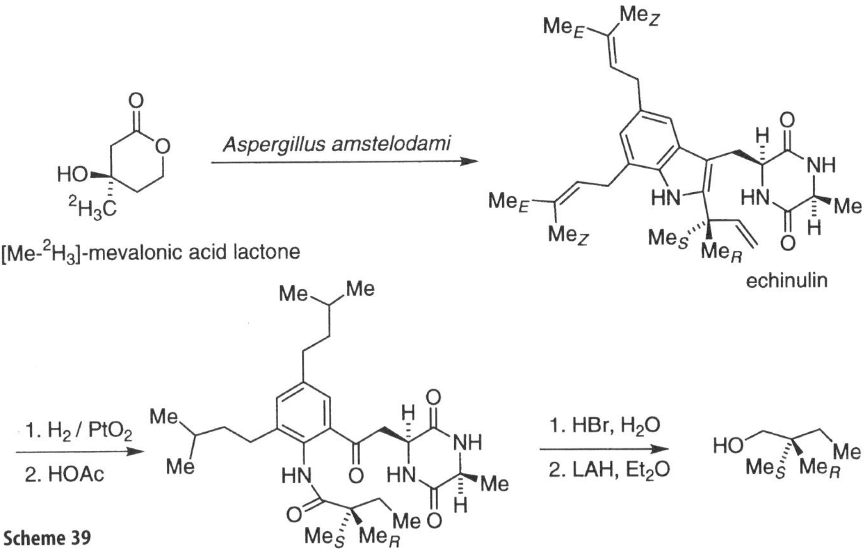

Scheme 39

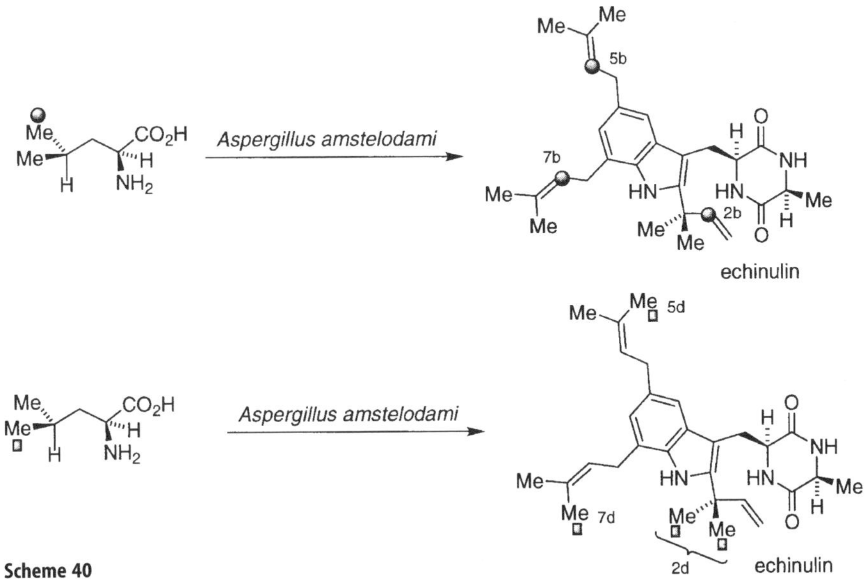

Scheme 40

This data was interpreted to indicate that, within the intrinsic limitations of this method, the *pro-R*-derived methyl group from leucine is incorporated into the *Z*-methyl group of the isoprenyl units (C-5 d, C-7 d) via the C-3 methyl group of mevalonic acid. The *pro-R*-derived methyl group is incorporated into the 2-position of the isoprenyl chain arising from C-4 of mevalonic acid. These workers interpreted this incorporation data via the pathway outlined in Scheme 41.

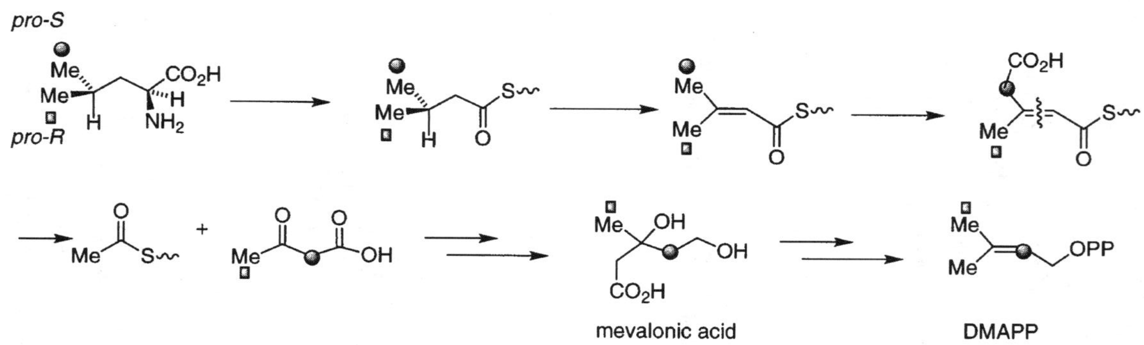

Scheme 41. Proposed mechanism of leucine catabolism to DMAPP [63]

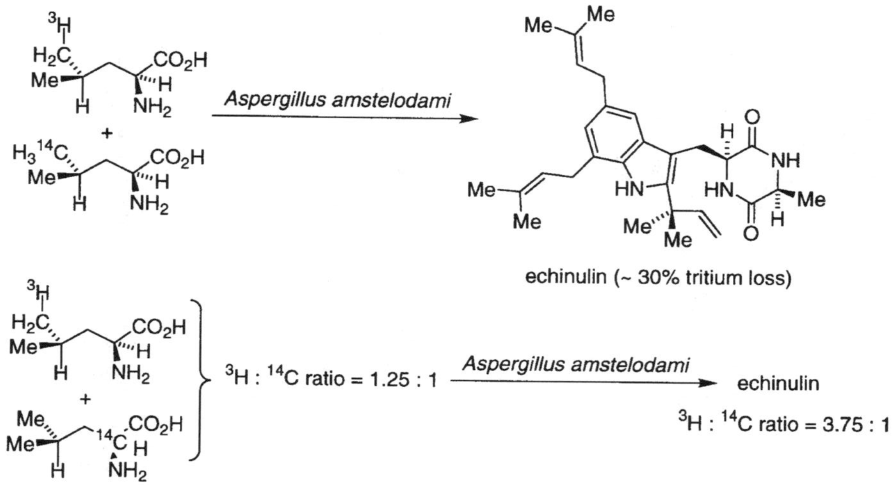

Scheme 42. Stereospecifically labeled leucine incorporation into echinulin [64]

Stereospecific α,β-desaturation of isovalerylcoenzyme A derived from the oxidative degradation of leucine would yield a β-methyl crotonylcoenzyme A with retention of the *pro-R*/*pro-S*-labeled leucine-derived stereochemistry as shown. Subsequent cleavage of β-hydroxy-β-methylglutarylcoenzyme A into acetyl CoA and acetoacetate followed by intact incorporation of the acetoacetate into mevalonate can account for the observed labeling propensities.

In order to confirm the above pathway, Fuganti et al. performed a second experiment using a doubly labeled leucine derivative to examine the fate of positions 2 and 3 of leucine (Scheme 42) [64]. In the first experiment, (4*S*)-[5-^{3}H]-L-leucine admixed with (4*S*)-[5-^{14}C]-L-leucine was fed to cultures of *Aspergillus amstelodami* yielding echinulin with ~30% loss of tritium. In a second experiment, (4*S*)-[5-^{3}H]-L-leucine admixed with [2-^{14}C]-L-leucine (^{3}H:^{14}C ratio = 1.25:1) was fed giving echinulin with a ^{3}H:^{14}C ratio = 3.75:1. This result clearly indicates significant loss of the C-2 carbon of leucine during the metabolic incorporation into the isoprenyl units of echinulin and provides additional, supporting evidence for the pathway depicted above in Scheme 41.

5 Verruculogen, Fumitremorgins, and Cyclotryprostatins

Verruculogen is a mold metabolite obtained from *Penicillium verruculosum* and is the prototypical structural example of a growing class of related secondary metabolites found in both the *Penicillia* sp. and *Aspergillus* sp. (Fig. 13). Verruculogen was isolated by Cole et al. from cultures of *Penicillium verruculosum* in 1972 and the relative structural assignment was determined in 1974 by X-ray crystallography [65]. The absolute stereochemistry was subsequently determined through the hydrolysis of (*S*)-proline from the related metabolite fumitremorgin A [65, 66]. Fumitremorgins A and B were isolated by Yamazaki et al. from cultures of *Aspergillus fumigatus* in 1971 [66]. Verruculogen and several related tremorgens in this family are mycotoxic alkaloids that are capable of eliciting a sustained tremoring response in vertebrates and has been the subject of investigation of animal neurological disorders. These substances are fashioned from L-tryptophan, L-proline, L-methionine (the source of the aromatic methyl ether carbon), and one or more isoprene moieties derived from mevalonate [67]. The co-existence of verruculogen and fumitremorgin B in cultures of *Aspergillus caespitosus* suggests a biosynthetic relationship between these substances [68].

More recently, the tryprostatins A and B [69], the spirotryprostatins A and B [70], and the cyclotryprostatins A–D [71] were isolated from *Aspergillus fumigatus* by Osada et al. and have been shown to be mammalian cell cycle inhibitors interfering with the cell cycle at the G2/M phase [72]. It should be noted that at least three different numbering systems have been used in the literature for this family of alkaloids and the numbering system adopted here is the one originally assigned by Yamazaki et al. [66] and most recently adopted by Osada et al. [69–71].

Willingale et al. investigated the biosynthetic incorporation of the known metabolite TR-2, isolated from *Aspergillus fumigatus*, into verruculogen in cultures of *Penicillium raistrickii* [73a]. TR-2, a minor metabolite observed in cul-

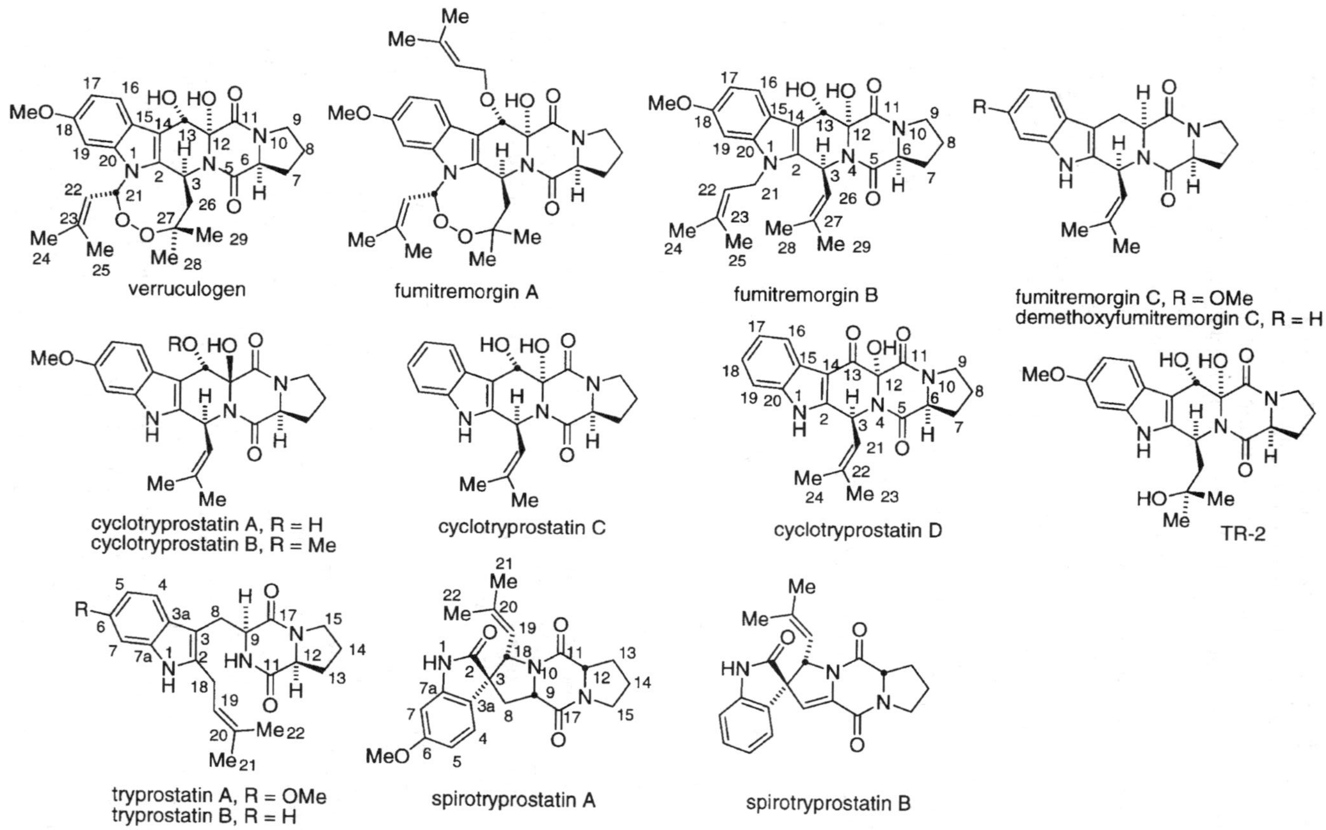

Fig. 13. Structures of verruculogen, the fumitremorgins, and tryprostatins

tures of *Aspergillus fumigatus* by Cole et al. [65c], was prepared biosynthetically using [U-^{14}C]proline and [2-^{14}C]mevalonic acid (organism vector not specified). Feeding of this substance to *Penicillium raistrickii* gave radioactive verruculogen and fumitremorgin B (co-migrating on PTLC). Further HPLC purification of each natural product was performed and the incorporation into verruculogen was estimated at 35% of the [^{14}C]TR-2 initially fed.

The authors propose a biosynthetic pathway as depicted in Scheme 43. It was noted that the yield of purified [^{14}C]fumitremorgin B was less than the usual ratio of these two metabolites produced by *Penicillium raistrickii* and suggested that fumitremorgin B spontaneously gives verruculogen upon addition of oxygen. The authors state: "This implies both that fumitremorgin B can be derived from TR-2 and that in aerobic culture some may become converted into verruculogen, mediated at least partly through a non-biological mechanism. Although it does not necessarily follow therefore that verruculogen is an artefact, it may not derive entirely from fungal biogenesis." Further work is needed to clarify the genetically pre-determined vs the adventitious production of verruculogen from fumitremorgin B or some other, as yet unidentified, metabolite.

A temporal picture of verruculogen biosynthesis was suggested by Mantle and Shipston in *Penicillium simplicissium* by feeding [2-^{3}H]mevalonic acid and [U-^{14}C]proline on days 2, 3, 4, 5, or 6 and monitoring the change in ^{3}H:^{14}C ratio [73b]. It was found that the ^{3}H:^{14}C ratio changed from 2.0:1 to 9.5:1 over the period of administration. This suggests that diketopiperazine formation precedes the isoprenylation steps although this study cannot discriminate between the isoprene moiety that is involved in the formation of the heterocyclic ring between the tryptophyl α-amino nitrogen and the peroxyl-containing ring. The findings of this study are consistent with the more penetrating TR-2 incorporation studies detailed in Scheme 43.

Horak and Vleggaar studied the stereochemical course of the peroxide ring-forming process as shown in Schemes 44 and 45 [74]. These workers first established that the oxygen atoms at C-18, C-12, C-22, and C-25 (Yamazaki numbering system) are derived from dioxygen through the biosynthetic production of verruculogen from *Penicillium verruculosum* grown under an atmosphere of $^{18}O_2$. The ^{18}O-labeling pattern at C-12/C-13 was rationalized through the agency of a (13*S*, 12*R*)-oxirane precursor that subsequently suffers ring-opening and attack by water on the *Re*-face of a conjugated iminium species at C-13 (Scheme 44).

Feeding of [1,2-$^{13}C_2$]acetate and examination of the one-bond ^{13}C-^{13}C coupling constants revealed that the 23-*pro-Z* methyl group (C-25) and the 27-*Re* methyl group (C-29) comprise intact acetate units [74]. Therefore, the 23-*pro-E* (C-24) and 27-*Si* methyl group (C-28) must be derived from C-2 of mevalonate. These workers found that the C-28 and C-24 signals were enhanced in the ^{13}C NMR spectrum (enrichment factors of 9.6 and 8.8, respectively) and that the signal for C-29 was also enhanced (enrichment factor of 3.3). A similar phenomenon was observed for fumitremorgin B isolated from the same [1,2-$^{13}C_2$]acetate feeding experiment. Feeding of (3*RS*)-[2-^{13}C]mevalonic acid lactone labeled the 23-*pro-E* methyl group (C-24; enrichment factor 6.8) of fumitremorgin B but, curiously, the 27-*pro-Z* methyl group (C-28) is enriched (enrichment factor

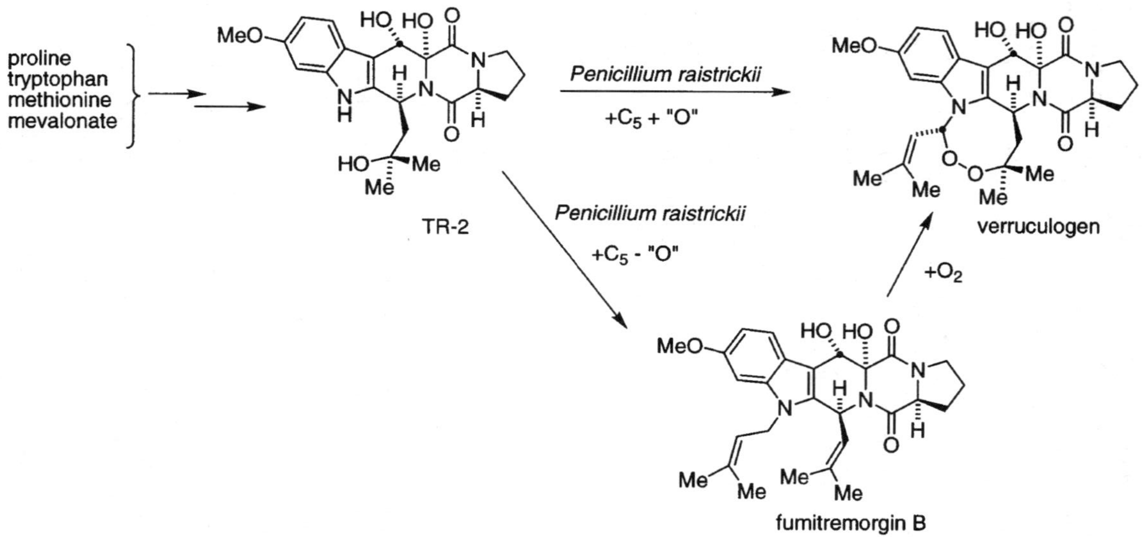

Scheme 43. Role of TR-2 in the biosynthesis of verruculogen and fumitremorgin B [73a]

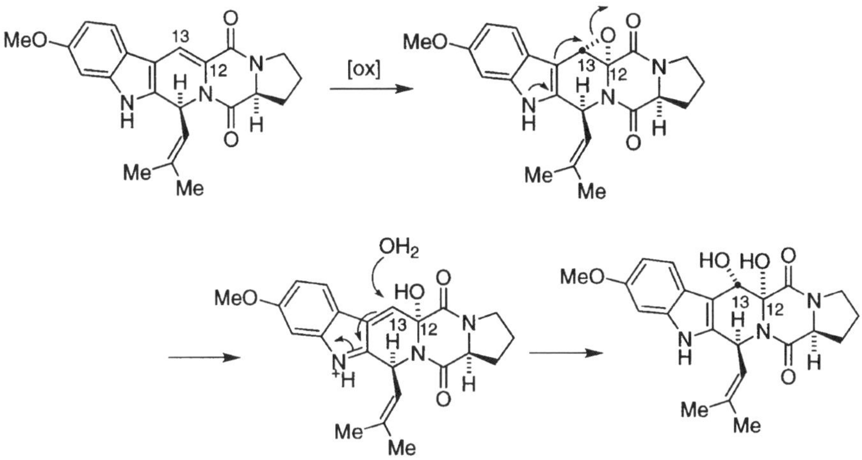

Scheme 44. Mode of dihydroxylation at C-12/C-13 as proposed by Horak and Vleggaar [74]

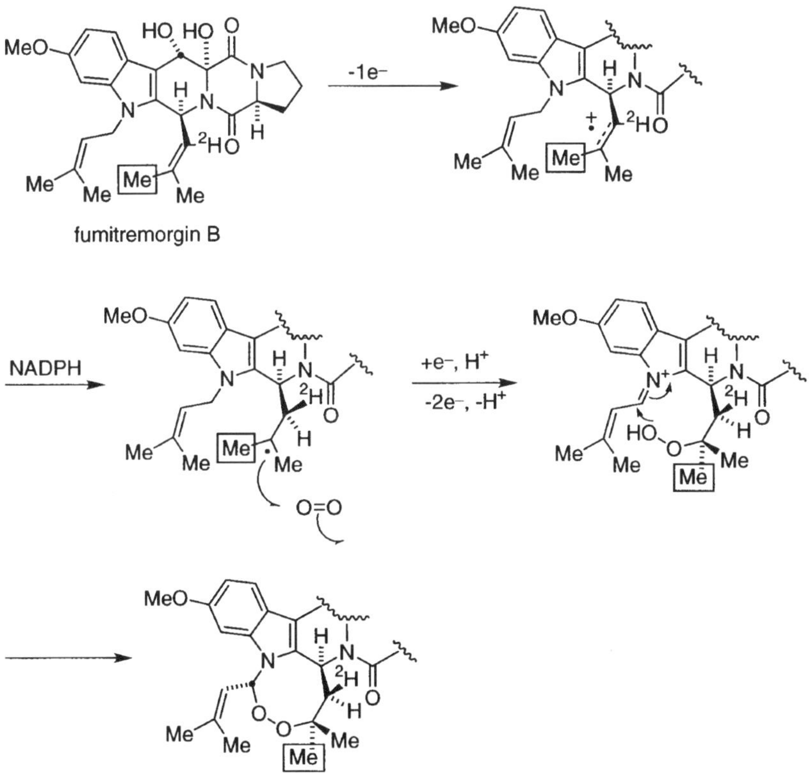

Scheme 45. Mechanism of peroxide ring formation as postulated by Horak and Vleggaar [74]

of 6.5). It was also observed that the signal for the 27-*pro-E* methyl group was enriched (C-29, enrichment factor of 2.0). Thus the stereochemical integrity of the diastereotopic methyl groups at C-27 (C-28 and C-29) is lost in the formation of fumitremorgin B and is inconsistent with direct S_N2 displacement on the methylene group of dimethylallyl pyrophosphate as the exclusive route for introduction of this isoprene unit.

In a second series of labeling studies, [1-^{13}C, 2-$^{2}H_3$]acetate was fed and incorporated into verruculogen. The deuterium atoms retained at C-26 and C-21 of verruculogen were assigned based on the characteristic β-isotope shifts of deuterium atoms two bonds removed from ^{13}C resonances and revealed in each case that the C-26 and C-21 hydrogen atoms are derived from the 4-*Re* position of mevalonate. Two-dimensional long-range (^{1}H, ^{13}C) chemical shift correlation spectroscopy was used to assign the relative stereochemistry of the deuterium atom at the diastereotopic C-26 position and it was found that the deuterium atom resided in the C-26-*Re* position. Based on these assignments, a biosynthetic pathway for the formation of the eight-membered ring peroxide in verruculogen was proposed as shown in Scheme 45. The authors propose a dioxygenase-mediated reaction on fumitremorgin B, envisioned to occur by one-electron oxidation of the C-26, C-27 olefin to form a radical cation. Subsequent NADPH reduction furnishes a C-27 radical that reacts with enzyme-bound O_2 to give, after a further one-electron reduction and protonation, the peroxide. Two-electron oxidation of the indolic *N*-prenyl group to the corresponding iminium species followed by nucleophilic addition of the peroxyl group furnishes the eight-membered ring peroxide. The isotopic labeling studies require that the overall addition of hydride ion and dioxygen must occur in an antarafacial manner; i.e., the 26*Si*, 27*Re*- and the 26*Re*, 27*Si*-face, respectively to yield the observed prochirality at C-26 and C-27 in verruculogen. The authors concede that the ring-closure step could, in principle, occur by either an ionic or a radical mechanism. The possibility that intramolecular transfer of a hydrogen atom from C-21 to C-26 at the 26*Si*, 27*Re*-face of the olefin in the ring-closure step was excluded based on the incorporation of (3*RS*)-[5-$^{2}H_2$]mevalonic acid into verruculogen.

In a subsequent study, Vleggaar et al. studied the process by which stereochemical integrity of the C-28/C-29 methyl groups are lost during the biosynthesis of fumitremorgin B and verruculogen [75]. These workers fed variously ^{2}H/^{13}C-labeled mevalonic acid lactones to *Penicillium verruculosum* to establish the origin and fate of the hydrogen atoms at C-29. The use of deuterium in conjunction with ^{13}C as a reporter nucleus utilizing both α- and β-isotope shifts in the ^{13}C NMR spectra provided the necessary accounting. As shown in Scheme 46, feeding of (3*RS*)-[6-$^{2}H_3$, 3-^{13}C]mevalonic acid lactone (>98 atom% ^{2}H, 99 atom% ^{13}C) to *Penicillium verruculosum* provided verruculogen whose NMR spectrum revealed that three deuterium atoms are retained at both C-29 and C-25.

It was observed that there was a single β-shifted ^{13}C signal for C-27 in the ^{13}C[^{1}H] NMR spectrum; this observation is inconsistent with a mechanism involving an sp^{2}-hybridized intermediate at C-29. This was further corroborated by an observed α-isotope shift in the ^{13}C[^{2}H, ^{1}H] NMR spectrum of verruculogen obtained from feeding (3*RS*)-[6-$^{2}H_3$, 6-^{13}C]mevalonic acid lactone (99 atom% ^{13}C) containing ^{13}C, $^{2}H_3$-(55 mol%) and ^{13}C, $^{2}H_2$-labeled (45 mol%) species at

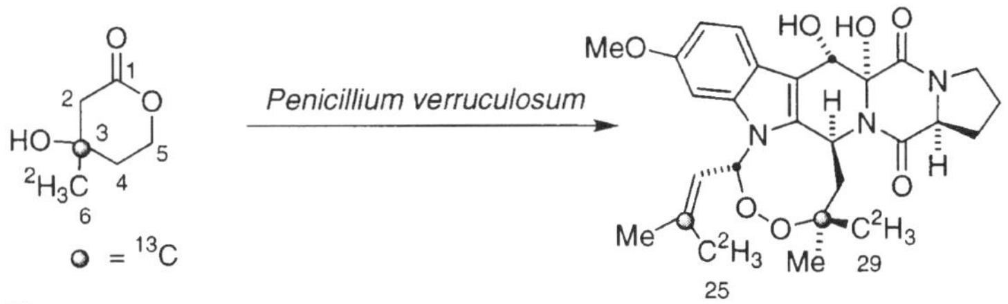

Scheme 46

C-6. It was reported that the relative intensities and magnitude of the two α-isotopically shifted signals for C-29 were basically the same as those for the C-25 resonance. Since no deuterium loss occurs from C-25 during the biosynthesis, this signal served as an internal reference for the above experiment confirming the β-isotope shift experiments. Similar α-isotopically shifted signals for C-28 were observed, also indicating that during the biosynthesis an sp^2-hybridized intermediate at C-28 can also be excluded. Confirmation of the above data was secured through the feeding of [2-$^2H_3$, 2-^{13}C]acetate which, upon incorporation into verruculogen, clearly revealed that C-24 and C-25 retain two and three deuterium atoms, respectively.

In order to explain the observed loss of stereochemical integrity of the diastereotopic methyl groups (C-28, C-29) at C-27, these workers proposed an intermediate allylic carbocation during the formation of the C-ring of fumitremorgin B as shown in Scheme 47. "Normal" prenylation of the indole N-1 and C-2 carbons with Walden inversion at the methylene carbon of DMAPP provides the doubly dimethylallylated, hypothetical substrate depicted in Scheme 47. Loss of one of the diastereotopic methylene hydrogens from C-3 provides the hypothetical allylic carbocation which allows for free rotation around the C-26/C-27 bond, thus scrambling the C-28/C-29 methyl groups. Closure of the amide nitrogen atom from the *Si*-face by either a cationic or radical process provides the C-ring of fumitremorgin B and verruculogen with the observed stereochemistry.

Since this group had previously established that one deuterium atom is retained at both C-3 and C-21 of verruculogen produced with (3*RS*)-[5-$^2H_2$]mevalonic acid lactone [74], a second experiment with stereospecifically labeled mevalonate at C-5 was conducted to determine the stereochemistry of the oxidation process at C-3 and C-21. Thus, (3*RS*, 5*S*)- and (3*RS*, 5*R*)-[5-^{2}H, 4-^{13}C]mevalonic acid lactones were individually fed to *Penicillium verruculosum* and the verruculogen so produced was analyzed by ^{13}C NMR. It was found that deuterium was retained at C-3 in verruculogen derived from the 5*S* stereoisomer as evidenced by the β-isotope shift at C-26 whereas no isotope shift was observed at C-22. In a complementary fashion, the verruculogen derived from the 5*R* stereoisomer exhibited a β-isotope shift at C-22 whereas there was no shift observed at the C-26 resonance. These data clearly indicate that during the biosynthesis of verruculogen the 5*Si* proton of mevalonate is retained at C-3 and the 5*Re* proton of mevalonate is retained at C-21.

As of this writing, literature on the biosynthesis of the tryprostatins, cyclotryprostatins, and spirotryprostatins could not be found. It can be reasonably

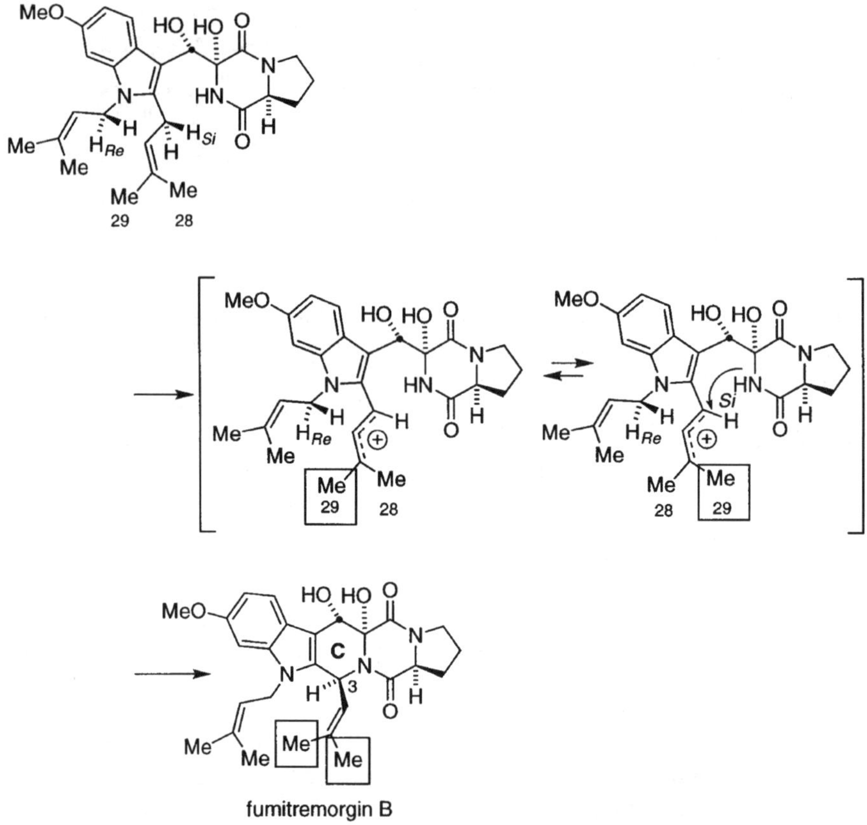

Scheme 47. Mechanism postulated by Vleggaar et al. for the loss of stereochemical integrity at the C-27 *gem*-dimethyl groups [75]

safely assumed that the biosynthesis of the tryprostatins and cyclotryprostatins closely parallels that of the fumitremorgins and verruculogen. Of particular interest will be elucidation of the spiro-ring-forming process in the biosynthesis of the spirotryprostatins. Based on metabolite co-occurrence and biogenetic and mechanistic considerations, the construction of the spirotryprostatins can be envisioned to arise via the pathway depicted in Scheme 48. Coupling of L-proline, and L-tryptophan followed by "normal" prenylation with DMAPP yields 6-demethoxytryprostatin. Divergence of this substance to spirotryprostatins A and B can occur via aromatic ring hydroxylation to tryprostatin B and S-adenosylmethionine (SAM) methylation to tryprostatin A. Oxidative ring closure, presumably analogous to that described by Vleggaar et al. for verruculogen [75] yields fumitremorgin C. Oxidation at the indole 2-position followed by a pinacol-type rearrangement would furnish spirotryprostatin A. A similar sequence of oxidative ring closure and spiro-ring-forming reactions from 6-demethoxytryprostatin and desaturation of the tryptophyl α,β-carbons would furnish spirotryprostatin B.

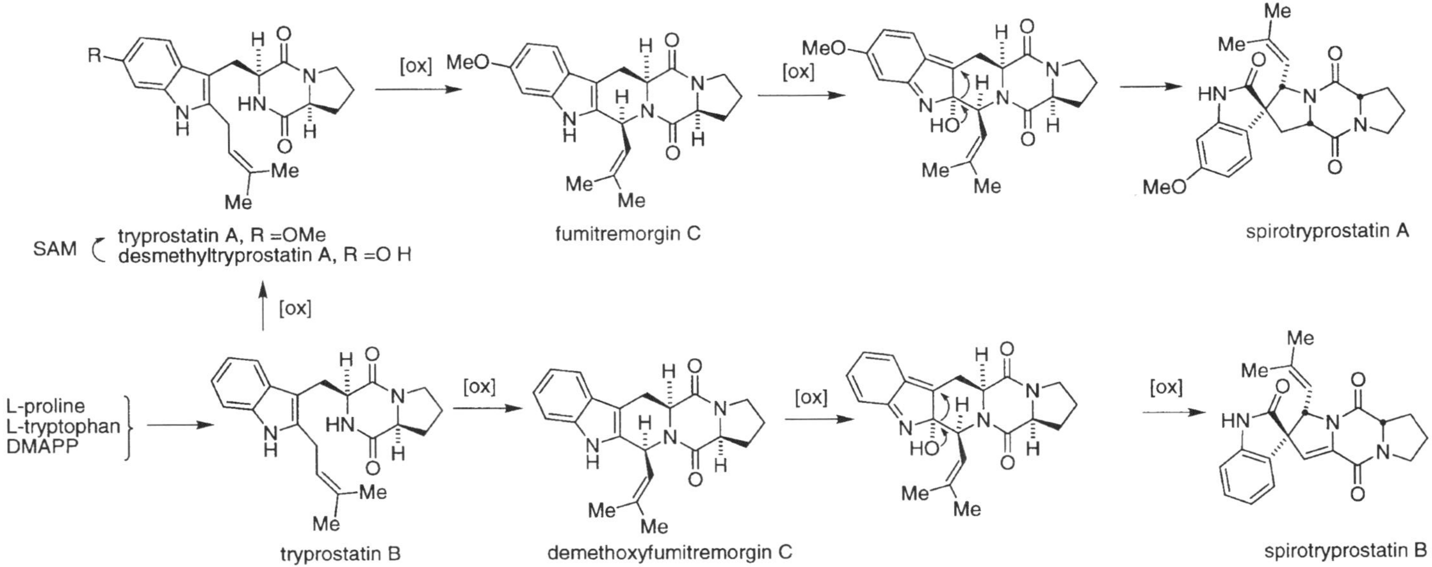

Scheme 48. Possible biosynthetic pathway to the spirotryprostatins

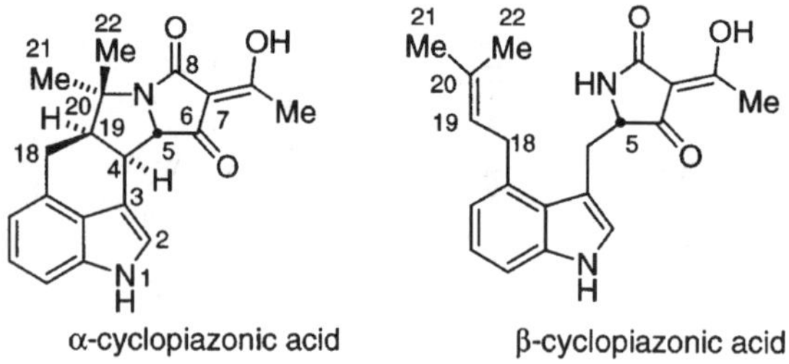

Fig. 14. Structures of α- and β-cyclopiazonic acids

6
α-Cyclopiazonic Acid

α-Cyclopiazonic acid (Fig. 14) is a mycotoxin produced by *Penicillium griseofulvum* Dierckx (formerly called *Penicillium cyclopium* Westling) as reported by Wilson et al. in 1968 [76]. A complete numbering scheme for α-cyclopiazonic acid could not be found and the one shown in Fig. 14 was taken from Gorst-Allman et al. (see below).

Holzapfel and co-workers carried out some preliminary biosynthetic work and identified via ^{14}C-labeled precursor feeding studies that this substance is fashioned from tryptophan, mevalonic acid, and acetate [77]. These workers were able to demonstrate further that β-cyclopiazonic acid is a biosynthetic precursor to α-cyclopiazonic via biosynthetic radiolabeling of β-cyclopiazonic acid and efficient incorporation of this precursor into α-cyclopiazonic acid. In addition, it was noted that β-cyclopiazonic acid is present in the mycelium at a stage when only traces of α-cyclopiazonic acid can be detected. The concentration of β-cyclopiazonic acid then rapidly decreases as the α-cyclopiazonic acid concentration increases and seems to indicate clearly a temporal separation of the biosynthetic steps.

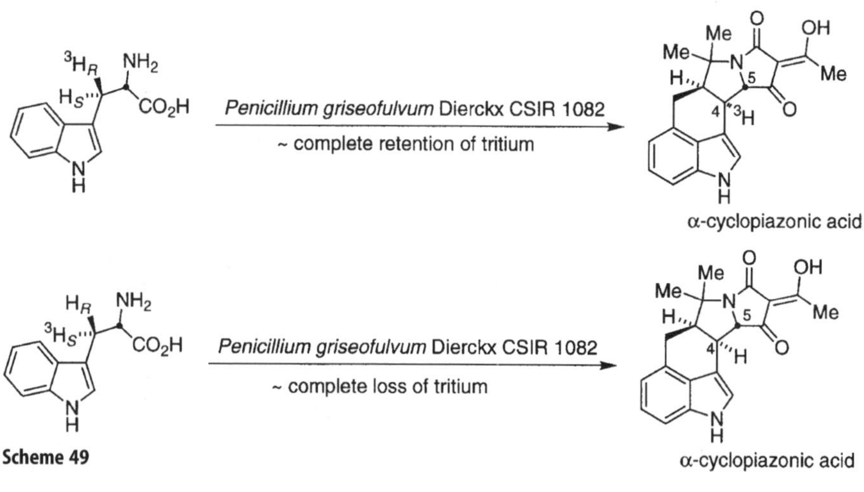

Scheme 49

Steyn et al. investigated the stereochemistry of proton removal from the methylene position of tryptophan culminating in closure of the D-ring [78]. To this end, (3*R*)-[3-^{3}H]tryptophan and (3*S*)-[3-^{3}H]tryptophan were synthesized and fed to *Penicillium griseofulvum* Dierckx. It was found that the 3*R* tritium was retained for both the (2*S*)- and (2*RS*)-forms of tryptophan and that for the (3*S*)-[3-^{3}H]tryptophan almost complete loss of the tritium was observed (Scheme 49, Table 6). A racemic mixture and the individual stereoisomers of the labeled tryptophan were efficiently incorporated into *β*-cyclopiazonic acid, thereby indicating the integrity of the methylene group during the early stages of the biosynthesis. Site-specific incorporation of the tritium atom at C-4 was secured indirectly by base-catalyzed exchange of the C-5 proton with no concomitant loss of tritium and through the feeding of (2*RS*)-[3-$^2H_2$]tryptophan which showed site-specific incorporation of the deuterium by NMR at C-4.

These workers [78] further established that the *α*-methine proton from tryptophan is not lost during the biosynthesis through feeding of (2*RS*)-[2-^{3}H]tryptophan admixed with (2*RS*)-[3-^{14}C]tryptophan as an internal standard. Subtracting loss of the (2*R*)-isomer via metabolism of the (2*R*)-isomer to indolylpyruvic acid gave 97.5% retention of tritium at C-5 in *α*-cyclopiazonic acid. These results are consistent with a mechanism for ring-closure that involves the formation of the C-C bond at C-4 occurring from the face opposite to the side that the proton is removed from.

Gorst-Allman et al. reported a study on additional stereochemical aspects of the D-ring formation via labeled mevalonic acid derivatives [79]. A ping-pong bi-bi mechanism was suggested for the cyclization reaction by Schabort and coworkers [80] and involves the loss of the *pro-S*-hydrogen from the methylene group of tryptophan (see Scheme 49). These workers prepared samples of stereospecifically labeled [4-^{3}H]mevalonic acid lactones and fed these with (3*RS*)-[2-^{14}C]- and (3*R*)-[3-^{14}C]mevalonic acid lactone to *Penicillium griseofulvum*

Table 6. Incorporation of labeled tryptophan derivatives into *α*- and *β*-cyclopiazonic acids [78]

Config. of Trp precursor	^{3}H:^{14}C ratio	^{3}H:^{14}C Ratios and retention of ^{3}H (%) in *α*-cyclopiazonic acid	and *β*-cyclopiazonic acid
(2*S*, 2*R*)-[3-^{3}H, 3-^{14}C]	4.53:1	4.44 (98.0)	4.32 (95.4)
	1.67:1	1.61 (96.4)	1.60 (95.8)
(2*RS*, 3*SS*)-[3-^{3}H, 3-^{14}C]	4.60:1	4.52 (98.3)	0.21 (4.6)
	3.99:1	–	0.20 (5.0)
	2.41:1	–	0.14 (5.8)
(2*RS*, 3*RR*)-[3-^{3}H, 3-^{14}C]	5.69:1	5.60 (98.4)	5.37 (94.4)
	2.16:1	–	2.09 (96.7)
(2 S, 3*R*)+(2*R*, 3*RS*)-[3-^{3}H, 3-^{14}C]	4.16:1	3.92 (94.2)	2.09 (50.2)
(2*RS*)-[2-^{3}H, 3-^{14}C]	7.32:1	–	3.57 (48.8)

Table 7. Incorporation of labeled mevalonic acid lactones into α-cyclopiazonic acid [79]

Mevalonic acid lactone mixture	$^{3}H:^{14}C$ Ratio of mevalonate precursor mixture	$^{3}H:^{14}C$ Ratio in α-cyclopiazonic acid
(3*S*, 4*S*)-[4-^{3}H]+(3*R*, 4*R*)-[4-^{3}H]+(3*RS*)-[2-^{14}C]	2.35	2.44
(3*S*, 4*S*)-[4-^{3}H]+(3*R*, 4*R*)-[4-^{3}H]+(3*R*)-[2-^{14}C]	1.67	0.83

Dierckx and examined the fate of these labels in α-cyclopiazonic acid. The results of these experiments are collected in Table 7.

From this data it was concluded that the (3*R*)-isomer of mevalonic acid lactone is incorporated into α-cyclopiazonic acid with retention of the 4-*pro-R*-hydrogen atom.

In an effort to establish the stereochemical origin of the geminal methyl groups in α-cyclopiazonic acid, Gorst-Allman et al. [79] assigned the diastereotopic geminal methyl groups by chemical degradation into a mixture of *O*-methyldesacetylcyclopiazonic acid and the corresponding C-5 epimer by treatment of α-cyclopiazonic acid with dilute sulfuric acid followed by etherification with diazomethane (Scheme 50). Based on a conformational analysis and chemical shift differences of the geminal methyl groups in the C-5 epimer, the assignment for the C-22 and C-23 methyl groups was made.

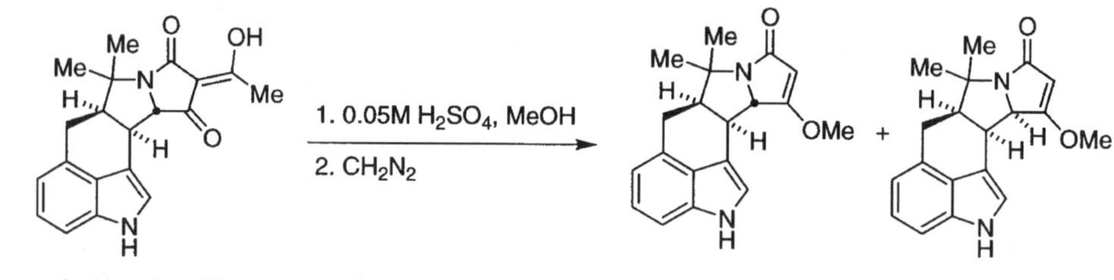

Scheme 50. Chemical degradation of α-cyclopiazonic acid into *O*-methyldesacetylcyclopiazonic acids [79]

Using the above assignments, these workers fed [1,2-$^{13}C_2$]acetate to *Penicillium griseofulvum* Dierckx and determined that the *Z*-methyl group (C-22) in β-cyclopiazonic acid becomes the β-methyl group (C-22) in α-cyclopiazonic acid (Scheme 51).

Although the detailed mechanism for the closure of the D-ring in α-cyclopiazonic acid via β-cyclopiazonic acid is not fully understood, it should be noted that the oxidocyclase that effects this transformation has been studied by Schabort and co-workers [80] and has been found to be a complex flavoenzyme system consisting of five isoenzymes. It was found that each isoenzyme contains

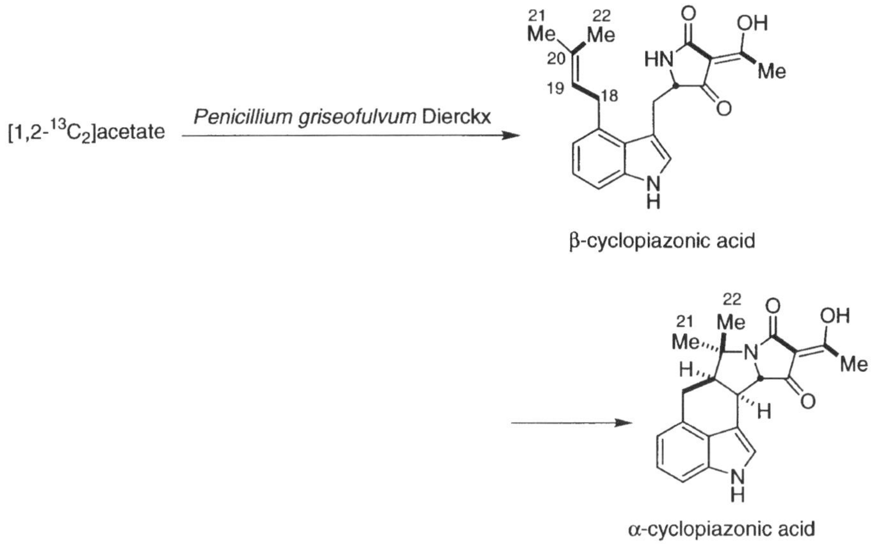

Scheme 51. Stereochemical origin of the geminal methyl groups in β- and α-cyclopiazonic acids [79]

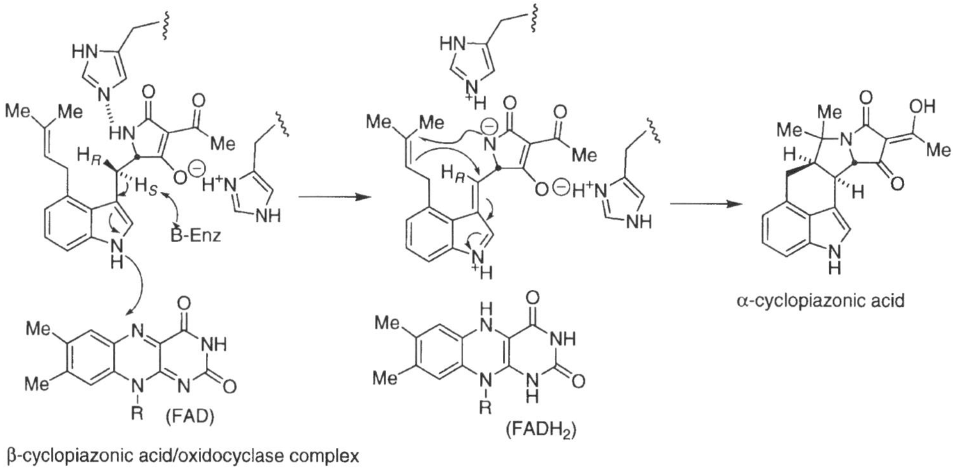

Scheme 52. Proposed mechanism for β-cyclopiazonic acid oxidocyclase [80]

one covalently linked flavin per molecule and requires O_2 or artificial electron acceptors for catalytic activity. Furthermore, pH-dependence studies suggest the presence of two histidine residues at the active site. These studies, reported by Schabort and co-workers have led to a mechanism for the ring closure as shown in Scheme 52 [80].

7
Ergot Alkaloids

The biosynthesis of the ergot alkaloids has been intensively studied over the past 40 years and comprises a large body of chemical literature. This family of alkaloids are secondary metabolites of the parasitic fungi *Claviceps* sp. (Clavicipitales) and display a wide array of interesting biological activities. The history and biosynthesis of the ergot alkaloids was reviewed in 1976 by Floss [81] and the material covered in this excellent review will not be elaborated on further here. In this section, only a few highlights of the initial prenylation of L-tryptophan and formation of the C-ring of the ergot alkaloid skeleton will be discussed. The biosynthesis of the D-ring system has already been extensively reviewed [81]. The later stages of ergot alkaloid biosynthesis and conjugation to the prolyl peptide moieties will also not be discussed.

7.1
Chanoclavine-I, Agroclavine, and Elymoclavine

Current knowledge of the biosynthesis of the ergot alkaloids appears to follow the pathway depicted in Scheme 53. Prenylation of L-tryptophan yields dimethylallyltryptophan (DMAT) which is N-methylated by *S*-adenosylmethionine (SAM) followed by oxidation, cyclization of the C-ring, and decarboxylation to give chanoclavine-I, the first isolable metabolite bearing the C-ring. In most of the studies discussed below on the early stages of ergot alkaloid biosynthesis, elymoclavine has been utilized to determine biosynthetic incorporation of fed precursors since this substance accumulates in sufficient quantities in most *Claviceps* spp. to allow for the determination of reasonably precise incorporation levels.

To accomodate the observed partial scrambling of the ^{14}C label between the two isoprenoid methyl groups of elymoclavine, agroclavine and lysergic acid (~90% at C-17 and ~10% at C-7) and the partial scrambling of the ^{3}H label of *pro-R* and *pro-S* hydrogens from C-5 of mevalonic acid into various clavines, Floss et al. proposed the mechanism shown in Scheme 54 [82]. In this mechanism, a double S_N' attack on DMAPP with partial rotation was invoked to account for the observed loss of stereochemical integrity of the labels.

To test this hypothesis, feeding experiments were performed with (5*S*)- and (5*R*)-[2-^{13}C, 5-$^{2}H_1$] mevalonate to cultures of *Claviceps* sp. strain SD58. Based on the mass spectra fragmentation patterns of the elymoclavine produced under these conditions the authors concluded that: "This experiment thus suggests that the scrambling of ^{13}C from C-2 of mevalonate between C-7 and C-17 of elymoclavine and that of ^{2}H from C-5 of mevalonate between the two allylic hydrogen positions of the the isoprenoid unit occur independently of each other (although not necessarily in different reaction steps)."

Arigoni et al. had proposed that in *Claviceps penniseti* partial loss of stereochemical integrity of the label from mevalonate must occur prior to or during the biosynthesis of DMAPP [83]. This was also confirmed by Floss et al. by

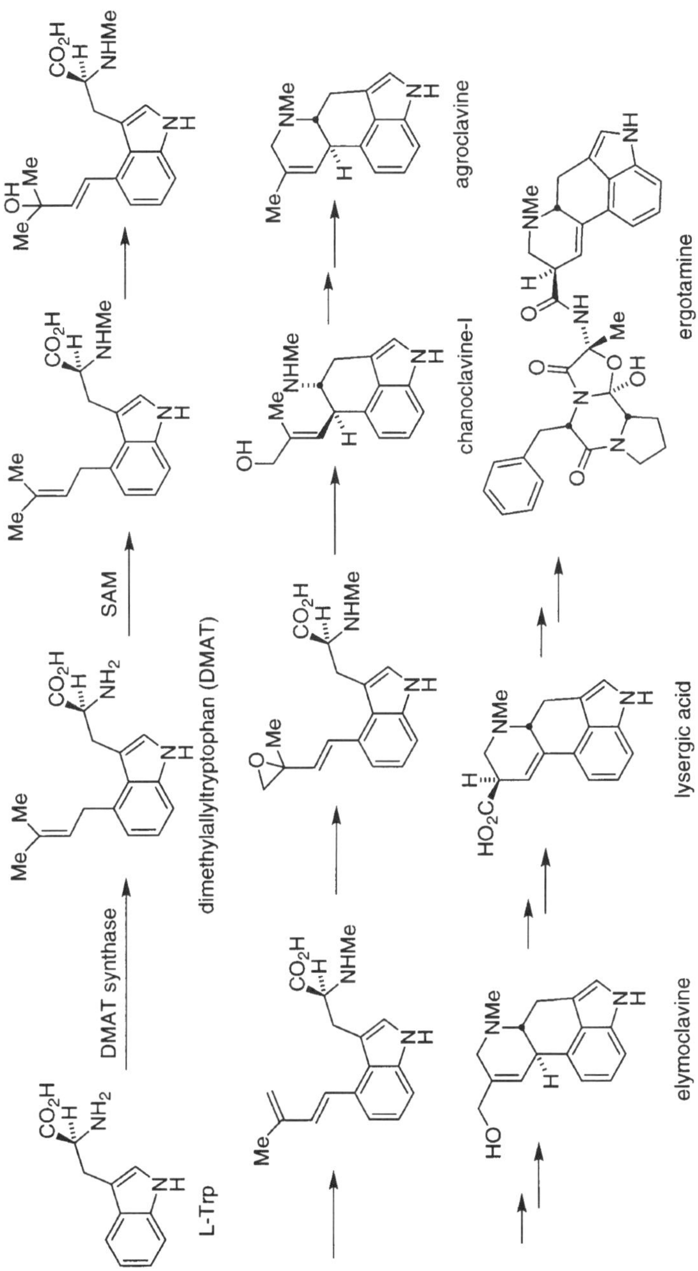

Scheme 53. Biosynthetic pathway to the ergot alkaloids such as ergotamine [82, 88]

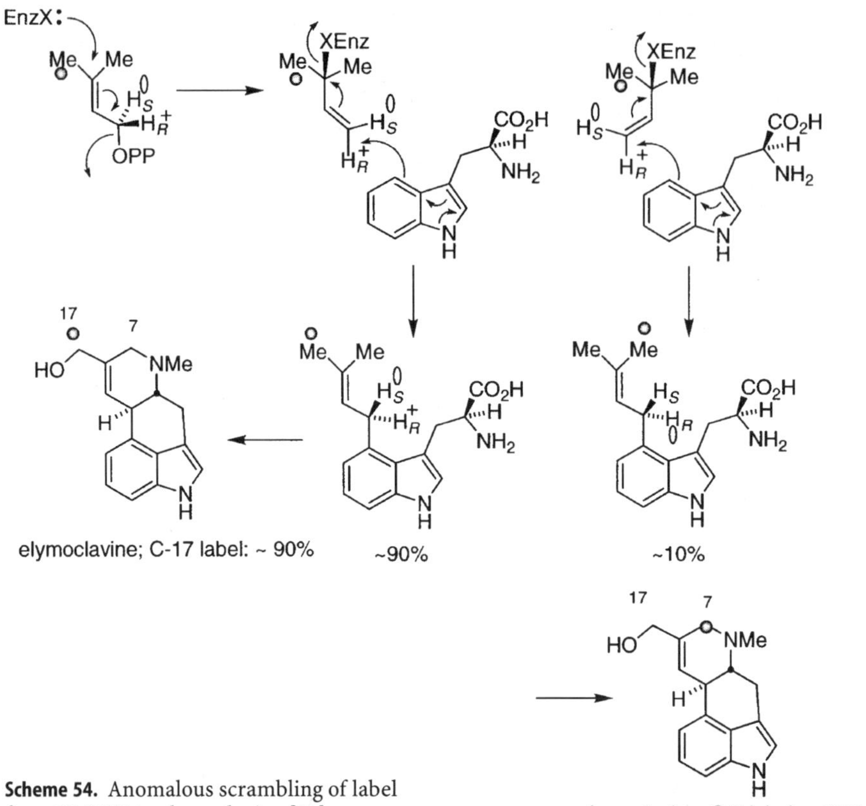

Scheme 54. Anomalous scrambling of label from DMAPP to elymoclavine [82]

showing that (*Z*)-[*methyl*-$^2H_3$]DMAT (97.2% Z, 2.8% E) generated from (*Z*)-[*methyl*-$^2H_3$]DMAPP with DMAT synthase was converted into chanoclavine and elymoclavine without loss of stereochemical integrity in *Claviceps* sp. SD58.

To rationalize these data, Floss et al. proposed the mechanism shown in Scheme 55 wherein a direct electrophilic aromatic substitution reaction from an enzyme-bound DMAPP ion pair species alkylates C-4 of the tryptophan nucleus [82]. The minor product, where partial loss of stereochemical integrity is sacrificed, was envisioned to occur via rotation around the C-1/C-2 bond of the allylic carbocation species as shown in Scheme 55. Poulter et al. subsequently published a mechanistic study on DMAT synthase that is fully consistent with this interpretation [84].

The second step in the pathway is believed to be N-methylation of the tryptophyl α-amino group by *S*-adenosylmethionine. Support for this notion was reported by Floss et al. as illustrated in Scheme 56 [85]. Synthesis and feeding of [α-^{15}N, N^{α}-*Me*-$^2H_3$]DMAT and [^{15}N, N-*Me*-$^2H_3$]dimethylallyltryptamine to cultures of *Claviceps* sp. SD58 was conducted to test this hypothesis. It was found that the labeled DMAT was efficiently incorporated into elymoclavine but, that the labeled tryptamine derivative was not incorporated. This demonstrates that

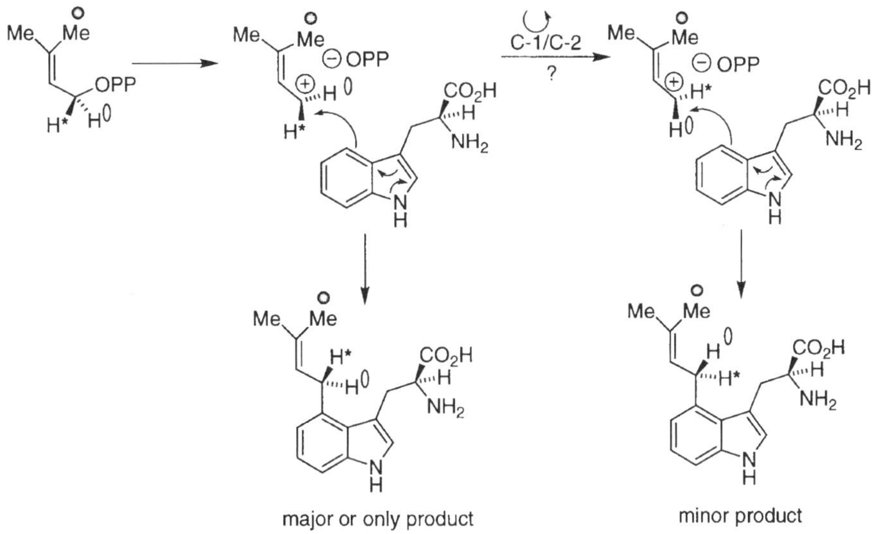

Scheme 55. Proposed ionization and C-1/C-2 partial rotation to accomodate observed scrambling [83]

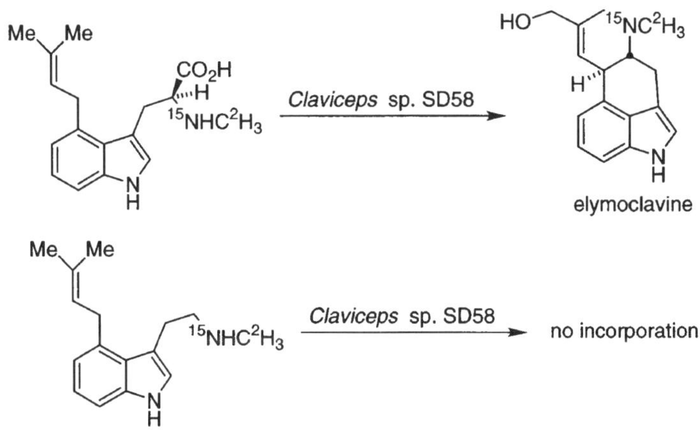

Scheme 56. Incorporation of *N*-Me-DMAT into elymoclavine [85]

N-methylation precedes the decarboxylation step and provides support for *N*-methyl-DMAT as a pathway intermediate.

The mechanism of cyclization to form the C-ring of the ergot alkaloid nucleus has been the subject of much speculation. Floss et al. proposed the dihydroxylation of the isoprenyl unit of N-Me-DMAT followed by an intramolecular S_N' cyclization coupled with a decarboxylation to yield chanoclavine-I (Scheme 57) [86]. On the other hand, Pachlatko and co-workers proposed the formation of an epoxide followed by a similar S_N' cyclization coupled with a decarboxylation [83]. To test the possible intermediacy of the diol, Floss et al. prepared [N$^{\alpha}$-*Me*-

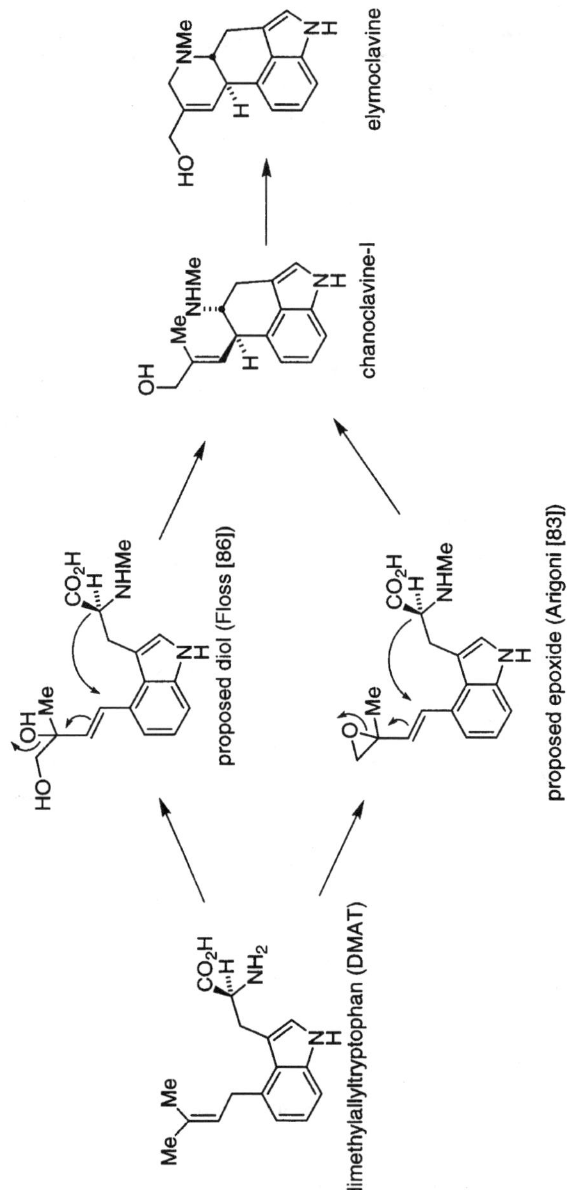

Scheme 57. Proposed modes of formation of the C-ring of chanoclavine [83, 86]

$^{2}H_{3}$] sample of the putative diol and fed this substance to *Claviceps* sp. SD58 and examined for incorporation into elymoclavine. These workers found no significant incorporation of the diol into elymoclavine lending credence to the intermediacy of Arigoni's proposed epoxide. Further support for this intermediate was obtained when Floss and Kobayashi incubated cultures of *Claviceps* sp. SD58 in an atmosphere of $^{18}O_2$ gas and obtained 76 atom% ^{18}O-enrichment of chanoclavine I [87]. Incorporation of molecular oxygen indicates a mechanism involving a possible carbocation at the benzylic position and a possible carbanion at C-α. The intermediacy of the epoxide and the mechanism shown in Scheme 57 are in accordance with the experimental data.

To elucidate further the mechanism of closure for ring C, Floss et al. synthesized the racemic amino acid shown in Scheme 58 [88]. Feeding experiments with the *N*-trideuteriomethyl labeled compound showed a 33% specific incorporation of the amino acid into elymoclavine. The high level and specificity of incorporation (no M+1 or M+2 species were detected) clearly point toward the intermediacy of the tertiary carbinol in the biosynthesis of the tetracyclic ergoline ring system. On the basis of these results, Floss et al. proposed the overall sequence shown in Scheme 53 for elaboration of the C ring system [88].

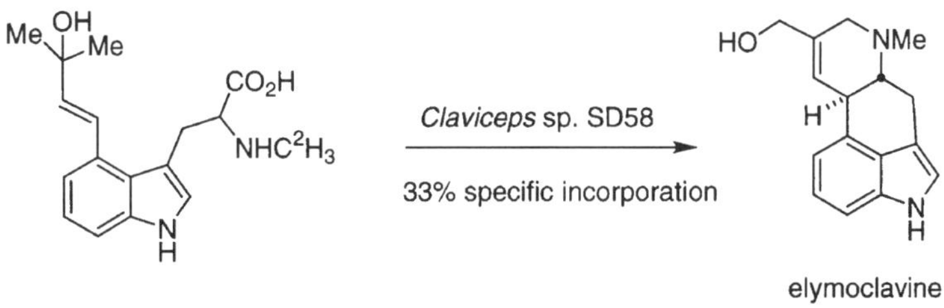

Scheme 58. Incorporation of the proposed tertiary carbinol into elymoclavine

7.2 Clavicipitic Acid

In the course of their studies on the biosynthesis of ergot alkaloids, Robbers and Floss isolated clavicipitic acid, a new indolic amino acid [89]. Based on the X-ray crystal data of the major isomer and the assumption that the stereocenter at C-5 retains its stereochemistry from L-tryptophan, the structure shown in Scheme 59 was assigned [90]. By performing feeding experiments on *Claviceps* sp. SD 58 with ^{14}C clavicipitic acid (biosynthetically labeled), the laboratories of Floss [90] and Anderson [91] independently determined that clavicipitic acid is not a precursor to elymoclavine. Instead, Floss argues that it arises from: "... a derailment of the metabolism leading to the tetracyclic ergolines between the first and second pathway-specific steps, the isoprenylation of tryptophan and the N-methylation of 4-(γ,γ-dimethylallyl)-tryptophan (DMAT)." Anderson and co-workers isolated an enzyme, DMAT oxidase, from *Claviceps* sp. which catalyzes the formation of clavicipitic acid from DMAT [92]. The enzyme, which requires

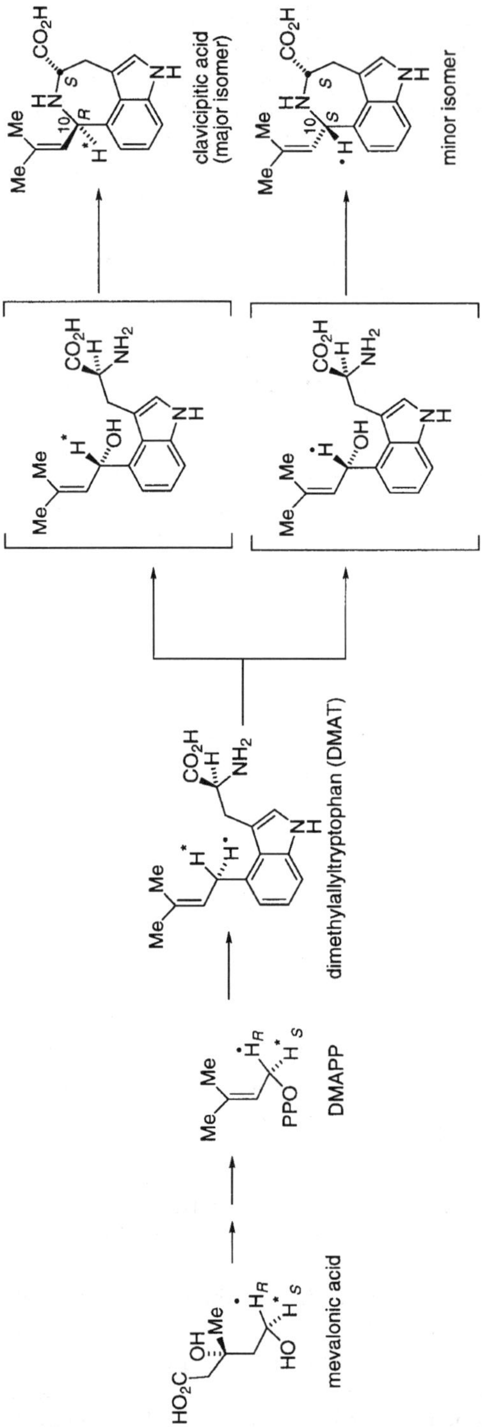

Scheme 59. Incorporation of labeled mevalonate into clavicipitic acid [90]

O_2 but no reducing agent, may catalyze oxidative cyclization directly or catalyze the 10-hydroxylation of DMAT which would then cyclize to clavicipitic acid (Scheme 59).

Floss et al. also performed feeding experiments with ^3H and ^{14}C labeled mevalonic acid to determine the fate of the C-5 *pro-R* and *pro-S* hydrogens of mevalonate (Scheme 59) [90]. There was complete retention of the *pro-S* hydrogen when (3*R*, 5*S*)-[5-^3H] mevalonic acid and (3*R*, *S*)-[2-^{14}C] mevalonic acid were fed to cultures of *Claviceps* sp. SD 58. However, these workers observed a peculiar result in the feeding experiment using non-stereospecifically C-5 labeled mevalonate wherein there was 75% retention of tritium at C-5 derived from mevalonate at C-10 in clavicipitic acid. Along with this result, they also found that feeding experiments with (3*R*, 5*R*)-[5-^3H] mevalonic acid and (3*R*, *S*)-[2-^{14}C] mevalonic acid retained half of the tritium from the C-5 *pro-R* hydrogen of mevalonic acid. Floss suggested the following to explain this unusual observation: "A possible explanation could be advanced if one assumes that the DMAT oxidase reaction discriminates only partially between the two heterotopic hydrogens at C-10, replacing one at a rate which differs from that of the other, and that the reaction involves a hydrogen isotope effect."

8
Addendum: Biosynthesis of the Primary Metabolites L-Tryptophan and Dimethylallyl Pyrophosphate

In this section, a brief and superficial overview of the biosynthesis of L-tryptophan and the key C_5 building block dimethylallyl pyrophosphate is described. The intent of this is to give a quick visual reference guide for the labeling patterns intrinsic to the primary metabolites that are then re-cast in the secondary metabolite structures that have been discussed throughout this article.

8.1
Biosynthesis of L-Tryptophan

The aromatic amino acids, including phenylalanine, tyrosine, and tryptophan are biosynthesized via the shikimate pathway as shown in Scheme 60 [93].

Tryptophan is constructed from a branch point in the shikimic acid pathway. Chorismic acid is transaminated with L-glutamine as the nitrogen donor to give anthranilic acid. Anthranilate is then enzymatically condensed with 5-phosphoribosyl-α-pyrophosphate to form *N*-(5′-phosphoribosyl)-anthranilate from which two carbon atoms of the indole ring will be constructed. An Amadori-type rearrangement on *N*-(5′-phosphoribosyl)-anthranilate catalyzed by *N*-(5′-phosphoribosyl)-anthranilate isomerase furnishes enol-1-carboxyphenylamino-1-deoxyribulose phosphate. Indole-3-glycerol phosphate synthase then effects the decarboxylation and dehydration of enol-1-carboxyphenylamino-1-deoxyribulose phosphate giving indole-3-glycerol phosphate.

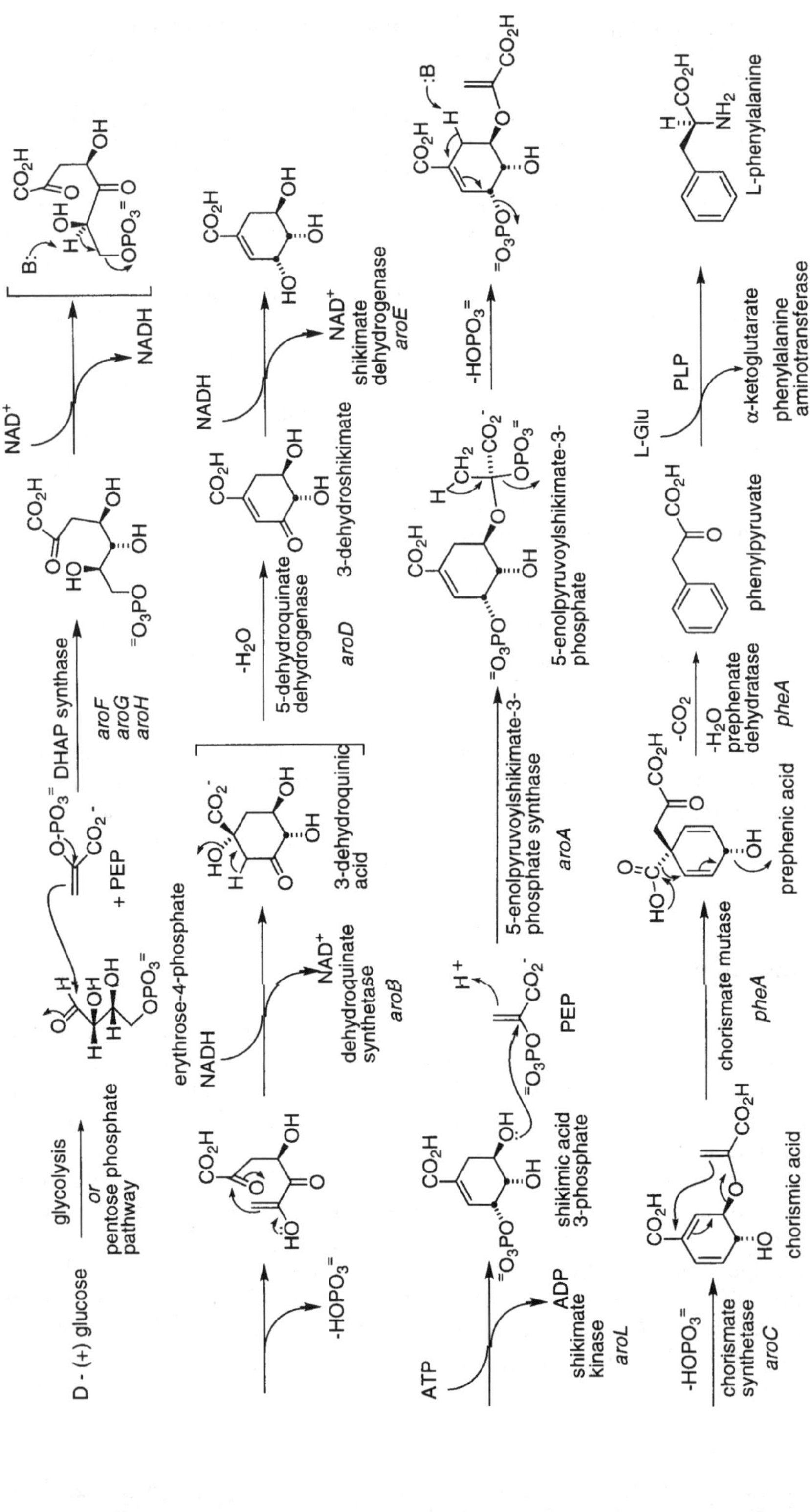

Scheme 60. The shikimic acid pathway

Indole-3-glycerol phosphate is the penultimate substrate which is fed into the tryptophan synthase complex. This enzyme is composed of two subunits (as an α_2- and β_2-tetramer) which utilizes substrate channeling to complete the construction of this amino acid. The α-subunit catalyzes the formation of indole from indole-3-glycerol phosphate and the β-subunit catalyzes the condensation of indole to a serine-derived aminoacrylate moiety bound to pyridoxal monophosphate (PLP).

The mechanism for cleavage of the C-C bond of glycerol to the indole ring is shown in Scheme 61.

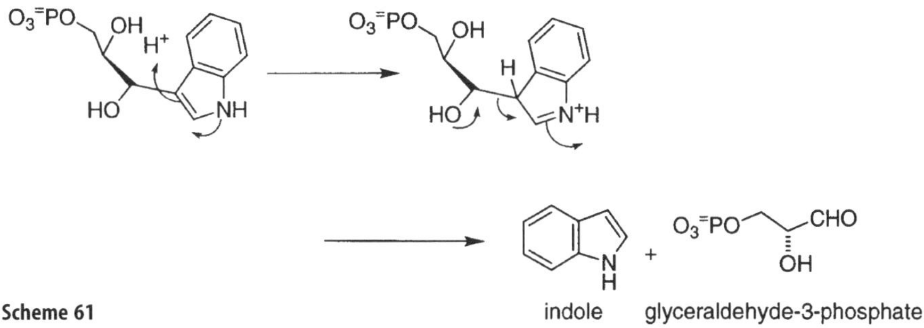

Scheme 61

The final assembly of tryptophan is shown in Scheme 62.

The biosynthesis of tryptophan poses an interesting challenge to the cell due to the lipophilic nature of the indole ring. The indole heterocycle formed in the first step would readily cross cell membranes and be lost to the cell if it were allowed to diffuse away from the enzyme. Nature has solved this problem by connecting the two functional parts together. Tryptophan synthase has a channel 25 Å long connecting the active sites of the two subunits. Thus, when indole is made in the α-subunit, it is "passed down" the channel to the β-subunit where the hydrophilic serine unit is connected. This is called "channeling" and increases the catalytic reaction rate more than tenfold. The crystal structure of tryptophan synthase, shown in Fig. 15, clearly reveals the two domains and the channel down the center of the enzyme [94].

8.2 Biosynthesis of the Isoprene Building Blocks

The five-carbon isoprene skeleton constitutes a ubiquitous biosynthetic building block in all cells. Nature constructs a daunting array of primary and secondary metabolites from the two basic precursors, *dimethylallyl pyrophosphate* (DMAPP) and *isopentenyl pyrophosphate* (IPP) (Fig. 16) [95].

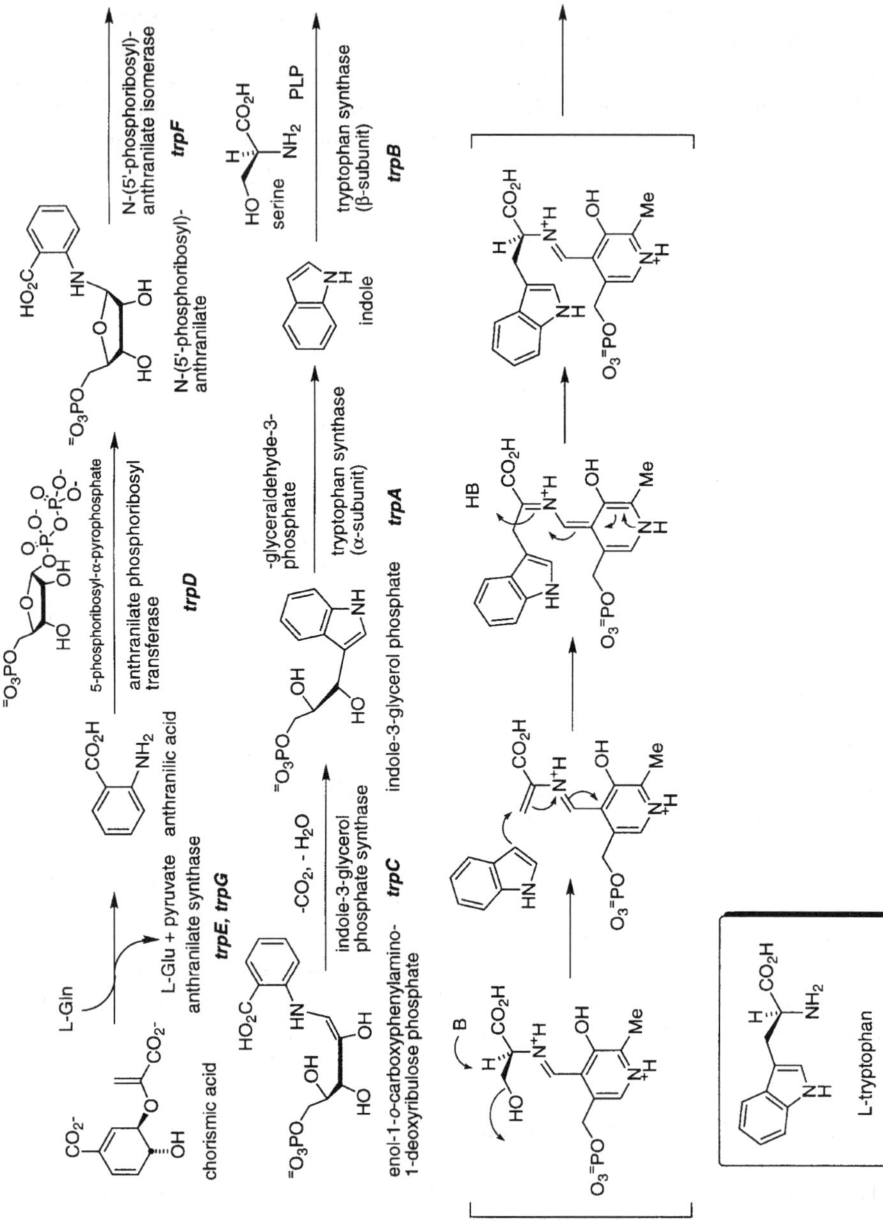

Scheme 62. The biosynthesis of L-tryptophan

Fig. 15. Two views of the crystal structure of tryptophan synthase. *Side view* showing both domains and *top view* showing the "channel" [94]

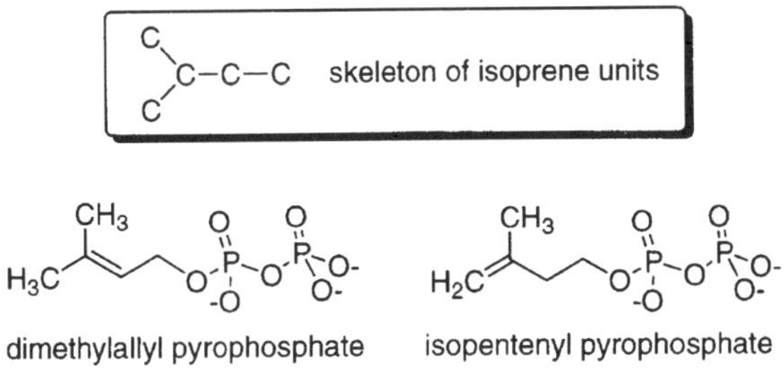

Fig. 16

8.2.1
The Mevalonate Pathway

The classical mevalonate pathway has long been recognized as one of the key primary metabolic pathways leading to the formation of the branched five-carbon skeleton of the isoprene building blocks, isopentenyl pyrophosphate and dimethylallyl pyrophosphate [95].

This biosynthetic path consists of the Claisen-type condensation of two acetyl CoA units to form the four-carbon substance acetoacetyl CoA; a third equivalent of acetyl CoA is then added in an aldol-type reaction giving, after hydrolysis of one of the thiol esters, hydroxymethylglutaryl CoA (HMG-CoA). HMG-CoA is then reduced by a net four electrons to mevalonic acid and is subsequently phosphorylated to mevalonic acid 5-pyrophosphate (MVA-5PP). This substrate is finally phosphorylated and decarboxylated with concomitant loss of inorganic phosphate to give the five-carbon isoprenoid isopentenyl pyrophosphate (IPP). IPP is then isomerized to DMAPP by an isomerase.

Scheme 63 illustrates the labeling pattern that is observed when doubly ^{13}C-labeled acetate is used as a precursor.

The alternate labeling pattern shown in Scheme 64 can be used as a guide when singly labeled acetate or glucose is utilized.

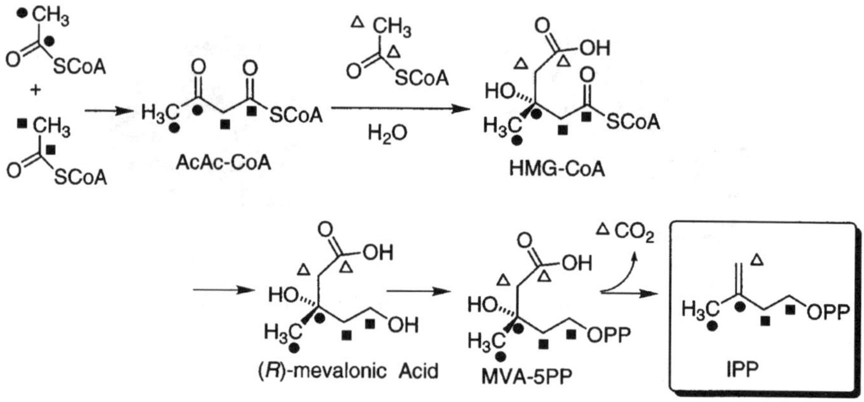

Scheme 63. The classical mevalonic acid pathway. Labeling pattern is via 1,2-doubly labeled acetate

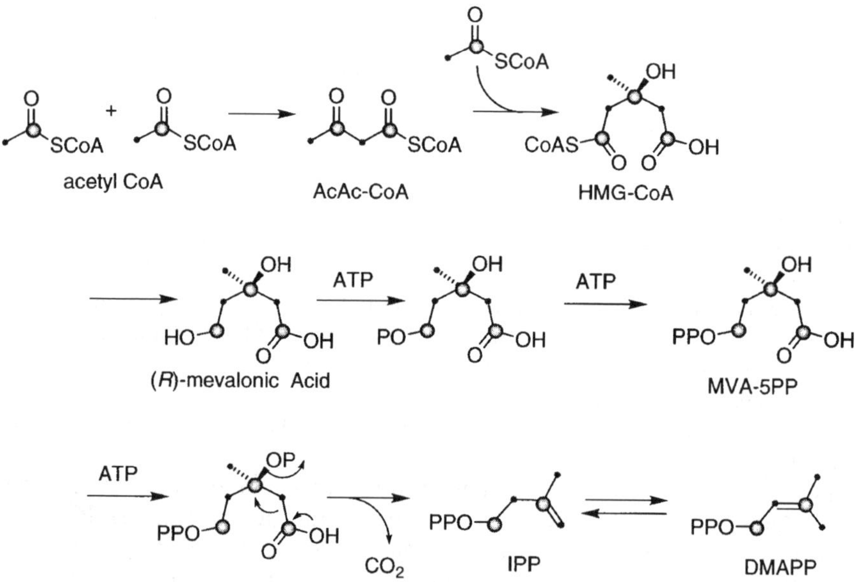

Scheme 64. The classical mevalonic acid pathway. Labeling pattern is via singly labeled acetate

8.2.2
The 1-Deoxy-D-xylulose Pathway

Recently, evidence has accumulated based on isotope labeling and feeding experiments that there must be an alternative pathway in some organisms for the formation of the isoprenoid building blocks [96–98]. Initially, it was proposed that distinct and compartmentalized acetyl CoA pools could account for the anomalous labeling patterns observed that were inconsistent with those expected from the mevalonate pathway. These results have recently been reinterpreted as resulting from a distinct biochemical pathway, named the deoxyxylulose pathway.

Based on a number of elegant experiments, Rohmer and co-workers suggested that dihydroxyacetone phosphate underwent an acyloin-type of condensation with an activated equivalent of acetaldehyde [96]. The resulting branched chain carbohydrate was postulated to undergo a skeletal rearrangement consistent with the "reshuffled" three-carbon fragment. On the other hand, Zenk and co-workers proposed a head-to-head condensation of glyceraldehyde 3-phosphate with the two-carbon unit resulting from the decarboxylative condensation of pyruvate with thiamine pyrophosphate (TPP) [97,98]. Experimental evidence for the latter pathway was secured through the incorporation of 1-deoxy-D-xylulose into ubiquinone in *E. coli*. The involvement of a reductoisomerase was recently reported by Seto et al. in the rearrangement of 1-deoxy-D-xylulose into the branched five-carbon aldehyde [99]. A mechanism for this pathway, based on currently available experimental data is illustrated in Scheme 65.

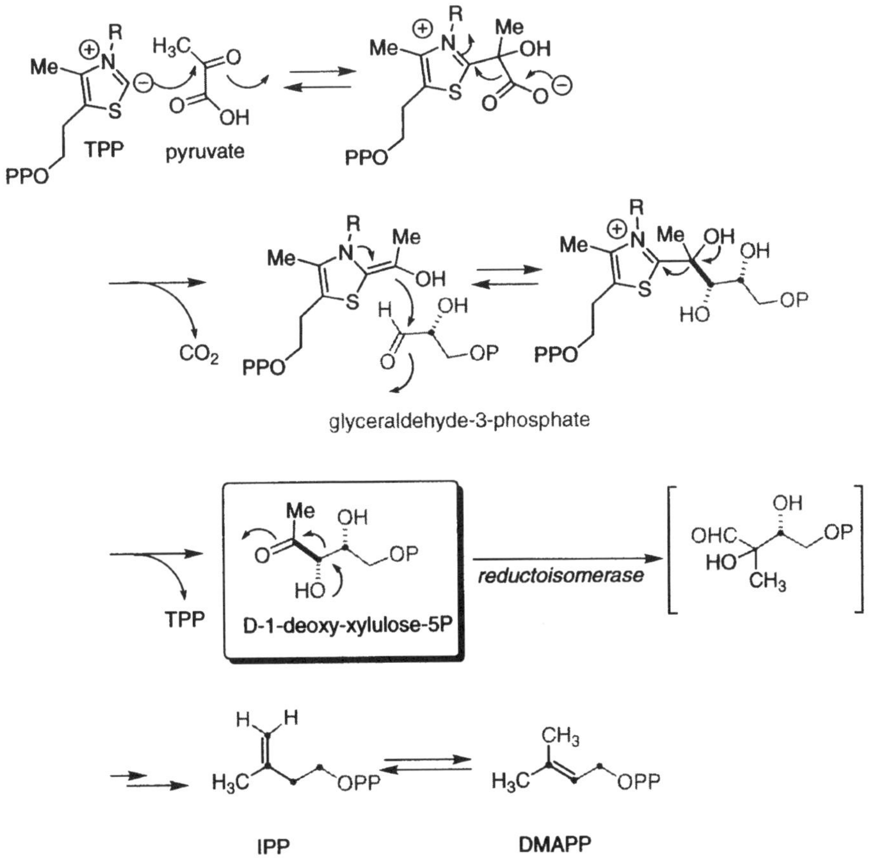

Scheme 65. The 1-deoxy-D-xylulose pathway to the isoprenoid building blocks

8.2.3
Evolutionary Origin of the Two Biosynthetic Pathways

The taxonomic distribution of the pathways as currently known in the limited number of organisms studied paints the following picture. Archaea appear to utilize the mevalonate pathway and includes such species as *Archaeoglobus fulgidus*, *Methanobacterium thermoautotrophicum*, and *Methanococcus jannaschii*. Orthologous genes for the mevalonate pathway have been found in the genome of *Borrelia burgdorferi*. The photosynthetic bacterium *Chloroflexus aurantiacus*, which represents the earliest known branch in the development of eubacteria, was demonstrated to produce the diterpene verrucosan-2β-ol via the mevalonate pathway.

On the other hand, the eubacteria *E. coli*, *Helicobacter pylori*, *Haemophilus influenza*, *Bacillus subtilus*, and the blue green alga *Synechocystis* sp. do not appear to have genes coding for the enzymes of the mevalonate pathway. These organisms contain putative orthologs of the *dxs* gene that presumably encode for the biosynthesis of 1-deoxy-D-xylulose. Taking these limited genetic data, it can be speculated that, in general, archaebacteria utilize the mevalonate pathway whereas eubacteria utilize the 1-deoxy-D-xylulose pathway and both pathways appear to operate, albeit in separate compartments, in higher plants. Thus, the mevalonate pathway in higher plants operates in the cytosol and the 1-deoxy-D-xylulose pathway appears to be compartmentalized in the plastids. There are no known cases of fungi that utilize the 1-deoxy-D-xylulose pathway and these organisms along with all higher animals are believed to utilize the mevalonate pathway exclusively.

The evolution of the plastid structure in higher plants is generally believed to have arisen by endocytosis and endosymbiosis of cyanobacteria. The genes coding for the enzymes of the 1-deoxy-D-xylulose pathway were presumably imported with the early endosymbiants and those genes relocated to the nuclear chromosome. It is interesting that these enzymes still retain a plastid-targeting sequence that has allowed the unusual compartmentalization of the two isoprene biosynthesis pathways in plants. Very recently, Martin and Muller proposed that eukaryotic cells originated from a methanobacterium that had incorporated a eubacterium [100]. This ancient biochemical progenitor of eukaryotic cells therefore gave birth to the cytoplasmic mevalonate machinery in plants and the 1-deoxy-D-xylulose pathway was later derived from the endosymbiotic cyanobacteria.

Acknowledgement. The authors acknowledge financial support from the NIH (Grand CA 70375 to RMW) and the DGICYT of Spain (Proyecto PB95-1089 to JFS-C). We also wish to acknowledge the ACS Division of Organic Chemistry Fellowship (sponsored by SmithKline Beecham) and the Pharmacia-Upjohn Company for Fellowship support for Emily M. Stocking.

9 References

1. Birch AJ, Wright JJ (1969) J Chem Soc Chem Comm:644
2. Wilson BJ, Yang DTC, Harris TM (1973) Appl Microbiol:633
3. Robbers JE, Straus JW (1975) Lloydia 38:355
4. Coetzer J (1974) Acta Cryst B30:2254
5. Steyn P (1971) Tetrahedron Lett 3331
6. Steyn P (1973) Tetrahedron 29:107
7. Coetzer J, Steyn PS (1973) Acta Cryst B29:685
8. Birch AJ, Wright JJ (1970) Tetrahedron 26:2329
9. Birch AJ, Russell RA (1972) Tetrahedron 28:2999
10. Ritchie R, Saxton JE (1975) J Chem Soc Chem Comm 611
11. Ritchie R, Saxton JE (1981) Tetrahedron 37:4295
12. Kametani T, Kanaya N, Ihara M (1981) J Chem Soc Perkin Trans I:959
13. Baldas J, Birch AJ, Russell RA (1974) J Chem Soc Perkin Trans I:50
14. Porter AEA, Sammes PG (1970) J Chem Soc Chem Comm 1103
15. Birch AJ (1971) J Agr Food Chem 19:1088
16. Williams RM, Glinka T, Kwast E (1988) J Am Chem Soc 110:5927
17. Williams RM, Glinka T, Kwast E, Coffman H, Stille JK (1990) J Am Chem Soc 112:808
18. Williams RM, Kwast E, Coffman H, Glinka T (1989) J Am Chem Soc 111:3064
19. Sanz-Cervera JF, Glinka T, Williams RM (1993) Tetrahedron 49:8471
20. Domingo LR, Sanz-Cervera JF, Williams RM, Picher MT, Marco JA (1997) J Org Chem 62:1662
21. Williams RM, Kwast E (1989) Tetrahedron Lett 30:451
22. Williams RM (unpublished results)
23. Harrison DM (1981) Tetrahedron Lett 22:2501
24. Shoop WL, Egerton JR, Eary CH, Suhayda D (1990) J Parasitol 76:349
25. Ostlind DA, Mickle WG, Ewanciw DV, Andriuli WC, Campbell WC, Hernandez S, Mochales S, Munguira E (1990) Res Vet Sci 48:260
26. Shoop WL, Haines HW, Eary CH, Michael BF (1992) Am J Vet Res 53:2032
27. Shoop WL, Michael BF, Haines HW, Eary CH (1992) Vet Parasitol 43:259
28. Schaeffer JM, Blizzard TA, Ondeyka J, Goegelman R, Sinclair PJ, Mrozik H (1992) Biochem Pharmacol 43:679
29. Yamazaki M, Okuyama E (1981) Tetrahedron Lett 22:135
30. (a) Ondeyka JG, Goegelman RT, Schaeffer JM, Kelemen L, Zitano L (1990) J Antibiot 43:1375; (b) Liesch JM, Wichmann CF (1990) J Antibiot 43:1380; (c) Blizzard TA, Mrozik H, Fisher MH, Schaeffer SM (1990) J Org Chem. 55:2256; (d) Blizzard TA, Marino G, Mrozik H, Fisher MH, Hoogsteen K, Springer JP (1989) J Org Chem 54:2657; (e) Blizzard TA, Margiatto G, Mrozik H, Schaeffer JM, Fisher MH (1991) Tetrahedron Lett 32:2437; (f) Blizzard TA, Margiatto G, Mrozik H, Schaeffer JM, Fisher MH (1991) Tetrahedron Lett 32:2441; (g) Blizzard TA, Rosegay A, Mrozik H, Fisher MH (1989) J Labeled Compounds and Radiopharmaceuticals 28:461
31. (a) Blanchflower SE, Banks RM, Everett JR, Manger BR, Reading C (1991) J Antibiotics 44:492; (b) Blanchflower SE, Banks RM, Everett JR, Reading C (1993) J Antibiotics 46:1355
32. Banks RM, Blanchflower SE, Everett JR, Manger BR, Reading C (1997) J Antibiot 50:840
33. Hayashi H, Nishimoto Y, Nozaki H (1997) Tetrahedron Lett 38:5655
34. Whyte AC, Gloer JB (1996) J Nat Prod 59:1093
35. (a) Polonsky J, Merrien M-A, Prange T, Pacard C (1980) J Chem Soc Chem Comm 601; (b) Prange T, Billion M-A, Vuilhorgne M, Pascard C, Polonsky J, Moreau S (1981) Tetrahedron Lett 22:1977
36. Fuchser J (1995) PhD thesis of Jens Fuchser "Beeinflussung der Sekundarstoffbildung bei Aspergillus ochraceus durch Variation der Kulturbedingungen sowie Isolierung, Strukturaufklarung und Biosynthese der neuen Naturstoffe", University of Goettingen. K. Bielefeld Verlag, Friedland (Prof. A. Zeeck).

37. (a) Kuo MS, Wiley VH, Cialdella JI, Yurek DA, Whaley HA, Marshall VP (1996) J Antibiot 49:1006; (b) Kuo MS, Yurek DA, Mizsak SA, Cialdella JI, Baczynskyj L, Marshall VP (1999) J Am Chem Soc 121:1763
38. Lee BH, Clothier MF (1997) J Org Chem 62:1795
39. Stocking EM, Sanz-Cervera JF, Williams RM, Unkefer CJ (1996) J Am Chem Soc 118:7008
40. Looser M (1989) PhD thesis. ETH (D. Arigoni)
41. Kellenberger JL (1997) PhD thesis. ETH (D. Arigoni)
42. Stocking E, Sanz-Cervera JF, Williams RM (1999) Angew Chem Int Ed Engl 38:786
43. Scott PM, Merrien M, Polonsky J (1976) Experientia 32:140
44. Barrow KD, Colley PW, Tribe DE (1979) J Chem Soc Chem Comm:225
45. Gorst-Allman CP, Steyn PS, Vleggaar R (1982) J Chem Soc Chem Comm 652
46. Bhat B, Harrison DM, Lamont HM (1990) J Chem Soc Chem Comm 1518
47. Bhat B, Harrison DM, Lamont HM (1993) Tetrahedron 49:10,663
48. Bycroft BW, Landon W (1970) J Chem Soc Chem Comm 967
49. Vleggaar R, Wessels PL (1980) J Chem Soc Chem Comm 160
50. Steyn PS, Vleggaar R (1983) J Chem Soc Chem Comm 560
51. (a) Mantle PG, Perera KPWC, Maishman NJ, Mundy GR (1983) Appl Environ Microbiol 45:1486; (b) For leading references on the biosynthesis of the penitrems and related metabolites, see Penn J, Mantle PG (1994) Phytochemistry 35:921
52. (a) Quilico A, Panizzi L(1943) Chem Ber 76:348; (b)Quilco A (1964) Res Progr Org Biol Medicin Chem 1:1964
53. (a) Birch AJ, Blance GE, David S, Smith H (1961) J Chem Soc 3128; (b) Birch AJ, Farrar KR (1963) J Chem Soc 4277
54. (a) MacDonald JC, Slater GP (1966) Can J Microbiol 12:455; (b) Slater GP, MacDonald JC, Nakashima R (1970) Biochemistry 9:2886
55. Marchelli R, Dossena A, Casnati G (1975) J Chem Soc Chem Comm 779
56. (a) Allen CM (1970) Biochemistry 11:2154; (b) Allen CM (1973) J Am Chem Soc 95:2386
57. Casnati G, Gardini GP, Palla G, Fuganti C (1974) J Chem Soc Chem Perkin Trans I:2397
58. Cardillo R, Fuganti C, Gatti G, Ghiringhelli D, Graselli P (1974) Tetrahedron Lett 3166
59. Cardillo R, Fuganti C, Ghiringhelli D, Graselli P (1975) J Chem Soc Chem Comm 778
60. Allen JK, Barrow KD, Jones AJ (1979) J Chem Soc Chem Comm 280
61. Grundon MF, Hamblin MR, Harrison DM, Derry Logue JN, Maguire M, McGrath JA (1980) J Chem Soc Chem Perkin Trans I:1294
62. Harrison DM, Quinn P (1983) J Chem Soc Chem Comm 879
63. Cardillo R, Fuganti C, Ghiringhelli D, Graselli P (1977) J Chem Soc Chem Comm 474
64. Fuganti C, Grasselli P, Pedrocchi-Fantoni G (1979) Tetrahedron Lett 2453
65. (a) Cole RJ, Kirksey JW, Moore JH, Blankenship BR, Diener UL, Davis NB (1972) Appl Microbiol 24:248; (b) Fayos J, Lokensgard D, Clardy J, Cole RJ, Kirksey JW (1974) J Am Chem Soc 96:6785; (c) Cole RJ, Kirksey JW, Cox RH, Clardy J (1975) J Agric Food Chem 23:1015; (d) Cole RJ, Kirksey JW, Dorner JW, Wilson DM, Johnson JC, Bedell DM, Springer JP, Chexal KK, Clardy JC, Cox RH (1977) J Agric Food Chem 25:826
66. (a) Yamazaki M, Suzuki S, Miyaki K (1971) Chem Pharm Bull 19:1739; (b) Yamazaki M, Sasago K, Miyaki K (1974) J Chem Soc Chem Comm 408; (c) Yamazaki M, Fujimoto H, Akiyama T, Sankawa U, Iitaka Y (1975) Tetrahedron Lett 27; (d) Yamazaki M, Fujimoto H, Kawasaki T (1975) Tetrahedron Lett 1241; (e) Yamazaki M, Suzuki K, Fujimoto H, Akiyama T, Sankawa U, Iitaka Y (1980) Chem Pharm Bull 28:861; (f) Yamazaki M, Fujimoto H, Kawasaki T (1980) Chem Pharm Bull 28:245; see also (g) Liu J, Yang Z-J, Meng Z-H (1996) Biomed Environ Sci 9:1
67. Day JB, Mantle PG (1982) Appl Environ Microbiol 43:514
68. Schroeder HW, Cole RJ, Hein H, Kirksey JW (1975) Appl Microbiol 29:857
69. (a) Cui C-B, Kakeya H, Okada G, Onose R, Ubukata M, Takahashi I, Isono K, Osada H (1995) J Antibiot 48:1382; (b) Cui C-B, Kakeya H, Okada G, Onose R, Osada H (1996) J Antibiot 49:527; (c) Cui C-B, Kakeya H, Osada H (1996) J Antibiot 49:534
70. (a) Cui C-B, Kakeya H, Osada H (1996) J Antibiot 49:832; (b) Cui C-B, Kakeya H, Osada H (1996) Tetrahedron 52:12,651

71. (a) Cui C-B, Kakeya H, Osada H (1997) Tetrahedron 53:59; (b) Kondoh M, Usui T, Mayumi T, Osada H (1998) J Antibiot 51:801
72. Osada H (1998) J Antibiot 51:973
73. (a) Willingale J, Perera KPWC, Mantle PG (1983) Biochem J 214:991; (b) Mantle PG, Shipston NF (1987) Biochem Int 14:1115
74. Horak RM, Vleggaar R (1987) J Chem Soc Chem Comm 1568
75. Vleggaar R, Horak RM, Maharaj VJ (1993) J Chem Soc Chem Comm 274
76. Wilson BJ, Wilson CH, Hayes AW (1968) Nature 220:77
77. (a) Holzapfel CW, Wilkins DC (1971) Phytochemistry 10:351; (b) Holzapfel CW, Schabort J (1977) S Afr J Chem 30:233
78. de Jesus AE, Steyn PS, Vleggaar R, Kirby GW, Varley MJ, Ferreira NP (1981) J Chem Soc Perkin Trans I:3292
79. Chalmers AA, Gorst-Allman CP, Steyn PS (1982) J Chem Soc Chem Comm 1367
80. (a) Schabort JC, Potgieter JJ (1971) Biochem Biphys Acta 250:329; (b) Steenkamp DJ, Schabort JC, Holzapfel CW, Ferreira NP (1974) Biochem Biphys Acta 358:126; (c) Schabort JC, Marx M, Erasmus GL, Oosthuizen MJ, Pitout MJ, Ferreira NP (1978) Int J Biochem 9:427; (d) Schabort JC, Marx M, Erasmus GL, (1978) Int J Biochem 9:437; see also (e) Neethling DC, McGrath RM (1977) Can J Microbiol 23:856
81. Floss HG (1976) Tetrahedron 32:873
82. Shibuya M, Chou H-M, Fountoulakis M, Hassam S, Kim S-U, Kobayashi K, Otsuka H, Rogalska E, Cassady JM, Floss HG (1990) J Am Chem Soc 112:297
83. (a) Pachlatko P, Tabacik C, Acklin W, Arigoni D (1975) Chimia 29:526; (b) Pachlatko P (1975) PhD thesis. ETH, Zürich
84. (a) Gebler JC, Woodside AB, Poulter CD (1992) J Am Chem Soc 114:7354; (b) Tsai H-F, Wang H, Gebler JC, Poulter CD, Schardl CL (1995) Biochem Biophys Res Comm 216:119
85. Otsuka H, Anderson JA, Floss HG (1979) J Chem Soc Chem Chem 660
86. Kozikowski AP, Okita M, Kobayashi M, Floss HG (1988) J Org Chem 53:863
87. Kobayashi M, Floss HG (1987) J Org Chem 52:4350
88. Kozikowski AP, Wu J-P, Shibuya M, Floss HG (1988) J Am Chem Soc 110:1970
89. Robbers JE, Floss HG (1969) Tetrahedron Lett 1857
90. Robbers JE, Otsuka H, Floss HG (1980) J Org Chem 45:1117
91. Bajwa RS, Kohler R-D, Saini MS, Cheng M, Anderson JA (1975) Phytochemistry 14:735
92. Saini MS, Cheng M, Anderson JA (1976) Phytochemistry 15:1497
93. Herrmann KM, Somerville RL (1983) In: Amino acids biosynthesis and genetic regulation. Addison-Wesley, London, Chaps 18–20 and references cited therein
94. The crystal structure of tryptophan synthase was retrieved from the Brookhaven Protein data bank; see Schneider TR, Gerhardt E, Lee M, Liang P-H, Anderson KS, Schlichting I (1998) Biochem 37:5394
95. For reviews see: (a) Bloch K (1992) Steroids 57:378; (b) Banthorpe DV, Charlwood BV, Francis MJO (1972) Chem Rev 72:115; (c) Bach TJ (1995) Lipids 30:191
96. (a) Flesch G, Rohmer M (1988) Eur J Biochem 175:405–411; (b) Rohmer M, Sutter B, Sahm H (1989) J Chem Soc Chem Comm 1471; (c) Duvold T, Bravo J-M, Pale-Grosdemange C, Rohmer M (1997) Tetrahedron Lett 38:4679; (d) Putra SR, Charon L, Danielsen K, Pale-Grosdemange C, Lois L-M, Campos N, Boronat A, Rohmer M (1998) Tetrahedron Lett 39:6185
97. (a) Sagner S, Latzel C, Eisenreich W, Bacher A, Zenk MH (1998) J Chem Soc Chem Comm 221; (b) Arigoni D, Sagner S, Latzel C, Eisenreich W, Bacher A, Zenk MH (1997) Proc Natl Acad Sci USA 94:10,600
98. Eisenreich W, Schwarz M, Cartayrade A, Arigoni D, Zenk MH, Bacher A (1998) Chem Biol 5:R221–R233
99. Kuzuyama T, Takahashi S, Watanabe S, Seto H (1998) Tetrahedron Lett 39:4509
100. Martin W, Muller M (1998) Nature 392:37

Tropane and Related Alkaloids

Thomas Hemscheidt

Department of Chemistry, University of Hawai'i at Manoa, Honolulu, HI 96822, USA
E-mail: tomh@gold.chem.hawaii.edu

Tropane alkaloids comprise a large group of bases occurring predominantly in the family of the Solanaceae . Structurally they are esters of carboxylic acids with tropine (3-hydroxy-8-aza-8-methyl-[3.2.1]-bicyclooctane) and are biosynthetically derived from amino acid and acetate precursors. Despite the relative structural simplicity of the alkaloids, their biosynthesis is not well understood from a mechanistic point of view. In this article the available information pertaining to this question is summarized and discussed in context with the information that is available from the analogous pelletierine class of alkaloids. A new proposal for the mechanism of assembly of the acetate derived C_3 fragment of these alkaloids is introduced.

Keywords. Tropanes, Alkaloids, Biosynthesis, Mannich condensation

Topics in Current Chemistry, Vol. 209

1
Introduction

Some 80 years ago, Robert Robinson published his elegant synthesis of tropinone from succindialdehyde, methylamine and acetonedicarboxylate [1]. Prompted by his synthetic strategy he took up his speculations on the origin of natural products in Nature, among them the tropanes, which were published shortly thereafter [2]. A fascinating account of the history of these early ideas is presented by A.J. Birch [3]. Ever since, the biological origin of tropine has held an important position in our thinking about the biogenesis of alkaloids. Thus, when suitable experimental tools, i.e. radiotracers, for the investigation of the biosynthesis of natural products became available in the late 1940s and early 1950s, among the alkaloids it was again the origin of the tropane nucleus that was one of the first targets of investigations [4].

The current status of knowledge of the biosynthesis of the tropane alkaloids is almost paradoxical: our understanding of the mechanisms by which plants assemble the tropane nucleus (**1**) still leaves much to be desired whilst some of the late events in the formation of one particular tropane alkaloid, scopolamine (**2**), are understood in great detail. Applications of this knowledge in the sense of "metabolic engineering" of medicinal plants have even been described [5].

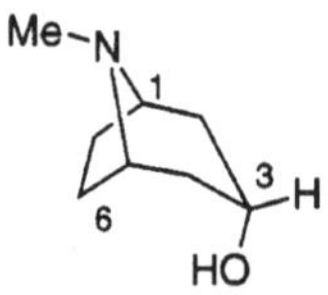

1

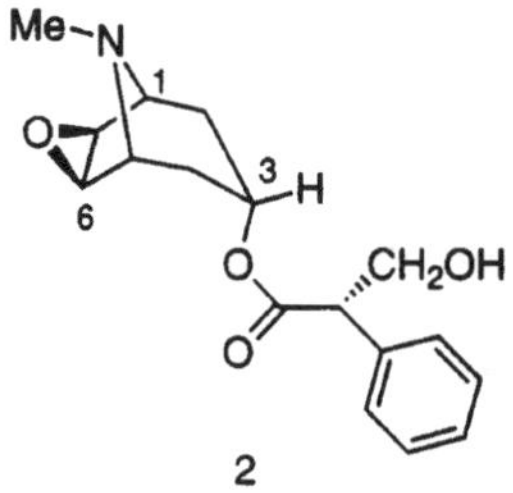

2

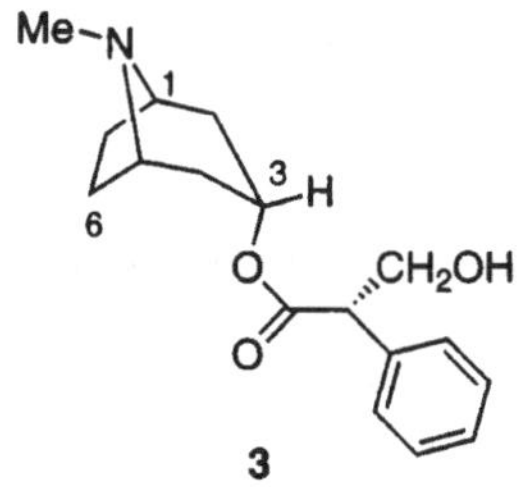

3

Apart from the sizable primary literature on the biosynthesis of the tropanes, a substantial number of reviews on this topic has been published. The two most recent of these describe the work from the laboratory of the late Eddie Leete [6] and the encouraging results that have been obtained on an enzymological level [7]. Experiments aimed at elucidating the biosynthesis of the tropic acid moiety present in many tropane alkaloids have been reviewed recently by workers in the field [8] and will not be covered here.

The present review will focus on progress made in the elucidation of the biosynthesis of tropane and related alkaloids from a more chemical perspective and will attempt to outline what in our current understanding is deficient and to identify the remaining problems. For this purpose, the discussion will be divided into two parts, dealing with the assembly of the amino acid derived C_4N portion, C-1,C-5,C-6,C-7,N-8, of the five-membered ring and then with the assembly of the acetate derived C_3 fragment, C-2,C-3,C-4, and the formation of the azabicyclooctane system. A liberal definition of the term tropane will be used, so that the biosynthesis of cocaine and some biogenetically related pelletierine-type alkaloids may be included in the discussion. In the view of this author, results obtained in the latter group may have a bearing on our thinking about the biosynthesis of the tropanes in the customary narrow definition.

2
The Amino Acid Derived Fragment

One of the first experimental results on tropane biosynthesis with radiotracer methodology in *Datura stramonium* [9] provided evidence that the C_4N fragment of the five membered ring of the tropane alkaloids is derived from ornithine (**4**) as had been envisioned by Robinson. This initial finding was subsequently confirmed by work from various groups in a variety of genera [reviewed in 10]. In more recent studies, employing inhibitors of ornithine and arginine decarboxylases, it has been suggested that arginine (**5**) rather than ornithine is the preferred source of the C_4N fragment C-1,C-5,C-6,C-7,*N*-8 [11, 12]. Because of the close biogenetic relationship between ornithine and arginine via the urea cycle this finding does not constitute a fundamental contradiction to the earlier results but rather a refinement of the model.

The next question addressed first by Leete [13, 14], concerned the occurence of a symmetrical intermediate on the pathway between ornithine and the alkaloids. The feeding of [2-^{14}C]ornithine to *Datura stramonium* led to specific incorporation of the label into only C-1, the stereocenter of (*R*) configuration in the tropine moiety of hyoscyamine (**3**), which was excised in unambiguous fashion. Leete concluded that an intermediate with C_{2v} symmetry could not be situated on the pathway between ornithine and the tropane alkaloids.

Several mechanistic interpretations have been put forth to explain this result. The first one of these envisioned the intermediacy of 5-*N*-methylornithine on the pathway from ornithine to the alkaloids. After decarboxylation of this nonproteinogenic amino acid *N*-methylputrescine (**6**) would be obtained in which the secondary amino group originates from N-5 of ornithine and the primary one from N-2 of the amino acid. No convincing experimental evidence could be accumulated to support this hypothesis and it was later abandoned in favor of a scheme adapted from one first proposed by Spenser to account for the nonsymmetrical incorporation of lysine into sedamine [15]. As formulated by Leete [16], this hypothesis requires the removal of the carboxyl group of the amino acid and the subsequent *N*-methylation to occur without a "free", i.e. not enzyme-bound, intermediate with C_{2v} symmetry. This proposal mandates that the *N*-methyl group of the resulting *N*-methylputrescine (**6**) is attached to the nitrogen atom originating from N-5 of ornithine and that N-5 of the amino acid is incorporated into the alkaloids (Path A1 in Scheme 1). An attractive variant of this proposal mentioned in passing by Leete, postulates that it may not be the *N*-methylation subsequent to decarboxylation that results in a nonsymmetrical intermediate. Instead, differentiation of the two ends of the bound putrescine on the putative decarboxylase enzyme would be achieved by transamination of the nitrogen atom from C-2 (ornithine numbering) subsequent to decarboxylation to yield Δ^1-pyrroline (Path A2). The usual mechanism of α-amino acid decarboxylases [17] involves binding of the α-amino group of the amino acid to the pyridoxal phosphate prosthetic group of the enzyme and delocalization of the negative

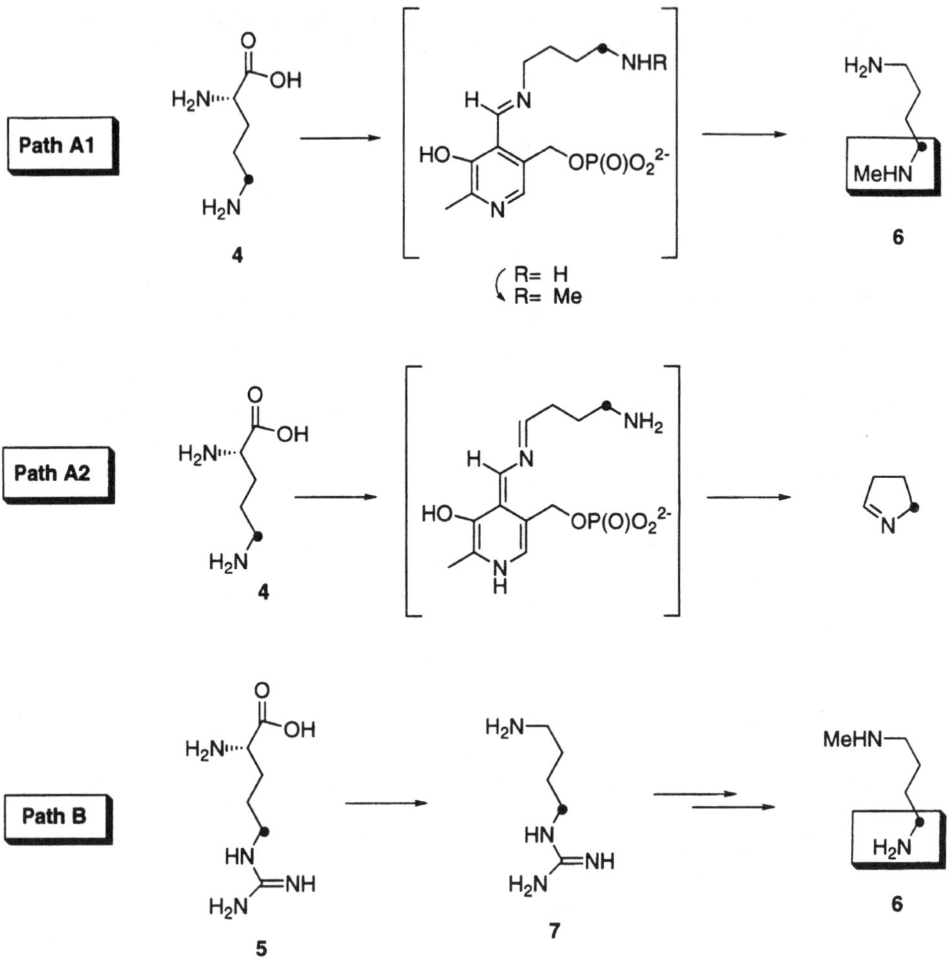

Scheme 1. Possible pathways for the incorporation of ornithine **4** or arginine **5** into tropane alkaloids

charge resulting from loss of C-1 into the aromatic ring of the cofactor (see the intermediate in Path A2). After tautomerization an enzyme-bound imine is obtained (see the intermediate in Path A1) which is then hydrolyzed to the biogenic amine. However, the first of these two intermediates can also be viewed as an intermediate in a transaminase reaction. Hydrolysis at this stage would result in formation of a carbonyl group at C-2 of the original amino acid and concomitant loss of the amine nitrogen. The amine nitrogen could subsequently be transferred via transamination to an α-keto acid to regenerate the aldehyde functionality of the cofactor. This variant is conceptually particularly attractive as it uses the pyridoxal phosphate prosthetic group common to amino acid decarboxylases and transaminases to achieve the generation of the nonsymmetrical intermediate. In any case, either version of this elegant proposal rationalizes the experimental observation concerning nonsymmetrical incorporation of ornithine into the tropane ring in *Datura stramonium*. It has remained experimentally unsubstantiated on an enzymological level to this day, however.

A third proposal to rationalize the nonsymmetrical incorporation of the amino acid precursor can be based on more recent experiments [11, 12] using mechanism-based enzyme inhibitors. The results suggest that L-arginine (**5**) rather than L-ornithine (**4**) is the central intermediate on the amino acid level in the formation of the alkaloids. Recognizing the fact that in (**5**) the two amino groups are differentiated, the model envisages that L-arginine is metabolized via its biogenic amine agmatine (**7**) in which the two nitrogen atoms N-2 and N-5 of the amino acid remain differentiated by the presence of the guanidino function on the latter. After methylation of the primary amine derived from N-2 of arginine to yield *N*-methylagmatine (**8**), the guanidino group could be removed liberating the primary amine (**6**) originating from N-5 of arginine (Path B in Scheme 1). In a subsequent step, this nitrogen atom would be removed by oxidation or transamination to yield aldehyde (**9**) in which the carbonyl carbon is derived from C-5 of arginine. Agmatine (**7**) is indeed incorporated into the alkaloids in root cultures of *Datura stramonium* [11,12]. Unfortunately, this study was restricted to measuring incorporation of [U-^{14}C]agmatine (**7**) into the alkaloids. It was assumed that (**7**) would be metabolized via putrescine, i. e. an intermediate with C_{2v} symmetry, without rigorous experimental evidence to that effect.

The experimental verification of this last proposal by means of radiotracers would require the use of a specifically labeled precursor and subsequent degradation of the labeled (**2**) to establish the distribution of label within (**1**). As the skills and patience required for such work are disappearing, it is unlikely that the solution to this problem will come from work with radiotracers. However, the two models still in contention for the rationalization of the incorporation of L-arginine or of L-ornithine via a nonsymmetrical intermediate differ by the origin of the nitrogen atom in the tropane nucleus as shown in Scheme 2. Either variant of Path A in Scheme 1 results in the incorporation of N-5 into the alkaloids and the loss of N-2. If desymmetrization proceeds via Path B on the other hand, N-5 would be lost and N-2 would be incorporated.

This could be established in a straightforward way by using [5-^{13}C,^{15}N]ornithine or [5-^{13}C,^{15}N]arginine and [2-^{13}C,^{15}N]ornithine or [2-^{13}C,^{15}N]arginine, respectively, followed by NMR analysis. It is most unfortunate that such a feeding experiment with successful incorporation via a nonsymmetrical intermediate into the tropane alkaloids has apparently never been reported. Yet it would be most desirable to be able to correlate the result from Leete's incorporation of [2-^{14}C]ornithine with our current stable isotope methodology. This would establish which of the two nitrogen atoms, N-2 or N-5, of the amino acid is retained in the alkaloids and differentiate between the two mechanistic proposals still in contention for the rationalization of incorporation of the amino acid into the alkaloid via a nonsymmetrical intermediate. A study reporting a feeding experiment with [2-^{14}C,5-^{15}N]ornithine in *Datura metel* would suggest that N-5 is retained in the alkaloids [18]. In this work incorporation of ^{14}C into the alkaloid via a nonsymmetrical intermediate was observed. Incorporation of ^{15}N was determined by mass spectrometry after degradation. An incorporation level of ^{15}N roughly in

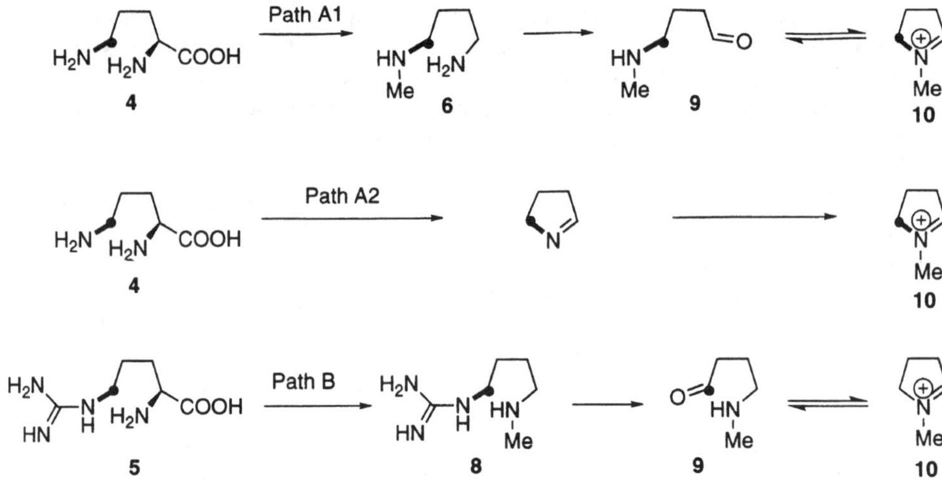

Scheme 2. The different isotopic labelling patterns resulting from the three possible paths in Scheme 1

accord with that of ^{14}C in this experiment was reported. It is not clear, however, whether incorporation levels at a fraction of a percent could be reliably determined by means of this methodology at that time. For instance, standard errors and detection limits were not provided or discussed. Moreover, some scrambling of the ^{15}N label was also evident. Consequently, a confirmation of this result using more contemporary methodology appears warranted, especially considering the confusion that exists in the literature over the involvement of a nonsymmetrical vs a symmetrical intermediate in tropane biosynthesis (see below).

Leete's result that described the incorporation of [2-^{14}C]ornithine into the tropane nucleus in *Datura stramonium* in a nonsymmetrical fashion had been obtained in a technically flawless manner and is beyond doubt correct. The result was reproduced also in a root culture of *Datura stramonium* [19]. Nevertheless, work in other organisms, e.g. *Hyoscyamus albus*, suggests that this pathway is not always operative and incorporation of ornithine into the tropane nucleus via a symmetrical intermediate has been recorded [20, 21]. Similarly, we have observed that [1,2-$^{13}C_2$]acetate is incorporated in *Datura stramonium* [22] not only into the C-2, C-3, C-4 fragment as expected (see below), but also into the ornithine/arginine derived portion C-1,C-6,C-7,C-5,N-8 albeit only at less than one tenth the specific incorporation observed for C-2,C-3,C-4 within the same sample. This labeling pattern arises probably in indirect fashion from acetate through the citric acid cycle via α-ketoglutarate and glutamic acid. The resulting glutamate receives bond-label at C-4 and C-5 and ornithine derived from it may be expected to be labeled in identical fashion. *N*-Methylputrescine derived nonsymmetrically from ornithine should thus only label the C-5/C-6 bond [23] after incorporation into tropine (**1**) since C-2 of (**4**) labels C-1 of (**1**) exclusively [13, 14]. A very recent reexamination of the observed pattern appears to suggest, however, that all four positions are labeled to an equal extent, indicative of a symmetrical intermediate between the amino acid and the alkaloid. This observation would not only be in direct contradiction to the result by Bothner-By et al. [23] but also to that of Leete [13, 14].

Leete has rationalized the discrepancies between the incorporation of ornithine via symmetrical and nonsymmetrical intermediates as metabolic differences between species [16], e.g. *Hyoscyamus albus* vs *Datura stramonium*. This may be the case since there is no published report which describes nonsymmetrical incorporation of an amino acid into (**1**) in *Hyoscyamus*. The available literature on work done in that species is thus internally consistent. In contrast, the situation in *Datura* species is more complicated. Some investigators have concluded based on isotope dilution experiments in root cultures of *Datura stramonium* and *Atropa belladonna* that the pathway resulting in stereospecific incorporation of (**4**) or (**5**) into the pyrrolidine ring of (**2**) is not of importance under so-called "physiological" conditions [11, 12]. Nothing could be further off the mark since the original observation of nonsymmetrical incorporation of [2-^{14}C]ornithine in *Datura stramonium* was made in experiments with intact plants, arguably the most "physiological" system imaginable. The differences can also not simply be ascribed to the experimental systems being used, i.e. "intact" plants vs in vitro material, as nonsymmetrical incorporation of ornithine into hyoscyamine has also been reported to occur in root cultures of *Datura metel* [19]. Thus, the available data for work in *Datura* species are not internally consistent. In the view of the author the discrepancies between the results concerning the symmetry of the intermediate in tropane alkaloid biosynthesis described in this species are real. Some data to be discussed below, in their current interpretation, cannot be reconciled with a pathway to (**1**) involving only nonsymmetrical intermediates. Yet the studies which invoke nonsymmetrical incorporation are numerous and appear to be solid. Unfortunately, we have so far failed to accept these discrepancies as a challenge for further study and for the design of experiments which explain the observations made in the early feeding studies. Our ability to describe, in mechanistic detail, the origin of the stereospecific incorporation of amino acid precursors into tropine in *Datura* species will be one of the measures of the maturity of our model of tropane biosynthesis. By comparison, the incorporation of ornithine via a symmetrical intermediate, i.e. putrescine (1,4-diaminobutane), as observed in a variety of other alkaloid classes, e.g. the pyrrolizidine alkaloids [24] (see T. Hartmann, p. 207–243), is the less interesting observation in that the enzyme steps required are well known. Yet, ironically, it is this latter branch of ornithine/arginine metabolism in tropane alkaloid-producing organisms that has received the attention of enzymologists to date.

Precedent for the incorporation of α,ω-diamino acids into alkaloids of higher plants via nonsymmetrical intermediates exists in some pelletierine type alkaloids which are derived from lysine (**11**). Thus, it has been reported that in *Sedum acre* label from DL-[2-^{14}C]lysine and from DL-[6-^{14}C]lysine is introduced in a regiospecific* fashion into the alkaloid sedamine (**12**). C-2 of (**12**) receives

* In the case of incorporation into tropine (**1**), which has a plane of symmetry, the word "stereospecific" has been used, but for a non-symmetrical alkaloid such as sedamine (**12**) the word "regiospecific" has to be used as the isotopically labeled species are no longer stereoisomeric.

label exclusively from the α carbon atom of (**11**) and incorporation of the ε carbon atom of lysine is restricted to C-6 of (**12**) [25]. These results led to the proposal that lysine was metabolized to Δ^1-piperideine (**14**) without an intermediate of C_{2v} symmetry [15]. It was this precedent on which Leete's proposal for the mechanism of incorporation of ornithine into (**1**) via a nonsymmetrical intermediate[16] is based. Interestingly, the lysine derived diamine [1-^{14}C]cadaverine (**13**) was also incorporated into sedamine (**12**) and *N*-methylpelletierine (**23**). Both carbon atoms α to nitrogen in (**12**) and (**23**) were labeled with equal efficiency as would be expected from the C_{2v} symmetry of the precursor [15].

It is generally assumed that cadaverine is metabolized to Δ^1-piperideine (**14**) by the action of a diamine oxidase and the incorporation of cadaverine into (**12**) and (**23**) suggests the presence of such an activity in *Sedum* species. Since the only known source of cadaverine is the decarboxylation of lysine catalyzed by lysine decarboxylase, evidence for the existence of the diamine oxidase suggests that lysine decarboxylase activity may be present in the plant as well. The regiospecific formation of *N*-methylpiperidinium ion by methylation of (**14**) rather than via enzyme-bound cadaverine(**13**) and *N*-methylcadaverine in analogy to Path A1 in Scheme 1 is favored. The available evidence suggests that formation of [*methyl*^{14}C]*N*-methylcadaverine from [*methyl*^{14}C]methionine is not taking place in plants which are actively producing [*methyl*^{14}C]*N*-methylpelletierine at the time of the experiment [15]. In any case, the available data suggest that the enzymatic machinery may exist within plants of the genus *Sedum* to effect incorporation of the amino acid precursor lysine (**11**) regiospecifically as well as non-regiospecifically. It is noteworthy, however, that in contrast to plants elaborating tropane alkaloids, no evidence has ever been obtained which would suggest that both symmetrical and nonsymmetrical pathways are actually used for the incorporation of lysine into alkaloids in *Sedum* species.

Thus in *S. acre*, for example, the rate of stereospecific formation of Δ^1-piperideine (**14**) from L-lysine (**11**) must far exceed the rate of its formation by decarboxylation of (**11**) to give the symmetrical diamine (**13**) followed by oxidation

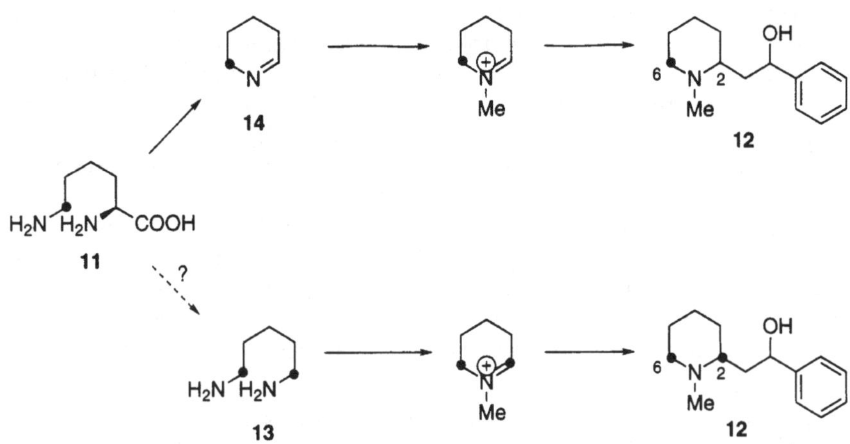

Scheme 3. Biosynthesis of sedamine **12**. Incorporation of lysine occurs by the upper pathway

by diamine oxidase, a pathway which would result in nonregiospecific labeling of (**14**) (Scheme 3). Given the sensitivity of radioisotope methodology when combined with rigorous degradation chemistry, a contribution from the latter pathway at even the 5% level to the overall incorporation of labeled lysine into sedamine (**12**) would have been easily detectable. Instead, complete regiospecificity within experimental error was observed.

It is peculiar that in *Datura*, on the other hand, experimental evidence for the operation of both symmetrical and nonsymmetrical pathways of metabolism of ornithine (**4**) exists. Furthermore, incorporation of (**4**) into (**1**) occurs with either apparent complete stereospecificity [13, 14, 18, 19, 23] or complete lack of stereospecificity [21, 22]. This suggests that either path may be followed but always at the exclusion of the other. Expression or operation of the enzymes of both paths in parallel should result in the observation of partial stereoselectivity, a result which has apparently never been reported.

As the preceding discussion shows, the literature which invokes the existence of an intermediate with C_{2v} symmetry between ornithine/arginine and the tropanes in *Datura* species commonly assumes that this intermediate is located on the pathway between the amino acids and *N*-methylpyrrolinium salt (**10**). From studies of the incorporation of more advanced precursors into tropine (**1**) to be discussed in detail below, it is becoming apparent that this assumption may not necessarily be warranted.

In plants such as *Hyoscyamus* species in which no evidence for stereospecific incorporation of the amino acids into the alkaloids exists, putrescine is the generally accepted intermediate with C_{2v} symmetry. A specific enzyme catalyzing the methylation of putrescine to *N*-methylputrescine (**6**) has been detected from plants of that species [26]. In *Datura* (**6**) appears to be an intermediate on the pathway to (**1**) as well. Thus, [1-^{13}C,^{14}C, *methylamino*^{15}N]*N*-methylputrescine (**6**) has been fed to intact plants of *Datura innoxia* and the scopolamine (**2**) isolated from this experiment was specifically labeled in the carbon atom giving rise to the downfield resonance corresponding to C-1/C-5 of the tropane nucleus in scopolamine [27] (Scheme 4). This result suggests that if (**6**) is formed regiospecifically from (**4**) or (**5**), this regiospecificity does not become scrambled in later steps. Observation of stereospecific incorporation of (**6**) into (**2**) by NMR is possible due to the presence of the chiral tropic acid residue of (*S*) configuration which is bound to the oxygen at C-3 of (**2**). The pairs of carbon atoms C-1/C-5, C-2/C-4 and C-6/C-7 in such esters are diastereotopic and their resonances have different ^{13}C chemical shifts. The interpretation of incorporation data, especially in a quantitative fashion, requires a great deal of care, however, owing to the facile racemization of the tropic acid moiety in scopolamine (**2**) and hyoscyamine (**3**).

Further metabolism of *N*-methylputrescine (**6**) involves the oxidative removal of the primary amino function with concomitant formation of an aldehyde. It is commonly assumed that this step is catalyzed by an amine oxidase activity and some experimental evidence is available that would support this hypothesis. Thus, an amine oxidase enzyme with favorable k_{cat} for *N*-methylputrescine vs putrescine and the homologous diamine cadaverine (**13**) has been detected and partially purified from root cultures of *Hyoscyamus niger* [28]. A role for this

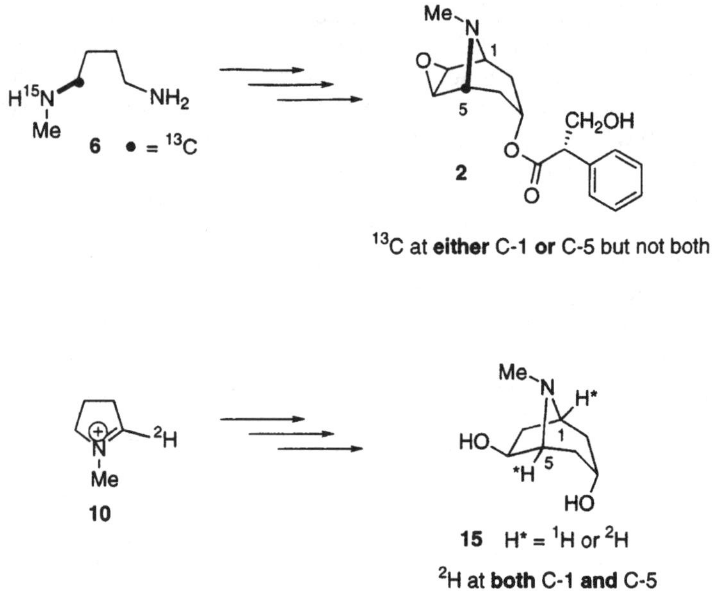

Scheme 4. Conflicting results from incorporation experiments in *Datura*

enzyme in tropane biosynthesis is postulated based on the observed substrate specificity. Moreover, it was observed that the specific activity of the enzyme when determined as a function of culture age, rises in parallel with the accumulation of the alkaloids by the culture. The aminoaldehyde product (**9**) expected from this enzymatic reaction is in equilibrium with the tautomeric carbinolamine and the iminium salt (**10**) derived therefrom by elimination of water (Scheme 2, Paths A1 and B).

We have observed non-regiospecific incorporation of deuterium from [2-^{2}H]pyrrolinium salt (**10**) into 6*β*-hydroxytropine (**15**) in *Datura stramonium*. This hydroxylated tropane derivative was chosen deliberately as the target alkaloid since the symmetry inherent in (**1**) is broken in this structure and the two bridgehead carbon atoms are different. Consequently, incorporation of label into these two sites can be established easily by NMR. In the event, we observed equal labeling of the two bridgehead positions H-1 and H-5 in this experiment [22] (Scheme 4). This result would suggest that an intermediate with C_{2v} symmetry is located on the pathway to (**1**) in *Datura* or, alternatively, that some other mechanism exists which results in loss of stereospecificity. Significantly, this symmetrization has to occur at some point on the pathway after the pyrrolinium salt (**10**), whereas the discussion until then had surmized that the symmetrical intermediate occurred between the amino acid precursors and the iminium salt. However, it should also be noted that this result cannot be reconciled with the observation of stereospecific incorporation of ornithine (**4**) and *N*-methylputrescine (**6**) into hyoscyamine (**3**) in *Datura* species, as observed by Leete [13, 14, 27].

All of the models discussed so far which attempt to rationalize the incorporation of precursors into the C-1,C-6,C-7,C-5,N-8 fragment of (**1**) with equal labeling of C-1 and C-5, invoke an intermediate with C_{2v} symmetry on the pathway before chiral centers are introduced into the skeletons of pathway intermediates. Observations concerning the incorporation of [1,2-$^{13}C_2$]acetate into the acetate derived portion C-2,C-3,C-4 of (**1**), to be discussed in more detail in the following section, suggest yet another explanation, namely the existence of one or more racemic intermediates on the pathway to the alkaloids. Thus, while the label from the amino acid precursor may well be introduced regiospecifically into the iminium salt (**10**) by any of the mechanisms in Scheme 1, this label might become scrambled further downstream in the pathway under some experimental circumstances.

3
The Acetate Derived Fragment

3.1
The Role of Hygrine

It has been known for a considerable time that the three carbon atoms C-2 to C-4 of (**1**) are derived from acetate [29, 30]. It is a testament to the power of Robinson's original hypothesis [2] that it was assumed, without much experimental scrutiny, that the introduction of these carbon atoms was proceeding via acetoacetate in a typical Mannich reaction with (**10**) in which C-1 of acetoacetic acid (or its CoA ester) is lost. The product of such a condensation would be the known alkaloid hygrine (**16**). Further transformation of this alkaloid was viewed as being straightforward involving oxidation of the remaining methylene carbon atom α to nitrogen to yield an iminium salt (**17**) which would cyclize by a subsequent intramolecular Mannich condensation (Scheme 5). A "biomimetic" model reaction forming tropinone (**18**) from (**16**) via (**17**) was interpreted as supporting this proposal [31].

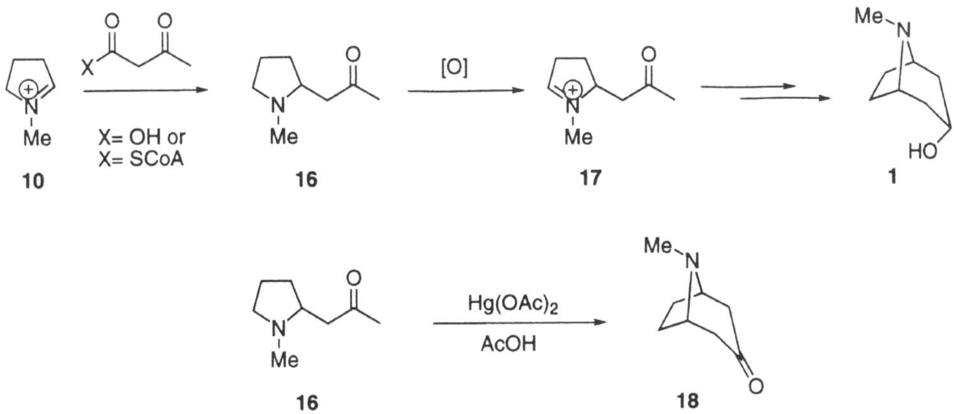

Scheme 5. The "classical" proposal for the biosynthesis of tropine **1** via hygrine **16** and a biomimetic reaction which supports it

Evidence in support of this scheme was obtained with the demonstration that [^{14}C]hygrine was incorporated into tropane alkaloids [32,33] and even a stereospecific incorporation of the (*R*) enantiomer of hygrine into (**3**) in *Datura innoxia* was reported [34]. However, evidence incompatible with the intermediacy of hygrine on the pathway to tropine has been accumulating over the last several years and it is fair to say that at present there is little support for this classical idea.

The first evidence that challenged the accepted scheme came from the investigations by Leete of the biosynthesis of methyl ecgonine (**19**), the base portion of cocaine (**20**). (**19**) had long been thought to arise from pyrrolinium salt (**10**) in a fashion analogous to that shown in Scheme 5 for the formation of tropine. It was assumed that the biogenetic differences between methyl ecgonine and tropine biosynthesis were, firstly, the retention of C-1 of acetoacetate after condensation with (**10**) in the case of (**19**) and, secondly, the different stereochemistry of the C-3 hydroxy group of (**19**) when compared to (**1**).

According to the classical hypothesis, the bridgehead carbon atom C-1 of (**19**) is derived from C-2 of the pyrrolinium salt (**10**) (Path C, Scheme 6) and C-2 of (**19**) originates from C-2 of acetoacetate. Incorporation experiments with [1-^{15}N,2-^{13}C]pyrrolinium salt (**10**) in *Erythroxylon coca* brought an unexpected result [35]. It was observed that the carbon atom derived from C-2 of (**10**) was not joined to C-2 of (**19**), located *α* to the carboxymethyl group, but was bonded instead to C-4, the methylene group distal to the ester function. Thus, C-2 of (**10**) becomes C-5 of (**19**). The authors deemed it unlikely that C-4 of acetoacetate (or its CoA ester) had attacked the iminium carbon atom of (**10**) resulting in 4-(1-methyl-2-pyrrolidin)-3-oxobutanoate (**22**). Instead they suggested that the observed regiochemistry of incorporation of 1-methyl-[1-^{15}N,2-^{13}C]pyrrolinium salt into cocaine was compatible with the stepwise introduction of acetate derived C_2 units into the ecgonine skeleton. Consequently, this hypothesis invokes

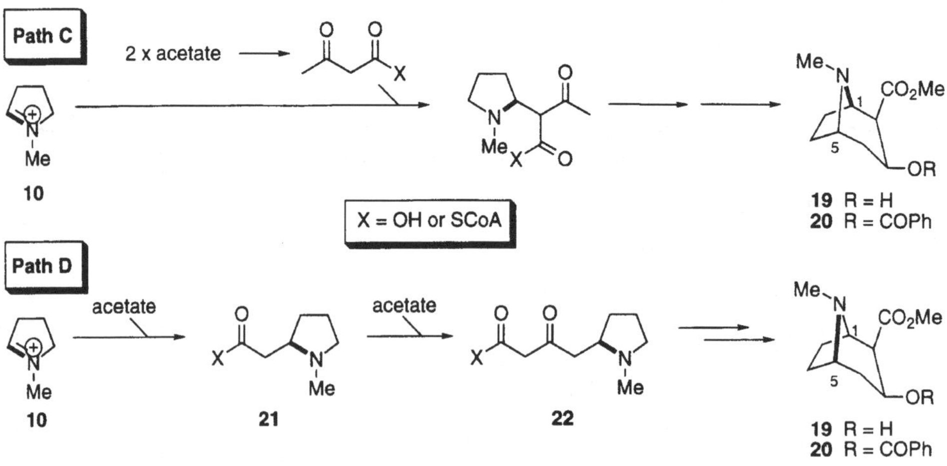

Scheme 6. Classical (path C) and revised proposals (path D) for the biosynthesis of methyl ecgonine **19** and thence of cocaine; the illustrated incorporation experiment with [^{15}N,^{13}C]**10** supported path D

the intermediacy of an equivalent of 1-methyl-2-pyrrolidineacetic acid (**21**) on the pathway, as shown in Path D, Scheme 6. This intermediate has been detected in several plant species [36,37,38] but attempts to effect incorporation of such an intermediate in ester or thioester form have been unsuccessful [39].

3.2 The Biosynthesis of N-Methylpelletierine

Prompted by Leete's finding which had significant ramifications for our thinking about the formation of the tropane class of alkaloids, we examined the formation of *N*-methylpelletierine (**23**) and of *N*-methylallosedridine (**24**) in *Sedum sarmentosum* [40]. We chose to analyze the incorporation of [1,2-$^{13}C_2$]acetate in admixture with unlabelled material into these alkaloids. The experimental system seemed ideal in that the alkaloids are not symmetrical and thus the regiochemistry of acetate incorporation could be readily analyzed.

In the event, we observed that the resonance assigned to the methylene group α to the ketone carbonyl, C-1', appeared as a singlet of increased intensity in the ^{13}C NMR spectrum while the resonances for the carbonyl carbon atom C-2' and C-methyl carbon atom C-3' were spin-coupled. This pattern was indicative, at first sight, of condensation of two acetate units to acetoacetate and subsequent incorporation of the C_4 compound (possibly as its CoA ester) via condensation of the carbon atom C-2 with the lysine derived imine (**14**) and subsequent or concomitant loss of C-1. The incorporation pattern into *N*-methylpelletierine (**23**) and its reduction product (**24**) appeared to be compatible with the classical Mannich scheme of Robinson, Path C1 in Scheme 8.

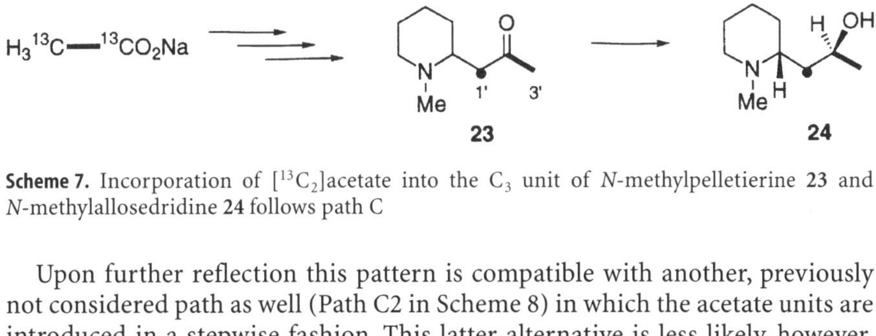

23 **24**

Scheme 7. Incorporation of [$^{13}C_2$]acetate into the C_3 unit of *N*-methylpelletierine **23** and *N*-methylallosedridine **24** follows path C

Upon further reflection this pattern is compatible with another, previously not considered path as well (Path C2 in Scheme 8) in which the acetate units are introduced in a stepwise fashion. This latter alternative is less likely, however, because it was found that the sodium salt of [1,2,3,4-$^{13}C_4$]acetoacetate is incorporated intact, yielding *N*-methylpelletierine carrying contiguous ^{13}C labels at the three carbon atoms of its side chain. Path C2 demands cleavage of the C_4 compound via a retroClaisen reaction and stepwise introduction of the resulting C_2 units.

Thus, the results from these two experiments support the conclusion that (**23**) is indeed assembled according to the classical Mannich scheme as envisioned by Robinson. There is one fundamental problem in arriving at this interpretation, however. It is well known that acetoacetate will condense with imines in aqueous

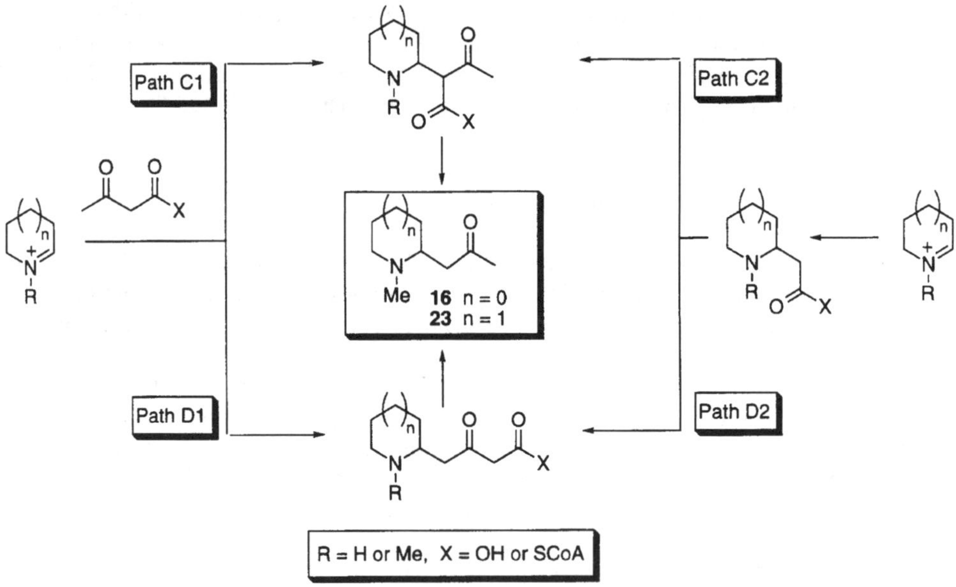

Scheme 8. Four possible paths for the introduction of the acetate-derived C_3 unit of *N*-methylpelletierine **23** and hygrine **16**

buffer to yield *β*-amino ketones without catalysis by enzymes. Thus, the question arises whether the labeling pattern observed in *N*-methylpelletierine (**23**) and *N*-methylallosedridine (**24**) after administration of [1,2,3,4-$^{13}C_4$]acetoacetate is a consequence of such a non-biological condensation of the acetoacetic acid supplied to the plants with an endogenous imine. Precedent for such an aberrant reaction does exist with the observation of hygrine formation upon administration of acetoacetate to *Nicotiana tabacum*, a plant which normally does not elaborate tropanes [41]. A variety of control experiments were considered which might support or refute such a notion but without success. For example, enantioselectivity in the formation of such a product, usually a reliable indication of enzyme action, may be expected to be wiped out completely before or during the isolation of the reaction product owing to the facile racemization of *β*-amino ketones by reversible retroMichael cleavage to yield an *α*,*β*-unsaturated *ω*-amino carbonyl compound (Scheme 9). Consequently, isolation of racemic product would not be a reliable indicator of nonenzymatic condensation of the administered acetoacetate with Δ^1-piperideine (**14**).

Whichever way the problem was approached, a completely conclusive experiment that would settle, unambiguously, the question of nonenzymatic condensation of *β*-ketoacids with imines under the conditions of in vivo feeding experiments in *Sedum* was not forthcoming. We eventually decided to interpret the results of the feeding experiments with [1,2-$^{13}C_2$]acetate described above as evidence that *N*-methylpelletierine and *N*-methylallosedridine were formed according to the classical scheme of Robinson (Path C1). This interpretation suggests that the acetate derived C_3 units of cocaine and the pelletierine alka-

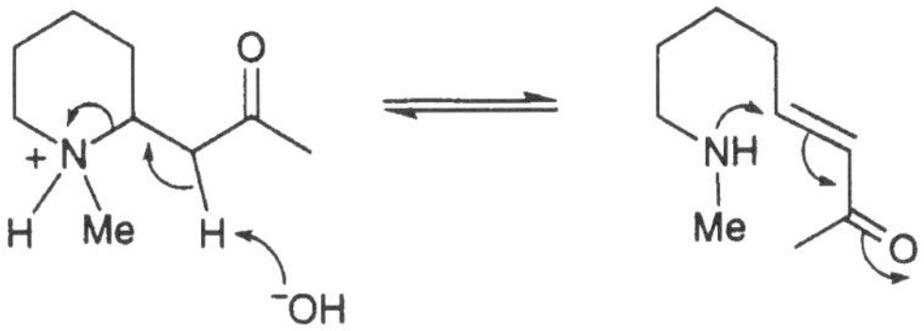

Scheme 9. Mechanism for the racemization of β-amino-ketones such as *N*-methylpelletierine

loids are assembled by two different mechanisms which accordingly we dubbed the "cocaine" (Path D) and the "pelletierine" (Path C) mechanisms. This discovery prompted us to investigate pelletierine biosynthesis in another plant in which pelletierine acts as an intermediate in the formation of a more complex alkaloid, lycopodine (**27**). These results will be discussed in Section 4.

3.3 Incorporation of Acetate into Tropine.

Four reports describe investigations of the incorporation of various isotopomers of acetic acid labeled with stable isotopes into the tropane alkaloids. The first of these, by Sankawa and Noguchi [42], reported the regiochemistry of $[1,2\text{-}^{13}C_2]$acetate incorporation into hyoscyamine (**3**) in *Hyoscyamus albus*. Surprisingly at the time, it was found that the ^{13}C resonances due C-2 and C-4 both showed coupling to the resonance assigned to C-3. The authors concluded that two labeled species were present in the sample, one in which an intact bond from acetate was introduced in such a fashion that C-2 and C-3 were bond-labeled and another one in which C-4 and C-3 were bond-labeled. The analysis of the ^{13}C NMR spectra of other alkaloids, the 6β- hydroxy derivatives of (**3**), isolated from this experiment indicated an identical pattern of incorporation of acetate into these substances as well.

We investigated the incorporation of $[1,2\text{-}^{13}C_2]$acetate into the tropane alkaloids independently and chose to work with *Datura stramonium* which produces 6β-hydroxytropine (**15**). In this alkaloid the symmetry of the parent azabicyclic system is broken and all carbon atoms are represented by well separated signals in the ^{13}C NMR spectrum. In the ^{13}C NMR spectrum of (**15**) isolated from this experiment a coupling pattern was observed which indicated that in this sample as well two labeled species were present which were bond-labeled between C-2/C-3 and C-3/C-4, respectively.

The most straightforward interpretation of this result invokes the formation of the tropane nucleus via both Path C and Path D, the former possibly occurring nonenzymatically as discussed previously. It should be noted that this rationalization of the observed labeling pattern still invokes a role for hygrine (**16**) in the biosynthesis of (**1**) as outlined in Scheme 10. In accordance with experimental observations, two labeling patterns in (**18**) and, after reduction, in (**1**) would result if (**18**) is formed by both pathways.

A second experiment with $[1,2,3,4\text{-}^{13}C_4]$acetoacetate in *D. stramonium*, performed just as in the investigation of the biosynthesis of *N*-methylpelletierine

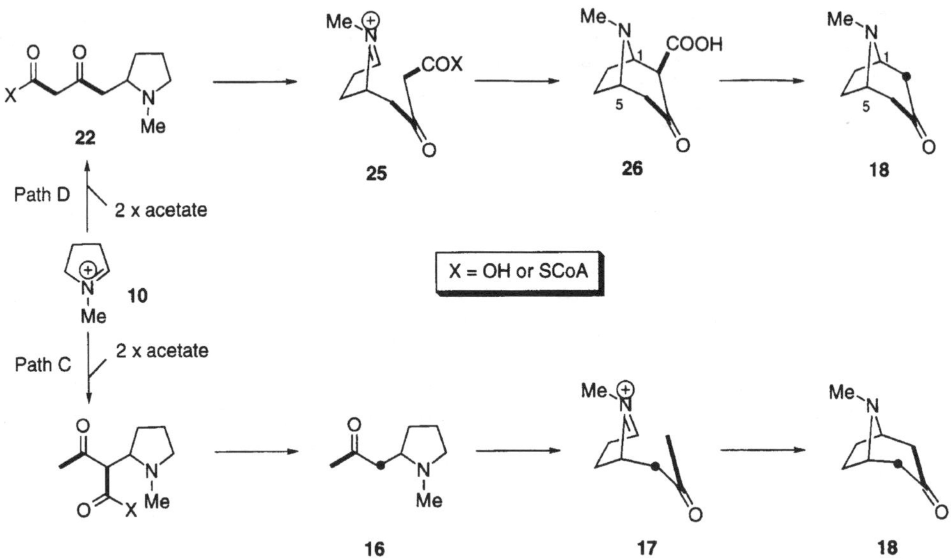

Scheme 10. One possible explanation for the scrambling of the label from [$^{13}C_2$]acetate: existence of two competing pathways

(**23**), led to an entirely different labeling pattern than had been observed in (**23**). No trace of intact incorporation of [1,2,3,4-$^{13}C_4$]acetoacetate was evident and all the label that ended up in (**15**) had been derived via [1,2-$^{13}C_2$]acetate after retroClaisen cleavage of the *β*-ketoacid. When taken at face value, this result suggests that the "pelletierine" pathway (Path C) is not involved in the formation of the tropane nucleus. However, intact incorporation of [1,2,3,4-$^{13}C_4$]acetoacetate into any alkaloid skeleton can only be observed if the rate of condensation of the intact C_4 unit with an imine partner is higher than the rate of retroClaisen cleavage of the C_4 compound. It is not unreasonable to assume that these two paths may be in competition for the precursor and the observed labeling pattern may be the fortuitous consequence of the relative rates of these competing reactions. Nonetheless, we deemed it unlikely that all of the $^{13}C_4$ precursor should be cleaved completely before being reassembled from the resulting acetate fragments and interpreted our observation as being of mechanistic significance. We concluded that the observed labeling pattern did not arise via contributions from both Paths C and D and considered other interpretations.

Another potential explanation for the equivalent labeling of the C-3/4 and C-2/3 bonds of (**1**) is the intervention, on the pathway, of a racemic intermediate. Thus it might be argued that either hygrine (**16**) on Path C or acid (**21**) or *β*-ketoacid (**22**) on Path D (Scheme 10) is formed stereospecifically but that it might accumulate for a sufficiently long time to allow racemization (as in Scheme 9). After oxidation to the iminium salt, *rac*-(**17**) or *rac*-(**25**), and intramolecular condensation of the two enantiomers, two regioisomeric labeling patterns in (**1**) would result; this is shown in Scheme 11 for the case of *β*-ketoacid (**22**).

Based on an additional experiment, discussed in detail above, in which [2-^{2}H]pyrrolinium salt (**10**) was administered and scrambling of label into positions H-1 and H-5 of (**15**) was observed, we came to favor the interpretation that a racemic intermediate on the pathway was involved. The same interpretation had been reached by Sankawa and Noguchi in the aforementioned, less extensive experiments on hyoscyamine biosynthesis [42]. Another explanation of the observed labeling pattern after administration of [1,2-$^{13}C_2$]acetate, not conceived of at the time of our publication, will be discussed later after a description of our results in a related system.

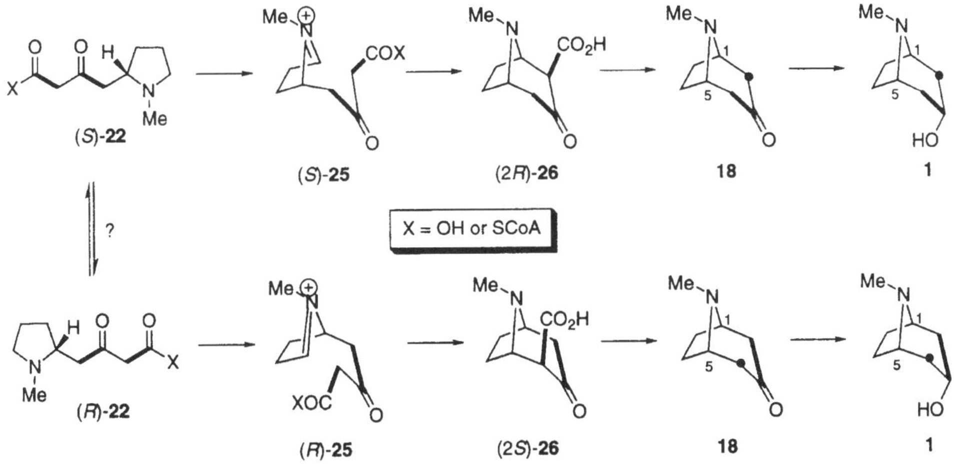

Scheme 11. A second possible explanation of the scrambling of the label from [$^{13}C_2$]acetate: existence of racemic intermediates, both enantiomers of which are utilized

A third study reporting a feeding experiment with acetate [43] attempted to refute the view put forth by Sankawa and Noguchi [42]. In this experiment [1-^{13}C,2-^{2}H]acetate was supplied to root cultures of *Hyoscyamus albus* and several tropane alkaloids were isolated. The analysis of the ^{13}C NMR spectrum of (**3**) indicated enrichment (approx. 3%) in only one carbon atom of the C_3 bridge, C-3, as would be expected. Oddly, the analysis of the incorporation of ^{2}H was performed in an indirect way via the integration of the signals due to H-2α and H-4α in the ^{1}H NMR spectrum of the sample rather than by acquisition of ^{2}H NMR spectra. An exact analysis of the data was not provided by the authors, but visual comparison of the integrals for H-2 and H-4 in the ^{1}H NMR spectra of hyoscyamine as reproduced in the publication suggests that they differ by about 30%. The authors imply, without stating so outright, that this observation indicates that H-4 receives more deuterium label than H-2 does. However, a quantitative analysis of the data raises serious doubts concerning the quality of the spectral data and the validity of their interpretation. A reduction of the ^{1}H integral for H-4α by 30% compared to H-2α implies that at least 30% of the molecules in the sample carry ^{2}H at H-4α. It is virtually impossible to propose a chemically rational mechanism to account for the purported observation that at least 30% of the molecules carry deuterium at H-4 when the specific incorporation of ^{13}C at C-3 is only 3%, approximately 10-fold lower. A higher specific incorporation of ^{13}C than of ^{2}H from such a doubly labeled precursor is a common occurrence and is most often interpreted as arising through chemical exchange of the "labile" deuterium atoms α to a carbonyl function. The reverse (higher specific incorporation of ^{2}H than of ^{13}C) is hardly explicable based on common chemical experience, however.

A fourth report on incorporation of [1,2-$^{13}C_2$]acetate into tropane alkaloids [44] also describes equal labeling of C-2 and C-4 in (**2**) isolated from *Datura stramonium* transformed root cultures to which this precursor had been applied. These latest results are qualitatively identical to the studies of incorporation of acetate discussed previously [22,42]. Thus, three investigations employing acetate labeled with stable isotopes agree in their experimental observations and the interpretation of the data. The postulation of a racemic intermediate on the pathway to (**1**) in *Hyoscyamus albus* [42] as an explanation of the labeling pattern can readily be accomodated in the model that has been developed for tropane biosynthesis in this species. The case is different, however, with the observations made in *Datura stramonium*. Stereospecific incorporation of labeled ornithine into (**1**) [13, 14, 19, 23] can only be explained if it is assumed that a single enantiomer of hygrine (**16**) or of alternative precursors such as (**21**) and (**22**) are involved in the biosynthesis. Racemization of these intermediates and incorporation of both enantiomers into (**1**) would result in labeling of both bridgehead carbons. Yet, in terms of the models discussed so far (Scheme 8), the labeling pattern from acetate of (**1**) and of (**15**) in *Datura stramonium* can be rationalized only if a racemic intermediate on the pathway is invoked.

The only obvious difference between these latest studies [22, 42, 44] which invoke racemic intermediates, and the earlier ones [13, 14, 19, 23] which report stereospecific incorporation of amino acid precursors is one of methodology. In the work reporting stereospecific incorporation, radioactive precursors were administered at "tracer" levels in the low microgram range. In the more recent work employing stable isotopes as analytical tools, substantial quantities of precursor were applied. This in itself is not unusual since it is almost an article of faith that because of the lower sensitivity of stable isotope methodology when compared to radiotracer methods, larger amounts of precursor need to be applied. The reasoning behind this standard procedure appears to be based on either one of two assumptions: firstly it is often assumed that the flux to product can be increased by providing more precursor; alternatively, it is also assumed that a larger fraction of the precursor molecules will be labeled if a larger amount of labeled precursor is applied externally. However, both lines of reasoning ignore the fact that one is dealing with a living organism with, among others, the capability to regulate metabolite flow. The dangers inherent in this simplistic view are illustrated in a recent study of galanthamine biosynthesis [45], following previous successful feeding studies employing radiotracers. It was observed that stable isotope-labeled precursors had to be administered in small amounts at high dilution to effect incorporation into the target alkaloid. Administration of precursor according to common practice resulted in inhibition of alkaloid formation under otherwise identical conditions. Thus, following blindly the prevailing custom may have unintended and detrimental consequences.

The observation of stereospecific as well as nonstereospecific incorporation, respectively, of precursors into (**1**) in *Datura stramonium* might similarly be rationalized as being a consequence of the techniques being used. It is striking that all bar one [27] of in vivo experiments employing stable isotopes have resulted in nonstereospecific incorporation into (**1**). All experiments in which stereospecific incorporation has been proved, on the other hand, were done using radiotracers (with the same one exception [27]). This leads to the admittedly speculative interpretation that, possibly, racemization of an intermediate on the pathway to (**1**) is observed only if this material accumulates in vivo. Such a scenario might play out only under the conditions of an experiment with stable isotopes in which an intermediate situated before the rate limiting step in the pathway accumulates for a sufficiently long time to allow for racemization before further metabolism occurs. Under true "tracer" conditions, on the other hand, homeostasis is not disrupted and the rates of the individual steps of the pathway are better matched. Accumulation of intermediates under these latter conditions would not be significant and consequently racemization is not observed.

3.4 Incorporation of Advanced Precursors

Feeding experiments have also been performed with more advanced putative precursors both for cocaine and tropane biosynthesis. In these experiments, the role of (**21**) and (**22**) in cocaine and tropane biosynthesis was examined. No incorporation of the ethyl ester derivative or of the *N*-acetylcysteamine thioester derivative of (**21**) into either of the alkaloids, cocaine or scopolamine, was observed [39]. However, isotope dilution experiments did provide evidence for the presence of (**21**) in *Erythroxylon coca* [46]. In the case of (**22**), fed as its ethyl ester, substantial specific incorporations into cocaine [47] and scopolamine [48] were recorded. These results were interpreted as corroborating evidence for the hypothesis that the acetate units in these alkaloids are introduced in a stepwise fashion via Path D2 in Scheme 8 with (**22**) as an intermediate.

One recent report [44], however, favors Path D1 in Scheme 8 as a mechanism for the introduction of the acetate derived C_3 unit, on the basis of the high specific incorporation of (**22**) into hyoscyamine (**3**) in *Datura stramonium*. It is, however, not obvious what the logical connection is between the level of specific incorporation of (**22**) into the alkaloids and mechanistic issues concerning the formation of (**22**).

A second argument that was furnished for the hypothesis that (**22**) is formed via the direct condensation of C-4 of acetoacetate with (**10**) stemmed from a feeding experiment with [1,2-$^{13}C_2$]acetate in *D. stramonium* in which a small proportion of molecules of (**3**) carried contiguous ^{13}C label at C-2, C-3 and C-4. However, this is simply due to the incorporation of two labelled acetate molecules into the same molecule of (**3**), which could happen whichever pathway is followed. The extent to which such multiple labeling occurs is solely determined by the degree of dilution of the labeled precursor by natural abundance material before it is incorporated into the product. This may happen either by the investigator diluting the labeled sample with natural abundance material or by dilution with endogenous material present within the organism. It has been our experience that the dilution of the labeled precursor with two parts unlabeled carrier efficiently suppresses such intramolecular interunit coupling below the limit of detection, given the relatively low specific incorporations usually observed in plants.

Lastly, the postulation of a role of acetoacetate in the formation of (**22**) is not supported by our observation that [1,2,3,4-$^{13}C_4$]acetoacetate is cleaved completely prior to incorporation into (**15**) in the same organism [22].

The question of whether (**22**) is produced by Path D1 or D2 really amounts to the question of whether an enzyme can form an enol or enolate of a methyl ketone, e.g. at C-4 of acetoacetate or at C-3' of hygrine (**16**). Current thinking on the role of hygrine (**16**) in tropane alkaloid biosynthesis would suggest that this is not possible. If one considers all four pathways of Scheme 8 and the subsequent steps required to convert (**16**) or (**22**) into tropinone (**18**), shown in Scheme 10, one can see that pathways C1, C2 and D1, respectively, involve enolization of a methyl ketone, which is not nearly as easy as enolization of a β-dicarbonyl compound. Only Path D2, which invokes (**21**) and (**22**) as intermediates, postulates that all of the carbon-carbon bond-forming reactions receive the benefit of stabilization of the nucleophile via a β-dicarbonyl system (assuming the acetate units are introduced stepwise via malonate or its CoA ester). The only experimental fact not in accord with this last proposal is the failure so far to achieve incorporation of (**21**) into (**1**) or (**19**). A new proposal for the formation of (**22**) which circumvents this problem will be introduced in Section 5 of this contribution.

Under the same experimental conditions under which incorporation of (**22**) into the alkaloids at substantial levels was observed, no evidence could be gathered for the incorporation of (**16**) into (**2**) or (**3**) in several plant species in several laboratories participating in this study [ref. 23 in 48]. These experiments are the strongest, albeit negative evidence available to date for the hypothesis that hygrine (**16**) is not an intermediate in the formation of (**3**). If correct, these negative results would strongly suggest the mechanistic interpretation that a methyl ketone such as (**17**) is not sufficiently reactive to undergo an intramolecular Mannich reaction under physiological conditions. Instead, activation of the C-methyl group of (**17**) by an additional electron-withdrawing substituent, such as an ester, a carboxylate or a thioester, as in (**25**) is required to drive the Mannich reaction.

Mechanistic questions concerning the formation of (**22**) aside, inspection of the structure of the hypothetical intermediate (**25**) suggests that the observations concerning the incorporation of (**22**) should be viewed with a certain amount of caution. A β-ketocarbonyl system such as (**25**) must be regarded as being highly activated to undergo Mannich reactions spontaneously, particularly in an intramolecular fashion, to yield the bicyclic compound, tropinone (**18**) (Scheme 11) after decarboxylation. Any nonspecific monoamine oxidase might be expected to achieve the oxidation of (**22**) to (**25**). Subsequent isolation of compounds related to or derived from tropinone might be viewed as a consequence of such an adventitious reaction rather than the result of the action of enzymes specifically involved in tropane or cocaine biosynthesis. The hypothesis invoking a role of

(**22**) or its Coenzyme A thioester in the formation of (**1**) is consistent with the available experimental results but is accompanied by a sense of uneasiness owing to the lack of good enzymological data. At the same time it has also to be kept in mind that there is good evidence that in some biosynthetic pathways in higher plants spontaneous reactions do play a role in the formation of complex products such as ajmaline [49] and morphine [50]. Thus, the cyclization of (**25**) to give (**26**) and (**18**) might just be considered to be another example of such spontaneous reactions.

If, on the other hand, the ring closure of (**25**) is enzyme catalyzed, it is reasonable to assume that only one of the enantiomers would be used as substrate by the enzyme. Consequently, the acetate derived C_3 chain of the resulting tropinone (**18**) should show only one labeling pattern from the incorporation of [1,2-$^{13}C_2$]acetate either via (*S*)-(**25**) or via (*R*)-(**25**) (Scheme 11). Our observations [22], those of Sankawa and Noguchi [42] and those of Robins et al. [44] on the regiochemistry of incorporation of [1,2-$^{13}C_2$]acetate into (**1**) are not in agreement with this prediction. Moreover, recent observations [44] that incorporation of *rac*-[2,3-$^{13}C_2$]-(**22**) into (**3**) results in equal labeling of C-2 and C-4 do not conform to that prediction. With either acetate or (**22**) as precursor, two bond-labeled species are actually observed within (**1**) instead of only one. Based solely on the models discussed so far for the origin of the C_3 unit, these experimental observations would suggest that the enzyme catalyzing the oxidation of (**22**) to (**25**) can oxidize either enantiomer of the substrate. It is important to note that the existence of a racemic intermediate on the pathway as postulated earlier [22,42,44] is not in itself sufficient to explain the labeling pattern in (**1**) after incorporation of [1,2-$^{13}C_2$]acetate or of [2,3-$^{13}C_2$]-(**21**). This interpretation of the observed labeling patterns can hold only if it is additionally assumed that at least one step of the biosynthetic sequence lacks stereoselectivity with the result that both enantiomers are converted to product.

Both paths to (**1**), via either (*S*)-(**25**) or via (*R*)-(**25**), result in the formation of tropinone (**18**) as an intermediate which is a meso compound and not a suitable object for further study of these questions. However, the isolation of (**26**) from *Datura stramonium* has been reported recently [51]. An analysis of the stereochemistry of (**26**) isolated from such a plant and the demonstration that (**26**) serves as an intermediate in the formation of (**1**) would go a long way to putting many of these conjectures on to firmer grounds. Unfortunately no further details from this work have been published.

The situation is different in the case of methyl ecgonine (**19**) in which the carboxyl carbon of the putative intermediate (**25**) is retained. Methyl ecgonine isolated from *Erythroxylon coca* plants is optically active, it is the (–) antipode of (2*R*,3*S*) absolute stereochemistry, and for the purposes of this discussion we shall surmise for simplicity that it is optically pure. Cocaine isolated from plants to which *rac*-[1,2-$^{13}C_2$,1-^{14}C]-(**22**) had been supplied shows only one ^{13}C labeling pattern in its ecgonine portion [43] as would be expected.

This result suggests that only the (*S*) enantiomer of (**22**) serves as a precursor for (-) methyl ecgonine. Unfortunately, the authors did not investigate the fate of (*R*)-(**22**) in the plant during the feeding experiments. It remains unclear whether the latter simply accumulates within the plant or whether it is oxidized to (*R*)-

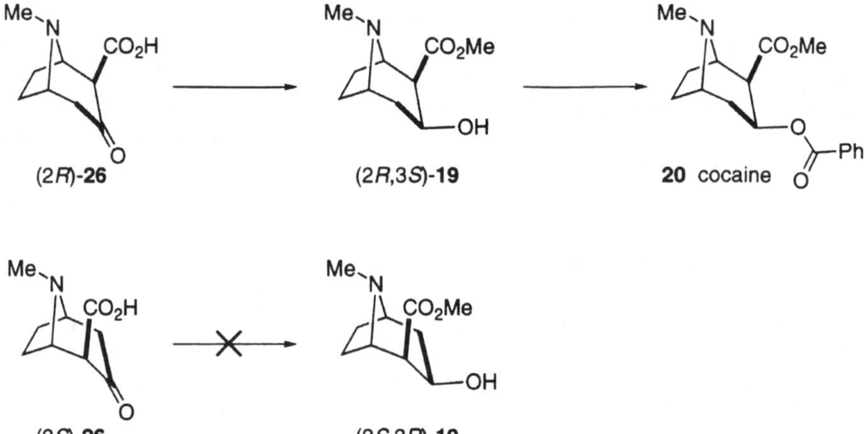

Scheme 12. Biosynthesis of cocaine via methyl ecgonine **19**; only the (2*R*,3*S*)-enantiomer is formed

(**25**) and cyclized to (2*S*)-(**26**). In any case, in view of the large doses of precursor (**22**) fed during those experiments and its facile racemization, this may technically not be trivial to establish.

Alternatively, the information might come from the seemingly trivial feeding experiment with [1,2-$^{13}C_2$]acetate in *E. coca* which has apparently not been performed. Thus, at present it is not clear that the labeling patterns from this precursor within (**19**) or (**26**) in *E. coca* and within (**1**) or (**26**) in *Datura* species will be identical. Nonetheless, such an experiment would be provide valuable experimental evidence for or against the presumed biogenetic analogy between (**1**) and (**19**). The results of a feeding experiment with acetate would also be highly desirable for the purposes of a new proposal for the formation of the acetate derived C_3 fragment of these alkaloids to be discussed below.

The discussion of the observed labeling patterns in the alkaloids so far has focused on the interpretation of the data according to either only one or a combination of several of the paths outlined in Scheme 8. Our investigation of the biosynthesis of a seemingly related alkaloid, lycopodine (**27**), uncovered yet another mechanism for the assembly of the C_3 fragments of the tropane- and the pelletierine-type alkaloids. These results will be discussed in the following section and the implications of this discovery for the biosynthesis of cocaine and tropane will be discussed subsequently.

4
The Biosynthesis of Lycopodine

In an effort to investigate the generality of the classical scheme for the generation of pelletierine-type compounds according to Path C1 in Scheme 8, we investigated the incorporation of acetate into lycopodine (**27**). Based on extensive work with radiotracers performed in the 1960s and 1970s predominantly by Spenser and

MacLean, a model had been developed which envisaged the carbon skeleton of lycopodine to arise via the dimerization of pelletierine (**28**) (Scheme 13 top), which was indeed proven to be a precursor for lycopodine [52]. Curiously, however, it served in this role only for the "left" half of the molecule C-9 to C-16. A rational explanation was subsequently presented which envisaged the intermediacy of 4-(2-piperidyl)-3-oxobutanoate (**29**), a molecule possibly biogenetically related to pelletierine, as the precursor for the "right" half of lycopodine, C-1 to C-8 [53] (Scheme 13 bottom).

When we initiated our work using stable isotope methodology, we postulated a variety of labeling patterns from [1,2-$^{13}C_2$]acetate in (**27**) which might be expected based on the models under discussion until then for the biosynthesis of the acetate derived fragments of tropane- and pelletierine-type alkaloids. Thus, for instance, the "left" half of (**27**) might be expected to show labeling according to the Path C whereas the "right" half might arise by Path D via (**29**). Alternatively, the pelletierine incorporated into the "left" half might arise via decarboxylation of (**29**) and thus to show a Path D labeling pattern (Scheme 14). An experiment with [1,2-$^{13}C_2$]acetate was deemed to be capable of deciding between these possibilities.

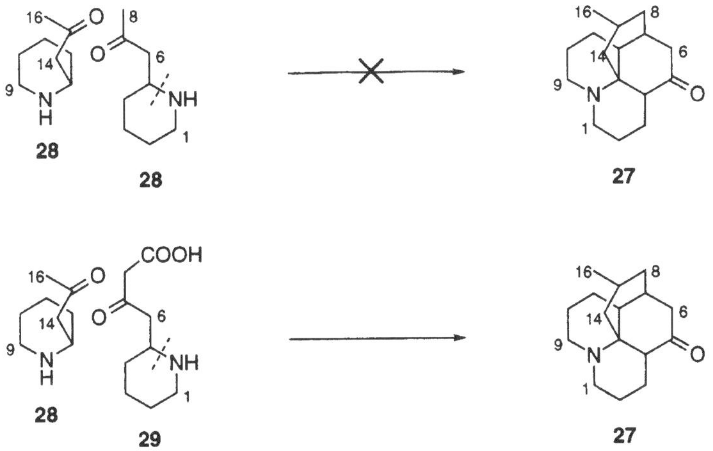

Scheme 13. Biosynthesis of lycopodine **27**; pelletierine **28** is only incorporated into the left half

On a practical level it required three years of effort to obtain the first usable incorporation result. This was for the most part due to the sharply reduced sensitivity of the stable isotope method in comparison to the radiotracer methodology employed in the earlier investigations, even if intramolecularly doubly labeled precursors are applied. Secondly, this work had to be done in the field as, to the best of our knowledge, nobody has succeeded in the culturing of any member of the family *Lycopodiaceae* in the greenhouse. Furthermore, the application of tracers to cuttings of plants, often a successful method, did not lead to any detectable incorporation in this instance. Thus, our work was confined to the months of July and August when blackflies and mosquitoes in the bush of

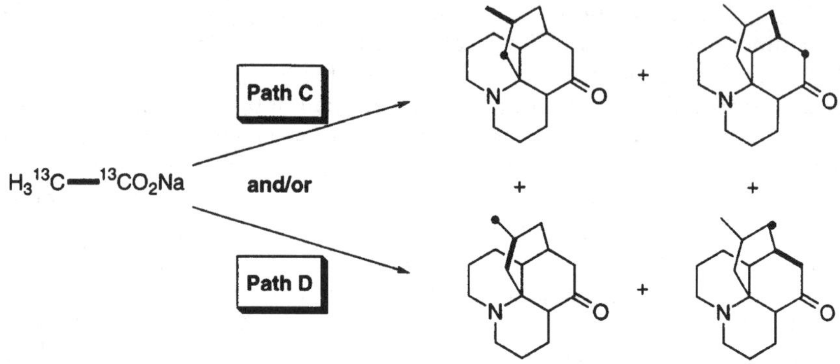

Scheme 14. Possible labelling patterns in lycopodine derived from [$^{13}C_2$]acetate; in fact all four labelling patterns were observed

Northern Ontario were tolerable, night temperatures did not fall below freezing and fresh growth at the tips of shoots was visible. Thirdly, the proper choice of plant material was crucial as well. Successful incorporations were observed only if young shoots at the ends of the above-ground rhizomes were used which showed a relatively high proportion of fresh growth. By using this selected plant material for the experiment, the small amount of labeled alkaloid formed during an experiment of one week's duration was not diluted by too large an amount of the endogenous material from previous growing seasons. In this way specific incorporations of 0.3 – 1.0% above natural abundance could be realized routinely.

In the event, we observed an entirely unexpected labeling pattern after incorporation of [1,2-$^{13}C_2$]acetate [54]. Analysis of the ^{13}C NMR spectra suggested that the patterns expected from Paths C and D were superimposed on each other in both acetate derived portions of the alkaloid. The two patterns were observed in a ratio of 1:1, which did not change over several repetitions of the experiment. In a separate experiment it was found that [1,2,3,4-$^{13}C_4$]acetoacetate was not incorporated intact but was instead first cleaved to acetate and then incorporated, resulting in an identical labeling pattern to that which had been observed when [1,2-$^{13}C_2$]acetate was fed. The results of these two experiments on (**27**) were thus qualitatively identical to our observations during the investigation of 6β-hydroxytropine (**15**) biosynthesis in *D. stramonium* [22].

In the interpretation of these experiments in *Lycopodium tristachyum* the argument could be made, as had been in the case of (**15**), that the preference for cleavage of the C_4 unit over intact incorporation was a consequence of higher reaction rates for the former rather than the latter process. We had rejected this explanation in the case of (**15**) on the basis of the argument that if an intact C_4 unit was a precursor for the acetate derived C_3 fragment, it would surely not all have been cleaved and at least some of the $^{13}C_4$ precursor should have survived. In the case of lycopodine, however, we did see an opportunity to probe the status of acetoacetate as a precursor which did not require the feeding of a C_4

precursor. We settled on an experiment with [1,2-$^{13}C_2$,2-$^2H_3$]acetate and decided to focus our attention on the acetate derived C_3 unit C-16,C-15,C-14 which originates from pelletierine [52, 53]. We reasoned that the labeling pattern in this portion of (**27**) suggested that half of the molecules were labeled according to Path C in Scheme 8 and half according to Path D. We expected that lycopodine molecules in which this portion is formed by Path C should retain at least some deuterium from [1,2-$^{13}C_2$, 2-$^2H_3$]acetate at the C-16 methyl group. Specifically, this methyl group is derived from the acetate starter unit used in the putative Claisen condensation of acetyl CoA with malonyl CoA to yield acetoacetyl CoA. Retention of deuterium at C-2 of acetate starter units of polyketides is well precedented. In the event, NMR analysis of a sample of lycopodine (**27**) from this experiment showed that it carried only ^{13}C and no trace of ^2H was observed. The ^{13}C NMR spectra were qualitatively identical to those obtained after feeding of [1,2-$^{13}C_2$]acetate. This unexpected result could not be reconciled with Path C1 and we focused our attention on finding an entirely different interpretation of our observations.

We concluded from the outcome of our double label experiment with [1,2-$^{13}C_2$, 2-$^2H_3$]acetate that the carbon atom destined to become the C-16 methyl group of lycopodine had to be activated at some time during the biosynthetic process to such an extent that facile deuterium/protium exchange was possible. A derivative which would fulfil such a structural requirement might be malonic acid (or its CoA derivative) or an analogous intermediate. Careful reexamination of the NMR data from all the experiments indicated that the ratio of the two labelling patterns in the acetate derived C_3 units of lycopodine was always exactly 1:1 (within the accuracy of ^{13}C integration). It was a conceptually crucial step when we came to realize that the observed 1:1 ratio was probably not adventitious. Instead, this ratio may be a necessary consequence of the mechanism by which the acetate derived fragments of (**27**) were assembled. If the mixed incorporation patterns had been due to competition between two paths, some variation in this ratio would have to be expected over several repetitions of the experiment.

The most straightforward explanation for the two labeling patterns occurring in a 1:1 ratio is the presence of a symmetrical intermediate between acetate and the immediate precursor for the C_3 unit, before the latter is joined to the imine (**14**). The structural prerequisites for such a putative intermediate were: firstly, a compound with C_{2v} symmetry capable of delivering a C_3 unit equivalent to acetone and, secondly, the two sites destined to become the methyl groups of this acetone equivalent had to be sufficiently activated to undergo ready exchange of the protons α to the keto group. These requirements are fulfilled by acetonedicarboxylic acid or its bisCoA thioester (**30**), Scheme 15.

Upon Claisen condensation of two molecules of malonyl CoA with one another, a molecule of acetonedicarboxylic acid CoA ester would be formed which could either be hydrolyzed to the free acid or further activated to its bisCoA thioester. It is important that the derivative (**30**) have C_{2v} symmetry to explain the labeling pattern in (**27**). The monoCoA ester would not fulfil this condition! After condensation of (**30**) with (**14**), the keto acid (**29**) would be obtained and would show two distinct labeling patterns depending on which of the methylene

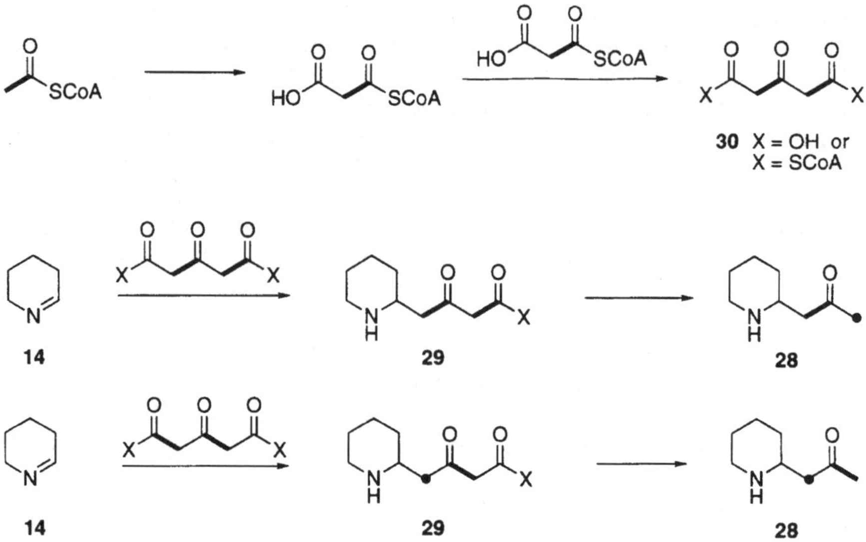

Scheme 15. New proposal for the biosynthesis of pelletierine in *Lyopodium tristachyum* via acetonedicarboxylic acid

carbon atoms of (**30**) formed the bond to C-2 of (**14**). After decarboxylation of (**29**), or hydrolysis/decarboxylation, pelletierine (**28**) would be obtained, which also would show two distinct labeling patterns. Combinations of the isotopomers of (**28**) and (**29**) with each other according to Scheme 13 would then yield the experimentally observed labeling patterns.

We were exceedingly fortunate that the quintessential proof of this idea, namely the direct incorporation of acetonedicarboxylic acid into lycopodine was indeed observed when [2,3,4-$^{13}C_3$]acetonedicarboxylate was fed to *L. tristachyum.* In fact, the experiment was so successful when it was first attempted that the resulting data were not easily interpretable. In order to achieve maximum sensitivity for the experiment, the precursor had not been diluted with unlabelled material. This precaution proved to be unnecessary as a 1% specific incorporation was observed. Careful analysis of the splitting patterns for C-15 and C-8 indicated that about 50% of the isotopically labeled molecules contained two labeled C_3 units and thus showed interunit coupling. A repetition of this experiment one year later with precursor properly diluted with unlabelled material, again yielded labeled alkaloid and this time the coupling pattern was readily analyzed and found to be consistent with our expectations.

The current proposal for assembly of the carbon skeleton of lycopodine (**27**) [55] is as shown in Scheme 13 bottom. 4-(2-Piperidyl)-3-oxobutanoate (**29**) is the central intermediate in this scheme. This material may suffer decarboxylation yielding pelletierine (**28**) which is incorporated into the C-9 to C-16 fragment. The keto group at C-15 (lycopodine numbering) of that fragment is then condensed in a Knoevenagel type reaction with the methylene group of the β-keto acid function of 4-(2-piperidyl)-3-oxobutanoate to form the C-15 to C-8 bond.

In an attempt to rationalize the underlying chemistry depicted in Scheme 13, one is bound to arrive at the same conclusion that was reached when the successful incorporation of (**22**) into tropane (**1**) was being considered. Apparently activation of the C-3' methyl group of pelletierine, destined to become C-8 of lycopodine (**27**), as a methylene placed between two carbonyl groups is required for successful formation of the C-15/C-8 bond.

5
A New Proposal for the Assembly of the Acetate Derived C3 Unit

At this point it is instructive to return to the discussion of the formation of the azabicyclooctane skeleton of the tropanes and the mechanism by which the acetate derived C_3 unit is assembled. The reader will have noticed that the labeling pattern we observed in the acetate derived C_3 units of lycopodine (**27**) and of tropane (**1**) after feeding experiments with $[1,2\text{-}^{13}C_2]$acetate is identical. In both cases two regioisomeric labeling patterns in a ratio of 1:1 are observed. Hitherto the pattern in the tropanes has been rationalized by invoking the existence of a racemic intermediate on the pathway between acetate and (**1**) [22,42,44]. However, our results on the biosynthesis of lycopodine and their interpretation do suggest another explanation for the labeling pattern from $[1,2\text{-}^{13}C_2]$acetate in (**1**): that in the tropanes the acetate derived C_3 unit C-2,C-3,C-4 arises via acetonedicarboxylic acid or a derivative thereof, Scheme 16, just as in Robinson's original biomimetic synthesis. Analogously it may be proposed that methyl ecgonine (**19**) is formed via condensation of (**10**)

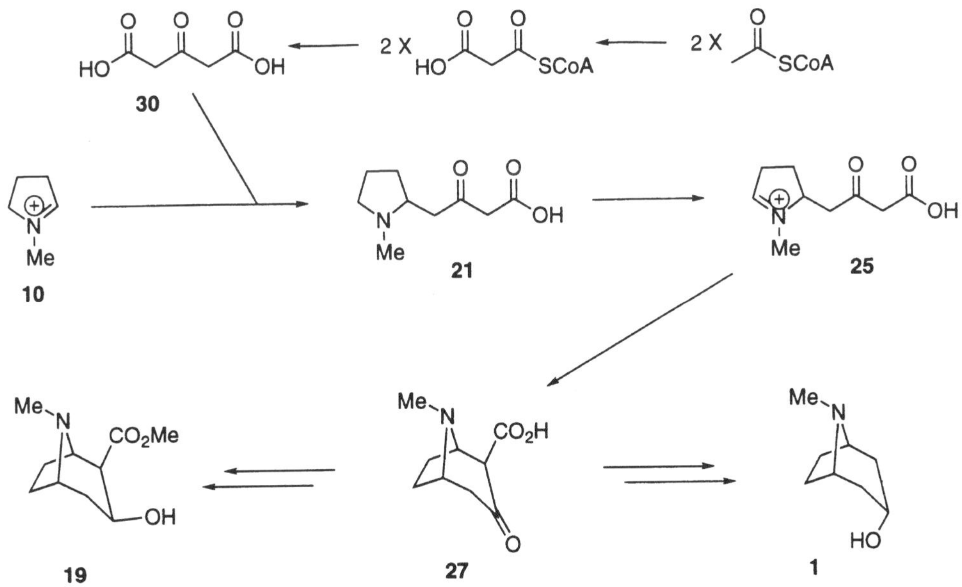

Scheme 16. New proposal for the biosynthesis of tropane alkaloids via acetonedicarboxylic acid

with acetonedicarboxylate followed by loss of only one of the pendant carboxyl groups. The resulting (**22**) could then be processed to (**19**) as shown (Scheme 16).

It is an attractive feature of this proposal that it establishes again a unified conceptual framework for the biosynthesis of tropane and pelletierine type alkaloids. According to this new hypothesis, tropane- and pelletierine-type alkaloids fall into two groups. Group A comprises alkaloids such as *N*-methylpelletierine (**23**), hygrine (**16**) and 2,4-dimethylindolizidine (**31**) in which the carbon atom corresponding to C-3' of (**16**) or of (**23**) is not bonded to another atom other than C-2'. These alkaloids would be made by reaction of acetoacetate with the appropriate cyclic imine or iminium ion (Path C1). Alkaloids such as ecgonine (**19**), tropine (**1**), cuskhygrine (**32**) or *ψ*-pelletierine (**33**), on the other hand, belong in group B. These latter alkaloids in which the analogous C-3' carbon atom is linked not only to C-2' but also to another carbon atom, would be formed via a double Mannich condensation of acetonedicarboxylate with (**10**) or (**14**). Ketoacids (**22**) and (**29**) are postulated as intermediates in the formation of alkaloids belonging to group B.

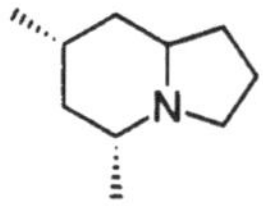

31

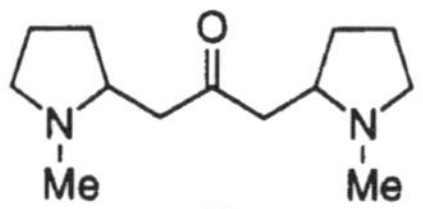

32

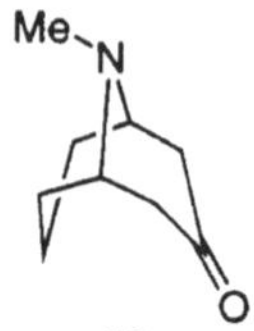

33

The experimental data accumulated so far for the origin of the acetate derived C_3 chain of (**1**) are in agreement with this proposal. The presence of two bond-labeling patterns in (**1**) after administration of [1,2-$^{13}C_2$]acetate has been interpreted hitherto as evidence for the presence of a racemic intermediate. This necessitated the postulation of a nonstereoselective enzymatic oxidation of (**22**) to (**25**) as discussed in detail above. However, this postulate in the face of the complete lack of enzymological data in its support, is not entirely satisfying. In contrast, the postulation of acetonedicarboxylic acid, or a derivative thereof, as a pathway intermediate does explain simply the labeling pattern in the C_3 sidechain of (**1**) and a precedent has now been found in the biosynthesis of (**27**). Moreover, the hypothesis also explains why (**21**) is not incorporated into tropane alkaloids.

Unfortunately, the new hypothesis does not eliminate all of the apparent confusion surrounding the question of a symmetrical vs a nonsymmetrical intermediate between ornithine and/or arginine and (**1**) in *Datura* species. The new model would, however, allow the occurrence of two labeling patterns from [1,2-$^{13}C_2$]acetate in the C_3 fragment of (**1**) to be explained even if nonsymmetrical incorporation of ornithine (**4**) or arginine (**5**) is observed. Thus, the apparent contradiction between the results of the early feeding experiments of Leete [13,14, 27] and of the Halle group [19] on the one hand and the results from the acetate feeding experiments [22,42,44] on the other can be resolved. The scrambling of the labeling pattern when (**22**) is incorporated into tropine (**1**) can still only be explained by supposing that both enantiomers of the administered (**22**) are converted into (**1**). Nonstereospecific labeling of (**1**) from the amino acids could still be rationalized either by invoking putrescine as an intermediate on the pathway or by postulating racemization and subsequent incorporation of both enantiomers of an intermediate.

This analysis suggests that experiments aimed at proving the acetonedicarboxylate route to (**1**) in *Datura* will have to assess the stereospecificity of incorporation of (**4**) and (**5**) into (**1**) and the pattern of incorporation of [1,2-$^{13}C_2$]acetate into (**1**) simultaneously. A further challenge will be to design and carry out successfully an experiment which will allow one to decide whether the C_3 unit in the tropanes is elaborated via Path D or the acetonedicarboxylate pathway even if racemic intermediates are involved. It will be crucial to avoid the use of precursors that are as highly activated as the β-keto acids fed by Leete [47,48] and Robins [44] and coworkers. This is because of the serious risk that subsequent incorporation of such putative precursors will be overinterpreted as being of mechanistic significance when in actual fact the result of an artefact is being observed. Thus for instance, feeding labeled acetonedicarboxylic acid to intact plants of the genus *Datura* or root cultures derived therefrom will not be the appropriate way to prove this notion. Under such circumstances we may expect that a Robinson type condensation may occur if a sufficient amount of (**10**) and a monoamine oxidase activity are present within the plant. Such a partially nonbiological sequence would be expected to yield (**18**) which is also an intermediate in the biological sequence to (**1**). A comparison of the specific incorporation of (**22**) into cuskhygrin (**32**) (9% above natural abundance) vs (**2**) (2% above natural abundance) during a recent feeding experiment in *Datura stramonium* [44] may serve as a cautionary tale.

The situation in our work on lycopodine (**27**) was different and a feeding experiment with acetonedicarboxylic acid could be more easily justified. Firstly, more meaningful tracer evidence from incorporation of acetate into (**27**) was available which was suggestive of a role of acetonedicarboxylic acid in the pathway to lycopodine. Secondly, a "biomimetic" formation of lycopodine from acetonedicarboxylic acid and imine (**14**) in the presence of ubiquitous enzymes such as a monoamine oxidase cannot be readily envisaged given the greater structural complexity of (**27**) when compared to (**18**). For this reason we are confident that the incorporation of acetonedicarboxylic acid into (**27**) reflects the biological process in *Lycopodium* species.

6 The Reduction of Tropinone

The reduction of tropinone (**18**) to tropine (**1**) has been subject to several studies in cell-free systems derived from intact plants and from root culture systems. In early studies of this step an enzyme was detected which reduced (**18**) to an alcohol with 3α-stereochemistry, namely (**1**) [56]. Subsequently, in *Hyoscyamus niger*, a plant which predominantly forms alkaloids with 3α-stereochemistry, an enzyme was found which reduces (**18**) to ψ-tropine (**34**) with 3β-stereochemistry [57] (Scheme 17). Other workers observed varying ratios of (**1**) and (**34**)

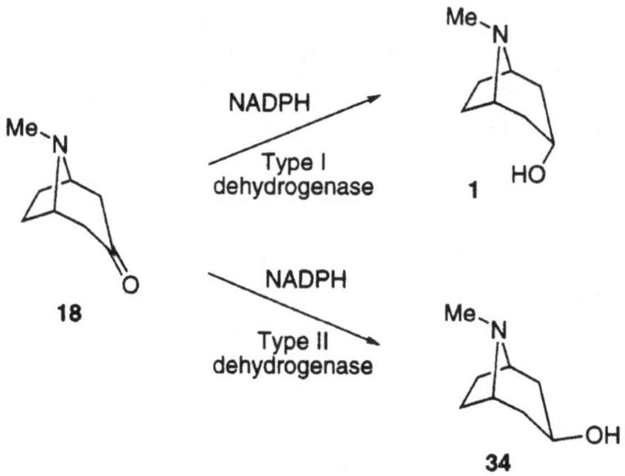

Scheme 17. The two dehydrogenases acting on tropinone **18** found in *Datura* and *Hyoscamus*

being formed in enzyme systems derived from roots of *D. innoxia* and concluded that two enzymes were involved, each of which stereospecifically formed one of the two products [58]. This view is supported by two studies which accomplished the separation of two such specific, NADPH dependent enzymes from *Datura stramonium* [59] and *Hyoscyamus albus* [60] root cultures. Type I enzymes produce (**1**) whilst Type II enzymes form (**34**) exclusively. Enzymes belonging to these two classes show marked differences in their physical properties such as pH optima and K_M values. The two enzymes from *D. stramonium* were studied in more detail on a genetic level [61] and were found, surprisingly in view of their kinetic differences, to be related to each other (64% identity on the amino acid level) and to be members of a family of short-chain dehydrogenases. Within this class of enzymes the cofactor binding site resides on the amino-terminal fragment of the protein and binding of the carbonyl substrate is thought to occur on the carboxy-terminal part. This suggests that the cofactor binding of enzymes of Type I and of Type II is identical and the two enzymes producing (**1**) and (**34**), respectively, from (**18**) would differ by the orientation of the substrate (**18**) within the binding site relative to the nicotinamide ring of the cofactor. This idea was put to the test by the construction, expression and assay of chimeric enzymes which resulted from shuffling of short stretches of genetic material between genes encoding Type I and Type II enzymes, respectively [62]. In this fashion it was possible to convert a Type I enzyme which forms (**1**) exclusively, into an enzyme yielding predominantly (**34**) as product. In accord with the domain model for this class of enzymes, it was observed that moving long stretches of DNA encoding the central and carboxy-terminal portion of the protein resulted in such a reversal of the stereoselectivity of reduction. Thus binding of the carbonyl containing substrate and its orientation within the active site are predominantly determined by the carboxy-terminal portion of the protein.

7 Conclusions

As the preceding discussion shows, our understanding of the biosynthesis of the tropane nucleus has advanced considerably in the past decade through molecular genetics and enzymology as well as more chemically oriented work. Despite these advances our model of tropane biosynthesis is still deficient in key aspects. Several reasons may be offered for the present state of affairs. Some of the problems are certainly due to the fact that different laboratories are using different species and genera of plants as experimental material. It is tempting to the organic chemist, post-Robert Robinson, to impose a unifying framework on biosynthetic pathways for a given natural product, regardless of producer species, and this point of view is fully justified in most groups of natural products. However, in the case of the biosynthesis of the tropane alkaloids Nature may not be cooperating and the chemist's simplification may be misleading. Thus, for instance, there remains little doubt that one key aspect of the biosynthesis of (**1**), stereospecificity of the incorporation of amino acids, is different in *Hyoscyamus* from *Datura*. To make matters worse, results concerning the same issue of stereospecificity in the latter genus are contradictory. What appears to be lacking first and foremost for the resolution of these inconsistencies is fruitful collaboration of chemists and enzymologists. Regretably, Leete's important early finding concerning the stereospecific incorporation of ornithine into (**3**) has not yet been taken up by enzymologists as an observation in want of a biochemical mechanism.

The central portion of the pathway to (**1**), the joining of the acetate derived and the amino acid derived portions of the alkaloid skeleton, remains an active area of investigation as well. The increased use of bond-labeled precursors with vicinal ^{13}C labels in the 1990s has brought significant progress. In retrospect it is surprising that this powerful method had not been brought to bear on the problem of tropane biosynthesis in a rigorous fashion earlier than it was. The origin of the acetate derived C_3 unit must be better defined than it is at present before enzymological work can be attempted with any hope of success. All of the competing laboratories appear to agree that (**22**) has a central role as an intermediate in the formation of (**1**) and of (**19**). In this contribution a new, experimentally testable hypothesis for the mechanism of assembly of the acetate derived C_3 fragment has been introduced and work is ongoing in this laboratory to substantiate it. The authors hopes that it will stimulate further experimental work in other laboratories as well.

8 References

1. Robinson R (1917) J Chem Soc 111:762
2. Robinson R (1917) J Chem Soc 111:876
3. Birch AJ (1993) Note Rec Roy Soc London 47:277
4. Diaper DGM, Kirkwood S, Marion L (1951) Can J Chem 29:964
5. Yun D-J, Hashimoto T, Yamada Y (1992) Proc Natl Acad Sci 89:11799
6. Leete E (1990) Planta medica 56:339
7. Robins RJ, Walton NJ (1993) The biosynthesis of tropane alkaloids. In: Cordell GA (ed) The Alkaloids, vol 44. Academic Press, San Diego, p 115
8. O'Hagan D, Robins RJ (1998) Chem Soc Revs 27:207

9. Leete E, Spenser ID, Marion L (1954) Nature 174:650
10. Spenser ID (1968) The biosynthesis of alkaloids and of other nitrogenous metabolites. In: Florkin M, Stotz EH (eds) Comprehensive biochemistry. Elsevier, Amsterdam, chap 6
11. Walton NJ, Robins RJ, Peerless ACJ (1989) Planta 182:136
12. Robins RJ, Parr AAJ, Walton NJ (1991) Planta 183:196
13. Leete E (1962) J Am Chem Soc 84:55
14. Leete E (1964) Tetrahedron Lett 1619
15. Leistner E, Spenser ID (1973) J Am Chem Soc 95:4715
16. Hashimoto T, Yamada Y, Leete E (1989) J Am Chem Soc 111:1141
17. Walsh C (1979) Enzymatic Reaction Mechanisms. Freeman, San Francisco, chap 24
18. Liebisch HW, Schütte HR (1967) Z Pflanzenphysiol 57:434
19. Liebisch HW, Ramin H, Schöpfinius I, Schütte HR (1965) Z Naturforschg 20B:1183
20. Leete E, Endo T, Yamada Y (1990) Phytochemistry 29:1847
21. Liebisch HW, Schütte HR, Mothes K (1963) Liebigs Ann Chem 668:139
22. Hemscheidt T, Spenser ID (1992) J Am Chem Soc 114:5472
23. Bothner-By AA, Schutz RS, Dawson RF, Solt MC (1962) J Am Chem Soc 84:52
24. Spenser ID (1985) Pure & Appl. Chem. 57:453
25. Gupta RN, Spenser ID (1967) Can J Chem 45:1275
26. Hashimoto T, Yamada Y (1989) Planta 178:123
27. Leete E, McDonell JA (1981) J Am Chem Soc 103:658
28. Hashimoto T, Mitani A, Yamada Y (1990) Plant Physiol 93:216
29. Kaczkowski J, Schütte HR, Mothes K (1961) Biochem Biophys Acta 46:588
30. Liebisch HW, Piesker K, Radwan A, Schütte, HR (1972) Z Pflanzenphysiol 67:1
31. Leete E, Kim SH (1989) Chem Commun 1899
32. O'Donovan DG, Keogh MF (1969) J Chem Soc (C) 223
33. McGaw BA, Woolley JG (1979) Phytochemistry 18:189
34. McGaw BA, Woolley JG (1978) Phytochemistry 17:257
35. Leete E, Kim SH (1988) J Am Chem Soc 110:2976
36. Bremner JB, Cannon JR, Joshi KR (1973) Aust J Chem 26:2559
37. Tomita H, Mitusaki H, Tamaki E (1964) Agric Biol Chem 28:451
38. Paßreiter CM (1992) Phytochemistry 31:4135
39. Huang MN, Abraham TW, Kim SH, Leete E (1996) Phytochemistry 41:767
40. Hemscheidt T, Spenser ID (1990) J Am Chem Soc 112:6360
41. Endo T, Hamaguchi N, Hashimoto T, Yamada Y (1988) FEBS Lett 234:86
42. Sankawa U, Noguchi H, Hashimoto T, Yamada Y (1990) Chem Pharm Bull 38:2066
43. Sauerwein M, Shimomura K, Wink M (1993) Phytochemistry 32:905
44. Robins RJ, Abraham TW, Parr AJ, Eagles, J, Walton, NJ (1997) J Am Chem Soc 119:10929
45. Eichhorn J, Takada T, Kita Y, Zenk MH (1998) Phytochemistry 49:1037
46. Leete E (1989) Heterocycles 28:481
47. Leete E, Bjorklund JA, Couladis MA, Kim, SH (1991) J Am Chem Soc 113:9286
48. Abraham TW, Leete E (1995) J Am Chem Soc 117:8100
49. Pfitzner A, Stöckigt J (1983) Chem Commun 459
50. Lenz R, Zenk MH (1995) J Biol Chem 220:31091
51. Bachmann P (1996) Abstracts, Biosynthesis of Secondary Products; Halle, Germany
52. Castillo M, Gupta RN, Ho YK, MacLean DB, Spenser ID (1970) J Am Chem Soc 92:1074
53. Castillo M, Gupta RN, Ho YK, MacLean DB, Spenser ID (1970) Can J Chem 48:2911
54. Hemscheidt T, Spenser ID (1993) J Am Chem Soc 115:3020
55. Hemscheidt T, Spenser ID (1996) J Am Chem Soc 118:1799
56. Koelen KJ, Gross GG (1982) Planta med 44:227
57. Dräger B, Hashimoto T, Yamada Y (1988) Agric Biol Chem 52:2663
58. Couladis MM, Friesen JB, Landgrebe ME, Leete E (1991) Phytochemistry 30:801
59. Portsteffen A, Dräger B, Nahrstedt A (1992) Phytochemistry 31:1135
60. Hashimoto T, Nakajima K, Ongena G, Yamada Y (1992) Plant Physiol 100:836
61. Nakajima K, Hashimoto T, Yamada Y (1993) Proc Natl Acad Sci USA 90:9591
62. Nakajima K, Hashimoto T, Yamada Y (1994) J Biol Chem 269:11695

Biosynthesis and Metabolism of Pyrrolizidine Alkaloids in Plants and Specialized Insect Herbivores

Thomas Hartmann · Dietrich Ober

Institut für Pharmazeutische Biologie, Technische Universität Braunschweig, Mendelssohnstrasse 1, 38106 Braunschweig, Germany
E-mail: t.hartmann@tu-bs.de; d.ober@tu-bs.de

Pyrrolizidine alkaloids (PAs) encompass a typical class of plant secondary compounds. During recent years PAs have proved to be an excellent choice to exemplify various mechanistic and functional aspects of plant secondary metabolism. PAs appear to play an important role in constitutive plant chemical defense, particularly, in plant-herbivore interactions. Biochemical and physiological aspects of PA biosynthesis, allocation, and accumulation are reviewed with particular emphasis on key enzymatic steps and chemical diversification of PA backbone structures. A number of taxonomically unrelated specialized insect herbivores sequester PAs from their food plants. The biochemistry of PA sequestration in lepidopterans and leaf beetles is outlined along with the mechanisms of PA absorption, storage, pheromone biosynthesis, and insect-specific PA transformations (i. e., formation of insect alkaloids). The unique feature of PAs, that they exist in two easily interchangeable forms, the pro-toxic free base and the non-toxic *N*-oxide, is stressed and related to the function of PAs as defensive compounds. The first molecular evidence concerning the phylogenetic origin of PAs is presented.

Keywords. Pyrrolizidine alkaloids (plants), Chemical diversification, Alkaloid sequestration (insects), Chemical defense, Chemical ecology

1 Introduction

Pyrrolizidine alkaloids (PAs) are typical constitutively produced plant secondary compounds. PAs exist in a great diversity of some 370 chemical structures [1–3]. They are assumed to have evolved as chemical defenses under the selection pressure of competing herbivores. This is evidenced by a number of insect herbivores from unrelated taxa which have developed adaptations not only to overcome PA-mediated plant defense, but also to sequester and utilize these alkaloids for their own defense against predators. PAs are an excellent choice to exemplify mechanistic and functional aspects of plant secondary metabolism [4, 5].

This chapter summarizes recent research in the field of PAs. It intends to combine chemical, biochemical, and functional aspects of these alkaloids not only in the producing plant species but also in the specialized insect herbivores which, in the course of their adaptation to PAs, have developed specific biochemical means to safely handle and utilize these defensive compounds for their own benefit.

There are a number of excellent recent monographs and reviews dealing with various aspects of PA research: chemistry and toxicology [1, 2, 6], biosynthesis [7–9], insect-plant interactions (chemical ecology) [5, 10–12] and general aspects [3].

2 Classification and Distribution of PAs

Pyrrolizidine alkaloids are ester alkaloids composed of a necine base (amino alcohol) mostly esterified with a branched aliphatic necic acid (Fig. 1). Extending Culvenor's chemosystematic suggestions [13] and including biogenetic aspects [3], PAs can be classified into five structural types (Fig. 1). Macrocyclic diesters of the senecionine type are abundant in species of the tribe Senecioneae (Asteraceae). More than one hundred structures are known and they all contain a branched C_{10}-necic acid, which is derived from the C_5-carbon skeletons of two molecules of isoleucine or rarely leucine. The related open-chain diesters of the triangularine type still possess two C_5-units but they are not linked. Triangularine-type PAs are found in species of the Boraginaceae and in a number of *Senecio* species. Macrocyclic diesters (senecionine type) and open-chain diesters (triangularine type) are never found together, but they sometimes occur in

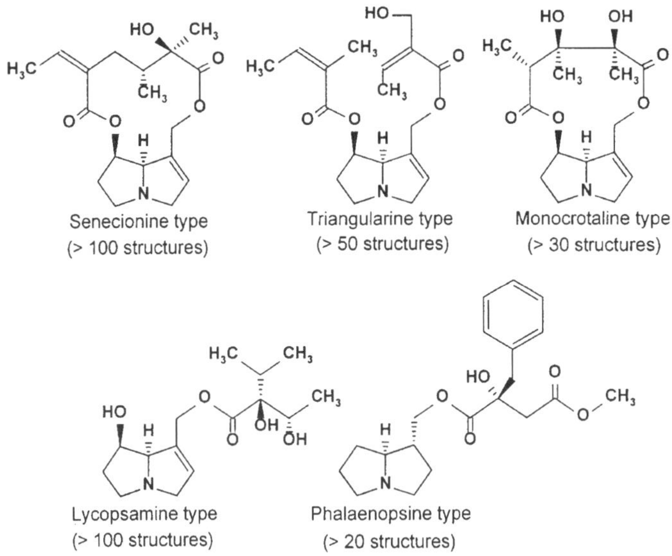

Fig. 1. The five major structural types of PAs

closely related species such as *S. nemorensis* (senecionine type) and *S. fuchsii* (triangularine type). Next to the senecionine-type PAs, alkaloids of the lycopsamine type with more than 100 structures represent the most diverse class of PAs. They are characterized by a unique branched C_7-necic acid with the carbon skeleton of 2-isopropylbutyric acid. PAs of the lycopsamine type occur in species of the tribe Eupatorieae (Asteraceae) and in the Boraginaceae. All lycopsamine-type alkaloids contain at least one C_7-necic acid and this is occasionally accompanied by a second necic acid. The few PA-containing species of the Apocynaceae typically produce lycopsamine derivatives in which two C_7-necic acids form macrocyclic triesters such as parsonsine [3]. A small group of closely related eleven-membered macrocyclic diesters are the monocrotaline type, produced by the alkaloid-containing species of the large genus *Crotalaria* (Fabaceae). The great majority of the individual structures belonging to one of the four PA types so far mentioned have two features in common: (i) they contain the 1,2-unsaturated retronecine as the most abundant necine base; (ii) they are present as alkaloid *N*-oxides. The importance of these two general features will be discussed later in more detail.

Simpler structures such as the orchid PAs are classified as the phalaenopsine type. They do not possess a 1,2-unsaturated necine base and they are often esterified with an aryl-, aralkyl-, or alkylnecic acid. The recently described ipangulines found in *Ipomoea hederifolia* (Convolvulaceae) also belong to this group of PAs [14, 15]. The orchid PAs are only partially *N*-oxidized [16] whereas the ipangulines are not at all [15].

Within the plant kingdom PAs are only found in the angiosperms. Here they occur quite scattered and often sporadically in phylogenetically unrelated taxa [3]. Major sources for PAs are the Asteraceae tribe Senecioneae (>230 species), the Asteraceae tribe Eupatorieae (>25 species), the Boraginaceae (>150 species), the Fabaceae genus *Crotalaria* (>81 species), and a few related genera (ca. 40 species), as well as the Orchidaceae (35 species). Sporadic occurrences of PA-containing species are known from the Apocynaceae (8 species), Celastraceae (1), Convolvulaceae (3), Ranunculaceae (2), Rhizophoraceae (2), Santalaceae (1), and Sapotaceae (4).

3 Biosynthesis and Maintenance of PAs in Plants

Pioneering tracer studies carried out in the laboratories of D. J. Robins and I. Spenser had demonstrated that the carbon skeleton of the necine base originates from ornithine or arginine via putrescine. Extension of these studies in the 1980s using precursors labeled with stable isotopes confirmed that retronecine is derived from putrescine via a symmetrical intermediate such as homospermidine [17–19]. The formation of the necic acids is less well understood. Early tracer-studies primarily carried out by D. H. G. Crout and collaborators revealed that the carbon skeletons of all aliphatic necic acids so far studied are derived from branched-chain amino acids, i.e., isoleucine, and less frequently leucine and valine [see 17].

The results of the tracer studies including the elucidation of the stereochemistry involved, provided a firm basis for a biochemical approach to PA biosynthesis, i.e., characterization of the enzymes that catalyze biosynthetic key steps and the specific mechanisms involved in translocation, subcellular accumulation, and metabolism of PAs. Early tracer work was carried out with intact plants to which tracers were applied for days or weeks. Meanwhile, in vitro plant systems, such as cell cultures and root-organ cultures of PA-producing plants are available. Root cultures were found to be excellent systems for biochemical and enzymatic studies of PA biosynthesis [20–22]. Dedifferentiated cell cultures do not synthesize PAs, but retain the ability to accumulate PAs. They are excellent systems to study the membrane transport of PAs and to identify the subcellular storage sites.

3.1 Biosynthesis of the Necine Base

Root cultures of *Senecio vulgaris* and related species such as *S. vernalis* and *S. erucifolius* were used to study PA biosynthesis. These cultures produce senecionine *N*-oxide as the major alkaloid [23, 24]. Tracer experiments with radioac-

Fig. 2. Biogenetic links between **A** primary metabolism and **B** the alkaloid-specific pathway. CAP, *N*-carbamoylputrescine; GABA, 4-aminobutyric acid. Inhibitors: *DFMA*, α-difluoromethylarginine, *HEH*, β-hydroxyethylhydrazine. Enzymes: *1*, arginine decarboxylase; *2*, agmatine iminohydrolase; *3*, *N*-carbamoylputrescine amidohydrolase; *4*, diamine oxidase; *5*, pyrroline dehydrogenase (NAD^+-dependent); *6*, spermidine synthase; *7*, a putrescine-producing polyamine oxidase (*HEH*-insensitive); *8*, homospermidine synthase; *9*, an *HEH*-sensitive polyamine oxidase

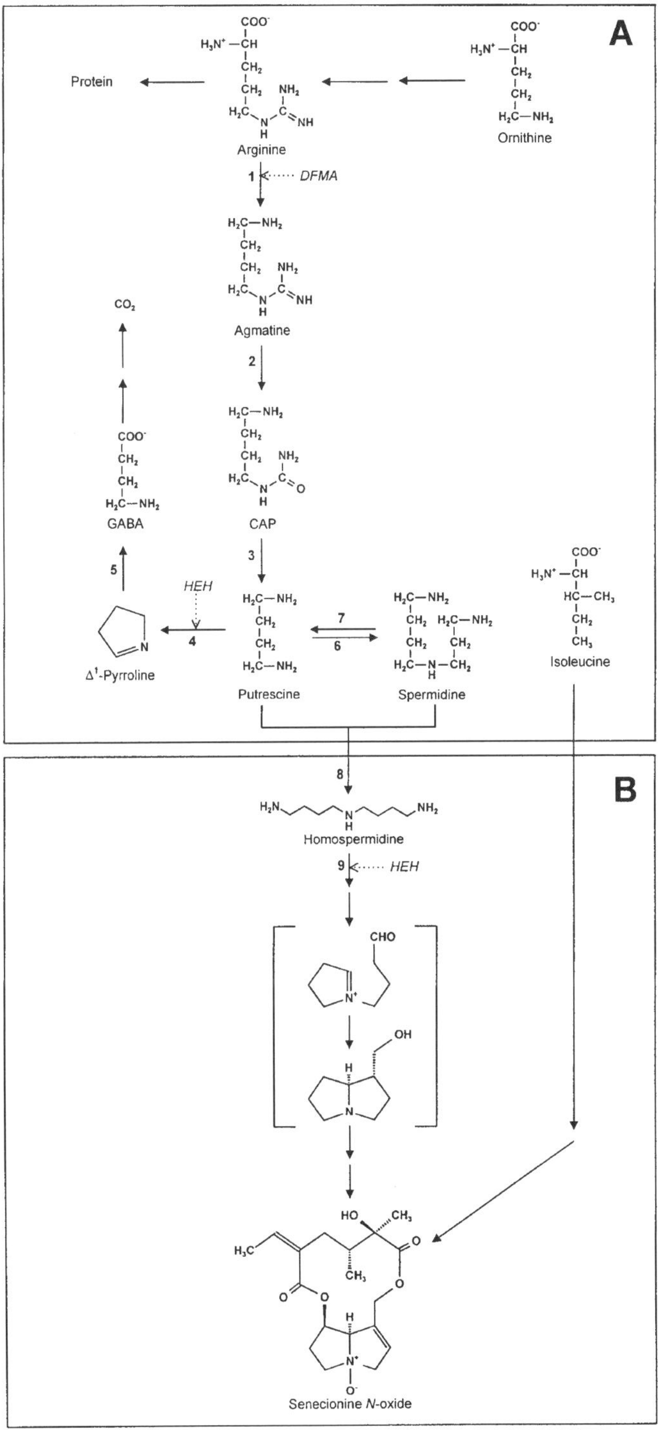
A
Protein
Arginine
Ornithine
1
DFMA
Agmatine
2
CO2
GABA
CAP
3
5
HEH
4
7
6
Δ1-Pyrroline
Putrescine
Spermidine
Isoleucine
B
8
Homospermidine
9
HEH
Senecionine N-oxide

tively labeled biogenetic precursors such as arginine, ornithine, putrescine, spermidine, and isoleucine yielded senecionine *N*-oxide as the only labeled biosynthetic product [23]. No biosynthetic intermediates were detected either with precursors of the necine base moiety or with isoleucine as a precursor of the necic acid moiety. In any case the alkaloids were found to be synthesized in the form of their polar *N*-oxides. The pre-alkaloid pathway (Fig. 2 A) which provides the ultimate alkaloid precursors was studied by feeding ^{14}C-labeled precursors and applying metabolic inhibitors. In the presence of α-difluoromethylarginine (DFMA), a suicide inhibitor of arginine decarboxylase, the incorporation of both arginine and ornithine was completely blocked, whereas the application of α-difluoromethylornithine (DFMO) did not affect the incorporation of the two amino acids [25]. *Senecio* root cultures are devoid of ornithine decarboxylase and the formation of putrescine from both ornithine and arginine proceeds via the arginine-agmatine path (Fig. 2). A participation of diamine oxidase in the formation of a symmetrical C_4-N-C_4 intermediate such as homospermidine from two molecules of putrescine as implicated by early tracer studies [17–19] could not be confirmed. Radioactively labeled Δ^1-pyrroline was not incorporated into senecionine *N*-oxide and in feeding experiments with [^{14}C]putrescine labeled homospermidine was trapped in the presence of β-hydroxyethylhydrazine (HEH), a potent inhibitor of diamine oxidase [26]. This confirmed the suggestion that a diamine oxidase may be involved in the oxidation of homospermidine [17–19] but makes participation of a diamine oxidase in the formation of homospermidine from putrescine unlikely (Fig. 2). Labeled homospermidine which accumulates in the presence of HEH is readily incorporated in senecionine *N*-oxide, when HEH inhibition is released. The only fate of homospermidine is its incorporation into the alkaloids. In contrast to putrescine and spermidine, this endogenously formed homospermidine does not undergo oxidative degradation [26]. Homospermidine is the first pathway-specific intermediate of PA biosynthesis (Fig. 2 B). This pathway leads to a product that, other than transformation of its backbone structure (Sect. 3.4), does not undergo turnover [24, 26]. Thus, formation of homospermidine is a key reaction in PA biosynthesis. It not only combines two building blocks of primary metabolism but also controls the substrate flow into the alkaloid-specific pathway. The focus of the subsequent work was on the elucidation of the biochemical and molecular mechanism of this key step.

The enzyme catalyzing the formation of homospermidine was identified as a homospermidine synthase [26], an enzyme previously described from two bacterial sources [27, 28] and seedlings of *Lathyrus sativus* [29]. None of these species produces PAs. One of the bacterial enzymes [28] has been purified. Bacterial homospermidine synthase was suggested to catalyze the following reaction:

$$\text{putrescine} + \text{putrescine} \xrightarrow{NAD^+} \text{homospermidine} + NH_3 \quad (1)$$

Homospermidine synthase (HSS) activity was detected in root cultures of four *Senecio* and two *Eupatorium* species. The enzyme was partially purified from root cultures of *E. cannabinum*, the source with the highest specific activity (i.e., 1.2 pmol · s^{-1} · mg^{-1}) in crude extracts [26]. A detailed characterization of the partially purified enzyme revealed that the enzyme also accepted spermidine as a substrate[30]. The same was demonstrated for the bacterial homospermidine synthase from *Rhodopseudomonas viridis* [30, 31]. The two enzymes are very similar in their molecular and kinetic properties and the following common reaction mechanism was suggested [30]: an aminobutyl group derived from putrescine or spermidine is transferred to a putrescine; ammonia or diaminopropane, respectively, are released; in a stoichiometric manner, NAD^+ functions as a hydride acceptor in the oxidative formation of the aminobutyl moiety and as a hydride donor in the reduction of the assumed imine intermediate (Fig. 3). Recently, the genes encoding bacterial homospermidine synthase from *Rhodopseudomonas viridis* and the plant enzyme from roots of *Senecio vernalis* were cloned, sequenced, and overexpressed in *Escherichia coli* [32, 33]. Surprisingly, no sequence homology was found between the two genes. The bacterial cDNA sequence has no detectable similarity with known sequences on the nucleotide or amino acid level [32]. On the contrary, the plant cDNA sequence showed a very high homology to genes encoding deoxyhypusine synthase, an enzyme that catalyzes the post-translational modification of the eukaryotic translation initiation factor eIF-5 A (Sect. 6). A reinvestigation of the kinetic properties of recombinant plant HSS revealed another essential difference between bacterial and plant HSS. In the plant enzyme spermidine was found to be the exclusive donor of the aminobutyl group that is transferred to putrescine [34]. The previous suggestion that either putrescine or spermidine may function as donor of the aminobutyl group [30] has to be revised for the plant enzyme. This has been overlooked in previous studies with the native enzyme because the very labile enzyme could only be stabilized in the presence of spermidine [26]. Even spermidine-free enzyme preparations may have contained nonspecifically bound spermidine in amounts sufficient to allow homospermidine formation with

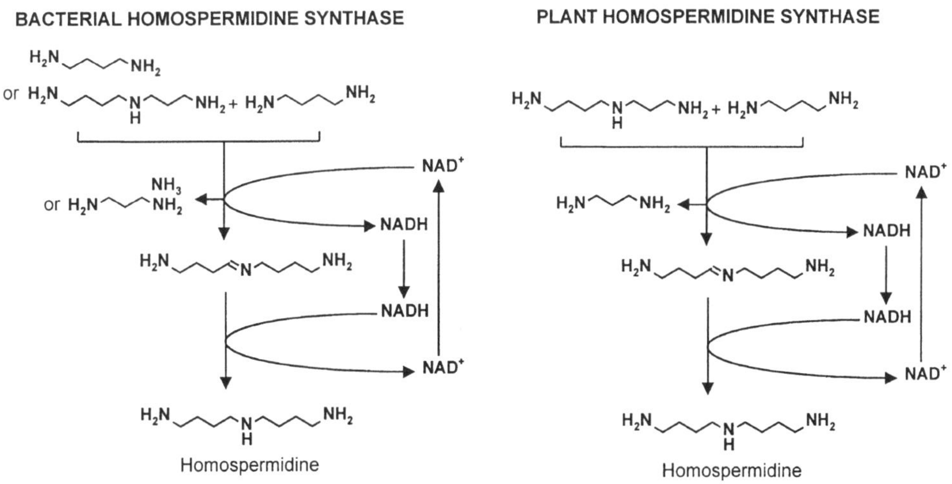

Fig. 3. Reactions catalyzed by bacterial and plant homospermidine synthase

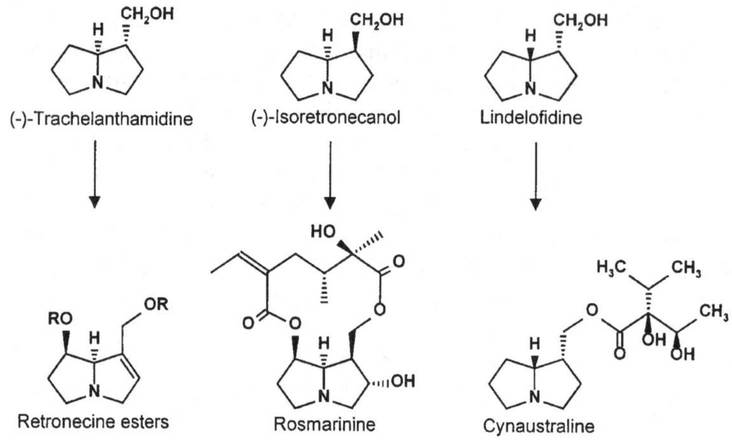

Fig. 4. Isomeric 1-hydroxymethylpyrrolizidines, the specific biogenetic precursors of necine esters. Most PAs are retronecine esters

putrescine as the only exogenous substrate. The specificity of homospermidine synthase for spermidine as aminobutyl donor is in accordance with the documented molecular relationship between the enzyme and deoxyhypusine synthase [33] (Sect. 6). In conclusion, homospermidine and thus the necine base moiety of PAs is half derived from spermidine (aminobutyl group) and half from putrescine.

The reaction step leading from homospermidine to the necine base moiety has not yet been characterized on an enzymatic level. Most likely an intermediate dialdehyde (Fig. 2) is formed by an enzyme with diamine oxidase activity as indicated by tracer and inhibitor studies [17–19, 26]. The cyclization of the carbon skeleton of the intermediate iminium ion may lead to four stereoisomers of 1-hydroxymethylpyrrolizidine; all of them have been identified as necines in naturally occurring PAs [1–3]. In isotope tracer experiments three isomers were shown to be formed as specific biosynthetic precursors of PAs (Fig. 4). (–)-Trachelanthamidine is the specific precursor of retronecine, by far the most widespread necine base of PAs [35, 36]. (–)-Isoretronecanol is specifically incorporated into the rare 2-hydroxylated PAs such as rosmarinine [36–39]. Orchids such as *Phalaenopsis* hybrids which contain both trachelanthamidine and isoretronecanol esters incorporate the two stereoisomeric necines into the respective alkaloids [16]. Cynaustraline one of the very rare examples of a PA with 8β-stereochemistry was shown to be formed via lindelofidine [(+)-isoretronecanol] in *Cynoglossum officinale* [40]. Epimerization of either trachelanthamidine or lindelofidine appears not to take place at the alkaloid level. Thus, the pathways probably diverge during cyclization of the iminium ions (Fig. 2), prior to the formation of the pyrrolizidine-alcohols, which in all the examples shown in Fig. 4 are efficiently incorporated into the PAs [41].

3.2 Biosynthesis of the Necic Acids

Complete biosynthetic labeling patterns have been obtained for the carbon skeletons of the necic acids of the two major classes of PAs, i.e., the senecionine type and the lycopsamine type. The biosynthesis of the senecic acid was studied by feeding [3,4-$^{13}C_2$]-2-aminobutanoic acid and [3,4-$^{2}H_5$]-2-aminobutanoic acid to hairy root cultures of *Senecio vulgaris* and plants of *S. pleistocephalus* [42]. The labeling patterns of the senecic acid moiety of senecionine isolated from *S. vulgaris* are shown in Fig. 5C. The same patterns were found in rosmarinine (1,2-dihydro-2-hydroxysenecionine) isolated from *S. pleistocephalus.* The labeling patterns confirm that in the two species senecic acid is formed from two units of 2-aminobutyric acid. The C-13 and C-20 positions of senecionine came from the C-3 of 2-aminobutanoic acid, and the C-19 and C-21 positions were originally C-4 of the amino acid. The labeling pattern is consistent with the metabolism of the labeled precursors via 2-oxobutanoic acid and the pathway of isoleucine biosynthesis, from which the necic synthesis branches off at the stage of 2-oxoisocaproic acid. The C_5-unit of senecic acid may be formed by oxidative decarboxylation of 2-oxoisocaproic acid (Fig. 5A, reaction 2).

A different technique was applied to establish the complete biosynthesis of 2-isopropylbutanoic acid, the carbon skeleton of the unique necic acids of the lycopsamine-type PAs [43]. Glucose containing 5% [U-$^{13}C_6$]glucose was fed to root cultures of *Eupatorium clematideum* which synthesize trachelanthamine as the major alkaloid. The ^{13}C-labeling patterns of the alkaloid and several biosynthetically related amino acids isolated from the cellular protein were quantitatively established by NMR techniques (Fig. 5D). The ^{13}C abundance of each carbon atom and the distribution of multiply ^{13}C labeled isotopomers were determined. The labeling patterns of central cellular metabolites, i.e., hydroxyethyl-TPP ("activated acetaldehyde"), pyruvate, and putrescine were reconstructed from the labeling pattern of amino acids on the basis of established mechanisms of amino acid biosynthesis in plants. The origin of the alkaloid carbon skeleton was then analyzed by a pattern-recognition approach using the labeling patterns of amino acids and central metabolites for comparison. This retrobiosynthetic approach has been successfully applied in biosynthetic studies of primary and secondary metabolites in a variety of organisms [44–46]. The ^{13}C-labeling pattern of the necic acid moiety of trachelanthamine is completely in accordance with the biosynthetic route outlined in Fig. 5B. Carbon atoms 1′, 2′, 5′, 6′, and 7′ reflect accurately the labeling pattern of valine and carbon atoms 3′ and 4′ reflect the labeling pattern of "activated acetaldehyde" (hydroxyethyl-TPP) (Fig. 5D). The labeling pattern of the necic acid is therefore best explained by the transfer of a hydroxyethyl moiety to the carbonyl group of the 2-oxoisovaleric acid yielding the branched C_7 carbon skeleton (Fig. 5B) which can be reduced to give the typical 2,3-dihydroxylated necic acids such as (+)-trachelanthic acid (2′*S*,3′*R*) or (–)-viridifloric acid (2′*S*,3′*S*). These two acids are the most prominent necic acids found in lycopsamine-type PAs [3]. This result confirms Crout's early idea [47] that the C_7-necic acids may be derived from valine and a two-carbon fragment possibly inserted as "active acetaldehyde".

From a biochemical point of view, the transfer of the hydroxethyl group to 2-oxoisovalerate (Fig. 5 B, reaction 3) should proceed analogously to its transfer to pyruvate and 2-oxobutyrate in valine and isoleucine biosynthesis (Fig. 5 A, B, reaction 1) which is catalyzed by acetolactate synthase. One just has to assume an extension of the substrate specificity of the enzyme [43].

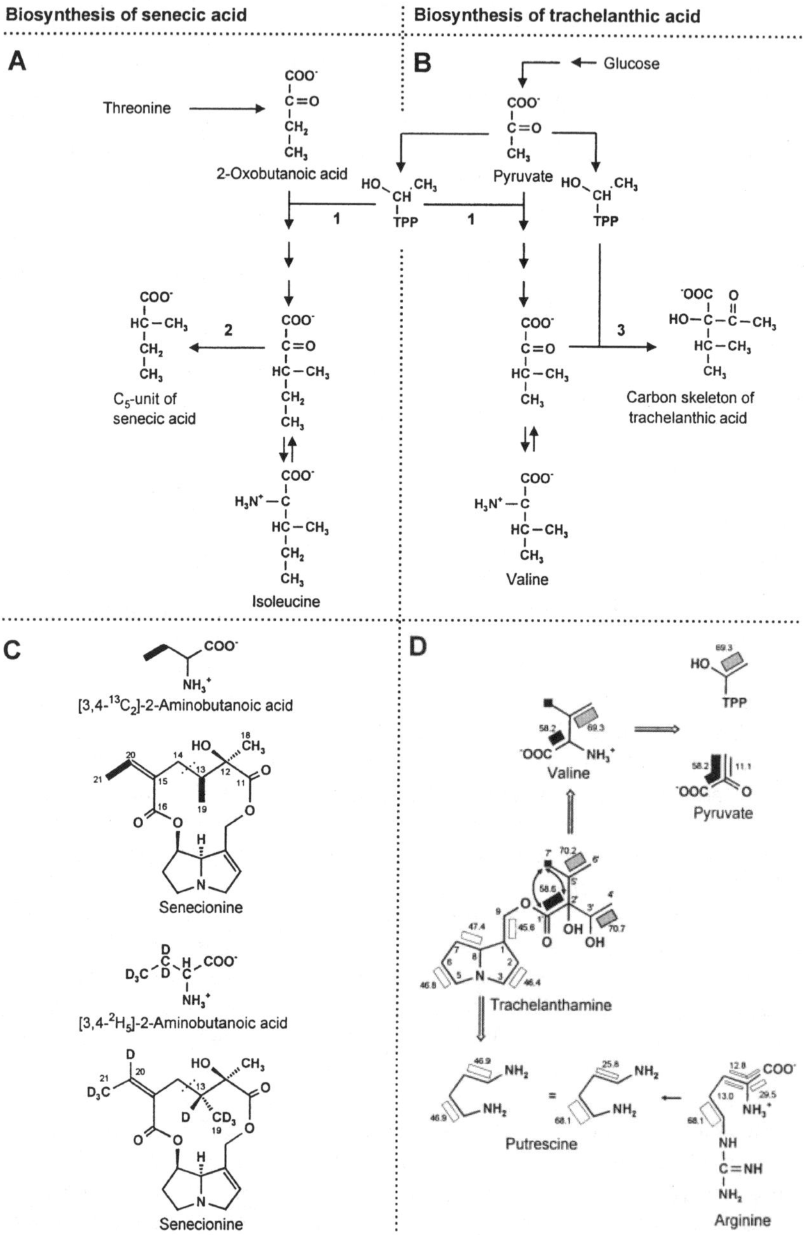

3.3
Site of Synthesis, Translocation and Storage of PAs

All PA-producing species of the family Asteraceae so far studied synthesize PAs in the roots [23, 25, 48]. In *Senecio* species which produce PAs of the senecionine type, senecionine *N*-oxide is the primary product of biosynthesis. This backbone structure is distributed within the plant and chemically modified to yield the species-specific PA pattern (Sect. 3.4). PAs are transported in the form of their polar *N*-oxides. Root-to-shoot translocation of PA *N*-oxides occurs exclusively via the phloem-path. This was demonstrated physiologically by stem-girdling experiments [48] as well as with the help of phloem-feeding aphids which store in their bodies PAs acquired with the phloem sap [49]. Long-distance transport via the living sieve-tubes requires specific phloem loading and unloading. Plants which do not produce PAs are incapable of translocating PA *N*-oxides via the phloem [48]. Within the shoots the PA *N*-oxides are channeled to the preferential sites of storage. PAs are found in all plant tissues, however, in a flowering *S. vulgaris* plant 60 to 80% of total PAs are located in the inflorescences (flower heads) and within the flower heads more than 90% are concentrated in the tubular flowers [50]. The subcellular localization of the alkaloids was studied with cell suspension cultures of various *Senecio* species. These cultures lost the ability to synthesize PAs but retained the ability to accumulate the alkaloids [51]. This PA accumulation is specific and cell cultures of plants that do not synthesize alkaloids are incapable of accumulating PAs. A membrane carrier was identified and characterized from *S. vulgaris* cell suspension cultures that specifically catalyzes the translocation of PAs via the tonoplast into the vacuole [52,53]. This carrier translocates both forms of the alkaloid, the tertiary alkaloid and its *N*-oxide, but storage of the *N*-oxide is more efficient than that of the free base. Vacuoles pre-filled with radioactively labeled PAs, easily lose the tertiary alkaloid during the isolation of vacuoles whereas the respective *N*-oxide is quantitatively retained within the vacuoles [53].

Fig. 5. Two approaches chosen to establish the biosynthesis of the necic acids of PAs of the senecionine type (**A**, **C**) and the lycopsamine type (**B**, **D**). **A** = Pathway of isoleucine biosynthesis; the C_5-unit of senecic acid branches off by oxidative decarboxylation of 2-oxoisocaproic acid (reaction 2). **B** = Pathway of valine biosynthesis; the carbon skeleton of trachelanthic acid is synthesized by transfer of an "activated acetaldehyde" to 2-oxoisovaleric acid (reaction 3). **C** = Labeling pattern of the necic acid moiety of senecionine isolated from a root culture of *Senecio vulgaris* after 14-days growth with ^{13}C- and ^{2}H-labeled aminobutanoic acids [42]. **D** = ^{13}C labeling pattern of trachelanthamine, valine and arginine from root culture of *Eupatorium clematideum* after 16-days growth with [U-$^{13}C_6$]glucose [43]. Isotopomers with 2 or 3 contiguous carbon atoms incorporated *en bloc* are indicated by *bold lines* (*squares*) adjacent to the structures. The *numbers* accompanying the lines indicate the fraction of the respective signal in % of the total signal intensity. The square widths reflect the relative contribution of the respective isotopomers. *Arrows* indicate multiple bond ^{13}C couplings. Isotopomer patterns which were transferred to trachelanthamine are differentially shaded. Due to the molecular C_2 symmetry of putrescine, its labeling pattern reflects the average of the contributing skeletal elements of arginine. The coupling patterns of pyruvate and hydroxyethyl-TPP were reconstructed retrobiosynthetically from the labeling patterns of valine on basis of established biosynthetic mechanisms. The labeling pattern of putrescine was predicted from that of arginine assuming established pathways

Not all PA-containing species synthesize the alkaloids in their roots. There is old evidence that *Crotalaria* species (Fabaceae) produce PAs in the shoots [54] and more recently it was shown that root cultures of *C. scassellatii* do not produce PAs [21]. Within the Boraginaceae the sites of PA biosynthesis may vary between species. *Heliotropium indicum* [16] and *H. spathulatum* [55] synthesize PAs in young shoots. *Symphytum officinale* synthesize PAs in the roots but not in shoots [16], whereas in *Cynoglossum officinale* both the roots and the shoots are able to synthesize PAs [16, 56]. In PA-producing orchids such as *Phalaenopsis* the alkaloids are synthesized in young root tips [16], whereas *Ipomoea hederifolia* (Convolvulaceae) again the shoots are the site of PA biosynthesis [57]. In any case PAs appear to be synthesized in very young growing tissues (i.e., root tips, shoots tips, and growing root cultures) [16, 24, 25].

3.4 Chemical Diversification of PA Backbone Structures

Senecionine *N*-oxides can be regarded as the backbone structure of many, if not most, macrocyclic PAs of the senecionine type [3]. Tracer studies with root cultures of *Senecio* species having different PA patterns, synthesize senecionine *N*-oxide as the first common alkaloid [21, 58]. To a limited extent roots are able to transform senecionine *N*-oxide into species-specific derivatives. These transformations proceed slowly and are separated in time and space from senecionine *N*-oxide synthesis [21, 24, 58]. In studies with five *Senecio* species shoots were found to be the major sites of species-specific alkaloid transformation [59]. In all species tested, leaves, stems, and inflorescences catalyze the transformation of senecionine *N*-oxide yielding the species-specific alkaloid patterns. The transformations proceed slowly and may vary between shoot organs in their efficiency and even specificity. All transformations established by tracer techniques represent specific reactions that proceed in a position-specific and stereoselective manner (Fig. 6). The stereospecificity and the fact that different *Senecio* species

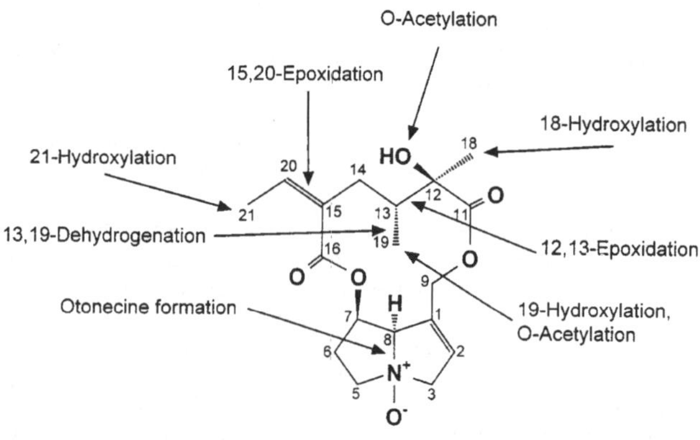

Fig. 6. Chemical diversification of the backbone structure (senecionine *N*-oxide) by species-specific transformations established for five *Senecio* species as documented in Table 1

Table 1. Species-specific transformations of radioactively labeled senecionine *N*-oxide in shoots of five *Senecio* species including two chemotypes. Nine steps of transformation create ten compounds and six species(population)-specific PA patterns [59]. The backbone structure of senecionine *N*-oxide is present in all species in proportions varying between trace-amounts and >50% of total alkaloids

Transformations	*S.vul*	*S.ver*	*S.ja*-A	*S.ja*-B	*S.eru*	*S.ina*
18-Hydroxylation	1	1				1
19-Hydroxylation				2	2	
21-Hydroxylation			3		3	
15,20-Epoxidation			4, 8			5, 10
12,13-Epoxidation				2	2	
12-*O*-Acetylation						5
19-*O*-Acetylation				6	6	
13,19-Dehydrogenation	7	7	7, 8	7		7
Otonecine formation		9				5, 10

S.vul = *S. vulgaris; S.ver* = *S. vernalis; S. ja*-A = *S. jacobaea* jacobine chemotype; *S.ja*-B = *S. jacobaea* erucifoline chemotype; *S.eru* = *S. erucifolius; S.ina* = *S. inaequidens.*
1 = retrorsine (1); **2** = erucifoline (2); **3** = eruciflorine (1); **4** = jacobine (1); **5** = florosenine (3); **6** = acetylerucifoline (2); **7** = seneciphylline (1); **8** = jacozine (2); **9** = senkirkine (1); **10** = otosenine (2). The numbers in parentheses give the steps of transformations needed to synthesize the compound.

transform senecionine *N*-oxide into unique products strongly indicate that the individual PA bouquets are brought about by genetically-controlled specific enzymes (Table 1). Transformation of senecionine *N*-oxide by indiscriminate enzyme activities as often postulated to be responsible for chemical diversification of secondary pathways [60] can be excluded. Most of the transformation products listed in Table 1 are formed from senecionine *N*-oxide by simple one-

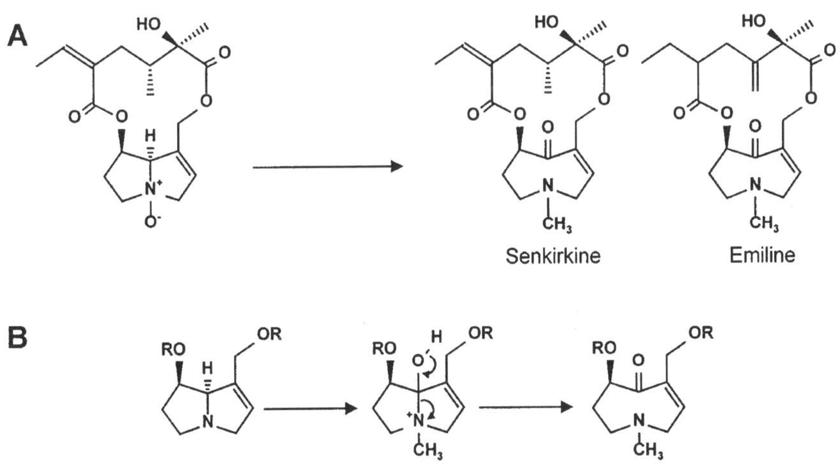

Fig. 7. Transformation of senecionine *N*-oxide into senkirkine [58] and emiline (**A**). Suggested mechanism for the conversion of the retronecine moiety into the otonecine moiety (**B**) [61]

step or two-step, rarely three-step, reactions. The only more complex transformation is the conversion of the retronecine moiety of senecionine *N*-oxide into the otonecine moiety of senkirkine. Root cultures of *S. vernalis* transform senecionine *N*-oxide into senkirkine (Fig. 7A) without occurrence of detectable intermediates [58]. In *Emilia flammea* retronecine is an efficient precursor of emiline (Fig. 7A) the major PA of the plant [61]. The formation of otonecine was suggested to proceed by hydroxylation at C-8 and *N*-methylation of a retronecine ester followed by ketone formation with concomitant cleavage of the C-N bond (Fig. 7B) [61]. The origin of the *N*-methyl group of senkirkine from *S*-adenosylmethionine has been demonstrated [62].

In comparison to the macrocyclic PAs of the senecionine type structural diversification of PAs of the lycopsamine type follows a different strategy. The backbone structure is represented by trachelanthamidine the simplest necine base (Fig. 2) esterified with stereoisomeric dihydroxylated 2-isopropylbutanoic acids (Sect. 3.2). The most abundant isomeric necic acids are (+)-trachelanthic acid (2′*S*, 3′*R* configuration) (Fig. 8) and (–)-viridifloric acid (2′*S*, 3′*S*). Necine esters with (–)-trachelanthic acid (2′*R*, 3′*S*) or (+)-viridifloric acid (2′*R*, 3′*R*) have been described but they are rare [3]. In PA-producing species of the family Boraginaceae [16, 56] and species of the tribe Eupatorieae [63] of the Asteraceae, diversification of the backbone structures is initiated by a common sequence of consecutive reactions:

Trachelanthamine *N*-oxide (2′*S*,3′*R*) → Supinine *N*-oxide → Intermedine *N*-oxide (7*R* OH) / Rinderine *N*-oxide (7*S* OH)

Viridiflorine *N*-oxide (2′*S*,3′*S*) → Amabiline *N*-oxide → Lycopsamine *N*-oxide (7*R* OH) / Echinatine *N*-oxide (7*S* OH)

(2)

The first step is the formation of the 1,2-double bond followed by a stereoselective hydroxylation at the C-7 converting the necine base moiety into either retronecine (7*R*) and heliotridine (7*S*) (Fig. 8). These structures may be further modified by introduction of a second necic acid (i.e., esterification at C-7) and additional modification of the necic acid moieties (i.e., hydroxylations, acetylations, etc.). The majority of the PAs of the lycopsamine type are monoesters or diesters of retronecine and heliotridine [3].

3.5 Plasticity and Variability of Individual Alkaloid Patterns

As discussed in the preceding section, the formation of unique species-specific PA patterns is achieved in two consecutive steps: (i) synthesis of the backbone structure and (ii) its structural diversification. From a biochemical point of view there are two additional important metabolic features: PAs do not undergo any degradation or turnover and they are slowly but steadily distributed within the plant. These features have been intensively studied with *Senecio* roots cultures

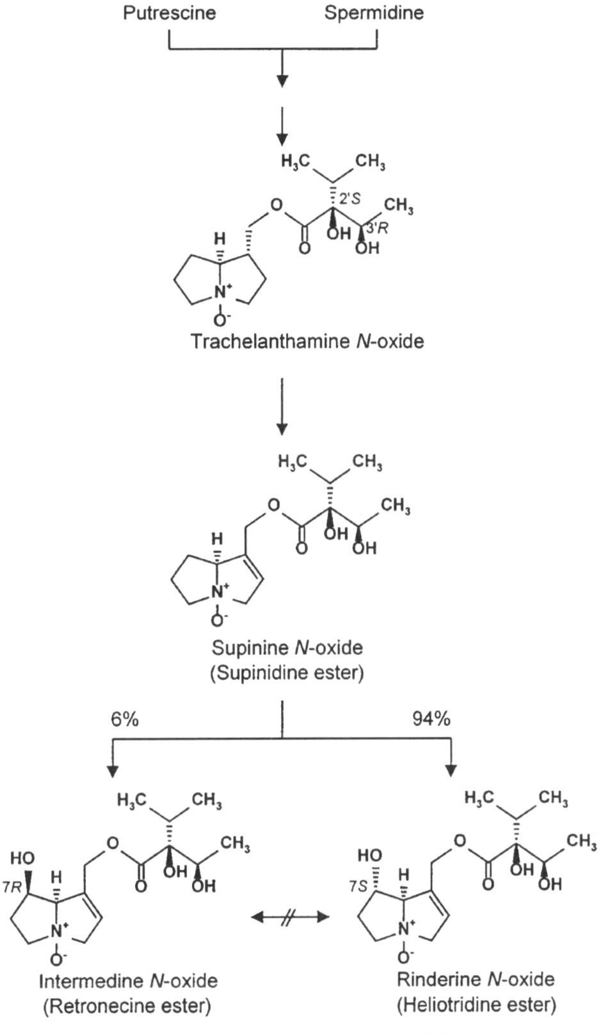

Fig. 8. Formation of the esters of supinidine, retronecine and heliotridine from trachelanthamine *N*-oxide, the backbone structure of the PAs in root cultures of *Cynoglossum officinale* [16]

[24, 25] and plants [59]. Radioactively labeled PAs obtained in pulse-feeding experiments with labeled senecionine *N*-oxide or distant biosynthetic precursors such as putrescine, remained entirely stable in root cultures over growth periods of up to three weeks [24]. In a typical experiment with *S. vulgaris* root cultures a "population" of radioactively labeled PAs was quantitatively recovered after 10 days of root growth, although during this period the root fresh weight and the amount of total alkaloid increased 6.1-fold and 6.4-fold, respectively [25]. In the same root cultures labeled PAs were found 10 days later distributed all over the root, even in tissues newly grown days after the labeled alkaloids had

been produced. The results obtained with the root cultures were confirmed in studies with whole plants. Again no substantial degradation or turnover was found over periods up to four weeks [59]. Moreover, PAs were slowly but steadily distributed within the plant independently of their "biogenetic age", i. e., labeled PAs were detected in roots, leaves, and flower heads that developed weeks after formation of the labeled alkaloids. Despite this spatial mobility PAs are not equally distributed within the plant; there are sites of preferential alkaloid accumulation as discussed in Sect. 3.3.

Taking all biochemical and physiological aspects of PA metabolism in a *Senecio* plant into account, we are facing an extremely variable and adaptive system, well suited to cope with the demands of a continuously changing environment (Fig. 9) [5]. The backbone structure of senecionine *N*-oxide is synthesized in the roots under a stringent metabolic regime that controls the plant's total level of PAs. This part of alkaloid metabolism appears to be highly conserved across species producing PAs of the senecionine type and thus apparently is under strong functional constraint. Genetic variability affecting the synthesis of senecionine *N*-oxide would alter PA biosynthesis quantitatively. Senecionine *N*-oxide, in turn, is subjected to chemical diversification to yield the alkaloid pattern unique to individual populations. This diversification brought about by specific one-step to three-step reactions (Table 1; Fig. 6) is highly plastic. It represents a dynamic equilibrium between a number of interacting processes such as the rate of supply of de novo synthesized backbone structure, the specificity and efficiency of the enzymatic transformations, continuous alkaloid distribution, and tissue specific alkaloid accumulation and vacuolar storage. Due to the lack of turnover, any genetic variation affecting this segment of alkaloid biosynthesis, par-

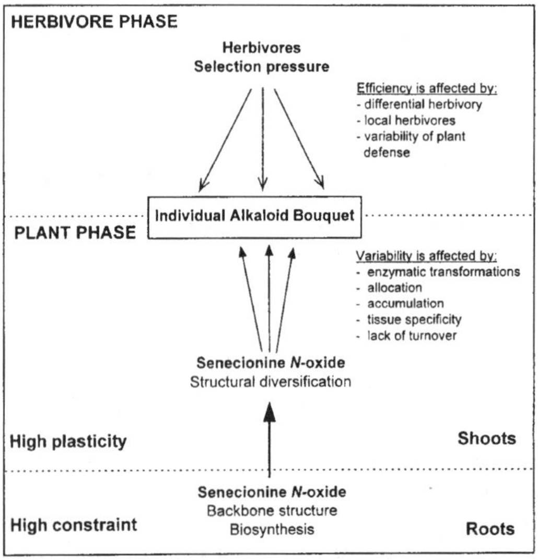

Fig. 9. General hypothesis of structural diversification of senecionine *N*-oxide under selection pressure of local herbivory

ticularly the transforming enzymes, would thus tend to modify the PA pattern without affecting its overall quantity. This view of strategy of PA biosynthesis predicts that the intraspecific PA patterns between populations should be extremely variable. This was recently demonstrated for *Senecio vulgaris* and *S. vernalis* [64] and *S. jacobaea* and *S. erucifolius* [65]. The relative abundance of the single components of PA patterns is highly variable between populations of species and their chemotypes. A genetic analysis of *S. jacobaea* populations revealed that the large phenotypic variation in total PAs contents and the relative amounts of the individual PAs are largely due to genetic variation [66]. Therefore, it was suggested that natural selection may undergo variation in PA concentration and composition in populations of *S. jacobaea*. Thus, biochemical and genetic evidence indicate that two discrete segments of PA biosynthesis exist with specific ecological roles: backbone supply and backbone derivatization [5, 59] (Fig. 9).

4 Biochemistry of Plant-Derived PAs in Specialized Insect Herbivores

Ecological evidence for a protective role of PAs in plant defense against herbivores is sparse. Young leaves of *Cynoglossum officinalis* which are rich in PAs are avoided by all generalist herbivores tested, whereas older leaves with low levels of PAs are eaten [67]. Almost all pure PAs were found to be strong feeding deterrents against many herbivorous insects [68–70]. It is well known that the majority of PAs causes serious diseases in domestic animals and humans through liver bioactivation [1, 6] (Sect. 5). Grazing animals, however, usually avoid PA plants unless there is shortage of other herbal food, apparently because of their deterrent taste [10]. The importance of such effects is difficult to assess empirically; often the most specific observation is that abundant generalist herbivores do not feed on PA plants. Convincing evidence favoring an important role of PAs in plant-herbivore interactions comes from specialized insect herbivores. A number of species from different insect taxa has developed adaptations not only to overcome the defensive barrier of PA-protected plants, but also to sequester and utilize these PAs for their own defense against predators [3, 10, 11]. Many butterflies and moths (Lepidoptera) and certain chrysomelid leaf beetles (Coleoptera) are known to store and utilize plant-acquired PAs for defensive purposes. All species which store plant acquired PAs in their bodies advertise their unpalatability to potential predators by conspicuous warning coloration (aposematic signals). Besides Lepidoptera and Coleoptera, individual species that sequester PAs from their host plants are known from the Orthoptera, i.e., the grasshopper *Zonocerus variegatus* [71–73] and the Homoptera, i.e., the aphid *Aphis jacobaeae* [49]. In the latter case ladybirds (*Coccinella septempunctata*) grazing on the aphids accumulate the plant-derived PAs from the aphids. The ingested PAs may reach 50% of the beetle's load of endogenously produced Coccinelidae alkaloids [49]. There is a flow of plant-produced defensive compounds through three trophic levels, i.e., from *Senecio* spp. via the herbivore to its predator. In the following sections mechanistic aspects of PA uptake, storage, metabolism in lepidopterans and leaf beetles are reviewed.

Alkaloid-storing Lepidoptera and leaf beetles are excellent models to study and compare the biochemical mechanisms developed most likely independently in the two taxa in the course of their adaptations to sequestration of PAs from plants.

4.1 Lepidoptera

Species which store plant-derived PAs are found among the moths of the families Arctiidae (12 genera) and butterflies of the subfamilies Danainae (4 genera) and Ithomiinae (38 genera) [3, 74].

4.1.1 *Sequestration and Defense*

PAs are taken up either by larvae, as in most Arctiidae, or by adults as in most Danainae and almost all neotropical Ithomiinae. All species that sequester PAs as larvae retain the stored PAs through to the adult stage. Larvae receive PAs with their herbal food whereas adult butterflies developed a special behavior. They are pharmacophagously attracted by PAs and ingest the alkaloid with nectar or, most frequently, from dead parts or withered twigs of PA plants. They wet the decaying plant material with fluid from their proboscis and then reimbibe it with the extracted PAs. By definition, pharmacophagous species search for plant secondary compounds (e.g., PAs), take them up, and utilize them for a purpose other than nutrient supply [10, 11, 75]. The average amounts of PAs sequestered pharmacophagously by Ithomiinae butterflies from their PA sources, i.e., Boraginaceae and *Eupatorium* sp. (Asteraceae), are 2–7% of dry weight but may reach 20% as in *Scada* spp. [74, 76]. All butterflies that sequester PAs seem to be well protected against predators. They are rejected, for instance, by the giant orb spider *Nephila clavipes* that cuts out any PA-storing butterfly unharmed from its web, whereas freshly emerged Ithomiinae butterflies that are still devoid of PAs are readily eaten [77–79]. Larvae of the arctiid *Utetheisa ornatrix* raised on a PA-free diet proved consistently palatable to wolf spiders, whereas larvae and adults containing PAs were rejected [80].

Pyrrolizidine alkaloids not only protect larvae and adults but also the eggs that are the perhaps most endangered stage in the life cycle of an insect. This is exemplified by the most complete study carried out by Eisner and Meinwald [81] with the arctiid moth *Utetheisa ornatrix*. Larvae of the species sequester PAs from their host plant *Crotalaria* (Fabaceae) and retain the alkaloids through metamorphosis into the adult stage. Both adult males and females provide PAs for egg protection. The female receives PAs from the male during copulation and transmits the alkaloids together with her own load to the eggs [82, 83]. The mating behavior is controlled by the PA-derived male courtship pheromone hydroxydanaidal (Sect. 4.1.3; Fig. 11), which signals the male's PA load to the female. The pheromone is emitted from a pair of androconial brush-like organs called coremata which are everted from the abdomen during close-range precopulatory interactions with the female. The amount of pheromone in the coremata of males correlates with his systemic level of PAs [83]. The female possesses antennal chemoreceptors highly sensitive to the pheromone [84]. Males reared on a laboratory diet devoid of PAs have fully developed coremata but produce no pheromone. These males are not able to stimulate female behavior and as a consequence are substantially less successful in courtship [85]. Field data indicate that *Utetheisa* males differ broadly in their systemic PA content, which ranged from 0 to 1010 μg per individual [83]. By mating preferentially with males endowed with the pheromone, females may ensure their acquisition of an alkaloidal gift. Laboratory studies with males raised on a diet containing monocrotaline and females raised on an usaramine-containing diet revealed that after mating and oviposition the eggs contained both PAs. The male's nuptial gift accounted for about 30% of total PAs [82]. Thus, both parents bestow upon eggs. The eggs are efficiently and long-term protected by their PA endowment (on average 0.8 μg per egg) against insect predators such as ants and coccinellid beetles [82, 86], but not against entomopathogenic fungi [87]. Similar biparental contributions to egg defense seem to exist in life cycles of the Asian arctiid *Creatonotos* [88], the queen butterfly *Danaus gilippus* (Danainae) [89], and most Ithomiinae [76, 90].

4.1.2
Mechanisms of PA Uptake and Storage

The uptake of dietary PAs, their storage and their tissue distribution have mainly been studied in larvae of arctiids such as the South Asian *Creatonotos transiens* and the European cinnabar moth *Tyria jacobaeae*. Tracer studies revealed that orally fed ^{14}C-labeled senecionine and its *N*-oxide are taken up with the same efficiency by larvae of the two species. With both tracers the ingested alkaloid accumulated exclusively as the *N*-oxide [91, 92]. Ingested PAs were found to be distributed in all larval tissues. The major proportion of total PAs was always associated with the integument [88, 93]. A comparison of the alkaloid concentrations revealed an almost equal distribution between hemolymph and integument, indicating a rather nonspecific tissue distribution of the accumulated PA *N*-oxides [92, 94]. In *Creatonotos* male and female adults translocate the stored PAs in a differential manner. In females, about 50–80% of total PA *N*-oxides are transferred to the ovaries, whereas in males ca. 40% are passed to the spermatophore and another 10–30% is allocated to the coremata where the alkaloids are utilized as pheromone precursor (Sects. 4.1.1 and 4.1.3) [88].

How the free bases and their *N*-oxides are taken up from the gut into the hemolymph? A first answer to this question was provided by uptake kinetics with *Spodoptera littoralis* (Noctuidae). Larvae of this lepidopteran feed freely on PA plants, they tolerate but do not sequester PAs. During feeding on radioactively labeled senecionine and its *N*-oxide substantial levels of senecionine, but never of the *N*-oxide, were temporarily built up in the hemolymph. The absorbed senecionine was rapidly excreted. If the two tracers were injected into the hemolymph, the two forms remained stable but again were rapidly excreted. These results suggest an efficient reduction of ingested PA *N*-oxide in the gut and passive uptake of the lipophilic free base [92]. The experiments were repeat-

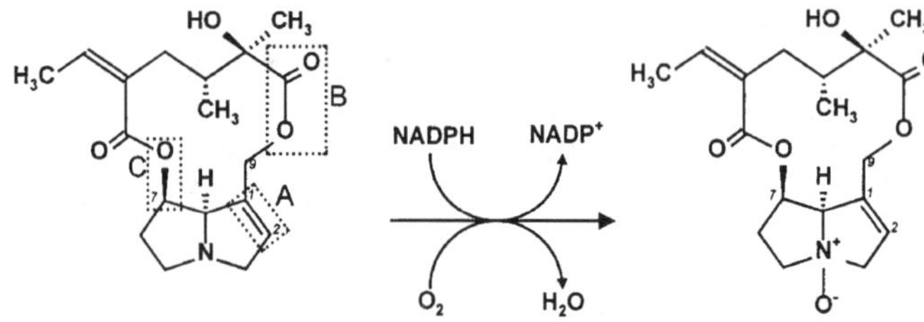

Fig. 10. Reaction catalyzed by senecionine *N*-oxygenase found in the hemolymph of arctiid larvae. The soluble enzyme is specific for PAs with the following structural features: *A*, 1,2-double bond; *B*, esterification of the allylic OH group at C-9; *C*, free or esterified OH group at C-7

ed with larvae of *C. transiens* applying doubly labeled [^{14}C]senecionine *N*-[^{18}O]oxide. FAB-MS analysis of the absorbed senecionine *N*-oxide isolated and purified from the hemolymph showed that the ^{18}O was completely replaced by ^{16}O. The same result was obtained with another arctiid (*Arctia caja*) and the taxonomically unrelated grasshopper *Zonocerus* [92]. Thus, in adapted PA-storing arctiids and *Zonocerus*, dietary PA *N*-oxide is reduced in the gut, taken up as free base, and again *N*-oxidized in the hemolymph. There is no evidence for a specific carrier-mediated uptake as claimed previously [95].

The enzyme that catalyzes the *N*-oxidation of senecionine was isolated from the hemolymph of *T. jacobaeae* larvae, purified, and characterized [92]. It is a soluble NADPH-dependent mixed function monooxygenase (senecionine *N*-oxygenase). The enzyme is a flavoprotein with a native M_r of 200,000 and a subunit weight of 51,000. It is specific for PAs with structural features that are essential for hepatotoxic or genotoxic PAs (Fig. 10). Twelve out of 20 PAs were found to be substrates. Non-toxic free necine bases, monohydroxylated and 1,2-saturated PAs as well as a great number of potential substrates other than PAs were not *N*-oxidized. The substrate specificity of senecionine *N*-oxygenase indicates that the enzyme must have been recruited and shaped from insect metabolism in the course of adaptation of the insect to PA sequestration. A comparison of the substrate specificity of the enzymes from three arctiids revealed a close relation to the host plant selection of the arctiids. The enzymes from the two generalists (i.e., *C. transiens* and *A. caja*) show a broader substrate efficiency than the enzyme from the specialist *T. jacobaeae* that feeds only on *S. jacobaea* [92].

4.1.3
Transformation of Plant-Acquired PAs

Specialized lepidopterans not only sequester PAs from their host plants but also transform them into insect-specific metabolites. The two best-studied examples are the PA-derived pheromones and the so-called insect PAs.

A number of male courtship pheromones are known to be synthesized from the necine base moiety of plant-acquired PAs by certain alkaloid sequestering Arctiidae, Danainae, and Ithomiinae [75, 96] (Fig. 11). The giant white danaine butterfly *Idea leuconoe* [97, 98] and most ithomiine butterflies even use lactones as pheromones derived from necic acids such as the (–)-viridifloric acid from

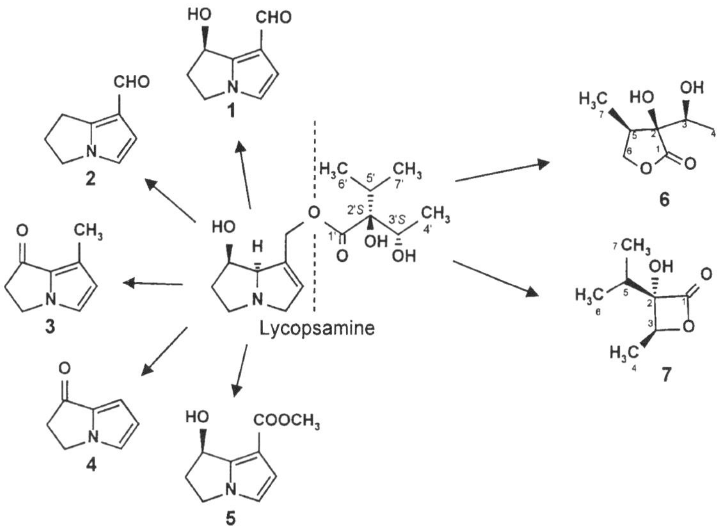

Fig. 11. Pyrrolizidine alkaloid-derived pheromones identified from the hairpencils (coremata) of PA-sequestering Lepidoptera. **1** = Hydroxydanaidal (Arctiidae, Danainae, Ithomiinae); **2** = danaidal (Arctiidae, Danainae); **3** = danaidone (Arctiidae, Danainae); **4** = nordanaidone (Danainae); **5** = methyl hydroxydanaidoate (Ithomiinae); **6** = ithomiolide A, derived from (–)-viridifloric acid (2'*S*, 3'*S*) (Ithomiinae); **7** = viridifloric β-lactone, (Danainae)

PAs of the lycopsamine type (Fig. 11) [96]. The behavior-modifying function of PA-derived male pheromones has so far only been established in a few cases (see Sect. 4.1.1). Male pheromone glands often contain complex mixtures of chemically diverse volatiles. The chemistry and behavioral biology of PA-derived pheromones produced in the androconial organs are discussed in detail in a number of excellent reviews [10, 11, 75, 96, 99].

Although it is well established that dietary PAs are transformed into the respective pheromones, until recently the biosynthetic sequence was obscure. Larvae of *Creatonotos transiens* easily convert monocrotaline with (*R*)-configuration at C-7 (Fig. 1) into the (7*R*)-configurated hydroxydanaidal (Fig. 12). Unexpectedly, feeding of the (7*S*)-configurated heliotrine also leads to hydroxydanaidal. Larvae fed with heliotrine always contain a mixture of the two 7-epimers, indicating that larvae are capable of epimerization at C-7 [100]. The whole biosynthetic sequence was established by feeding C-7 deuterated (7*S*)-heliotrine and (7*R*)-heliotrine (epiheliotrine) (Fig. 12) [101]. (7*S*)-Heliotrine is epimerized at C-7 by a stereoselective oxidation-reduction process with the C-7 ketone as an intermediate. As a consequence the deuterium is lost. Deuterated (7*R*)-heliotrine is incorporated into the pheromone with retention of configuration and the deuterium label is retained. Epimerization of heliotrine takes place already in the larval stage, whereas the further transformations (i.e., formation of the unstable pyrrole, followed by cleavage of the ester group, and the final oxidation) occur just prior to eclosion of the pupae. *Utetheisa ornatrix* which in nature appears to have dietary access to PAs with only the 7*R* configuration (e.g., monocrotaline)

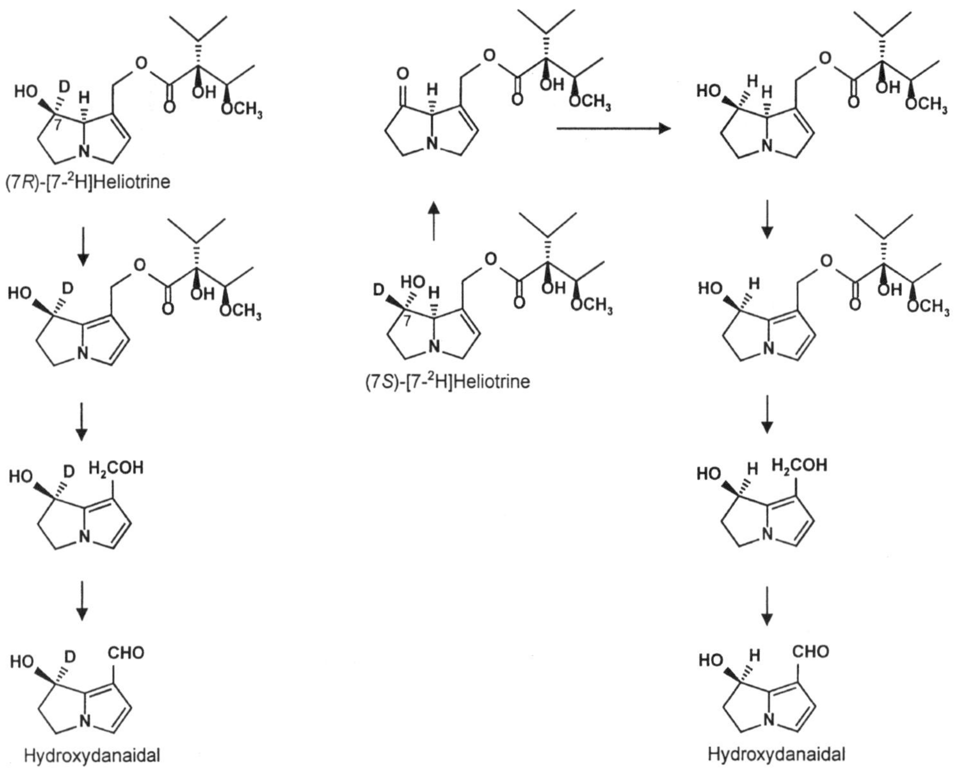

Fig. 12. Biosynthesis of the male courtship pheromone hydroxydanaidal from C-7 deuterated (*7R*)- and (*7S*)-heliotrine orally fed to larvae of *Creatonotos transiens*. Epimerization of the (*7S*)-configurated alkaloid occurs in the larvae, whereas the further steps occur just prior to eclosion of the pupae [101]

is not able to epimerize heliotrine [101]. Interestingly most Ithomiinae butterflies feeding on PAs of the lycopsamine type (Fig. 1) are able to epimerize PAs with (*S*)-configuration at C-7 (e.g., rinderine and echinatine) as well as those with (*R*)-configuration at the C-3′ of the necic acid moiety [102]. Thus, all Ithomiinae butterflies store lycopsamine as the major defensive PA, which has the correct stereochemistry for the formation of the necine base and necic acid-derived pheromones (Fig. 11).

Arctiids are able to esterify retronecine with insect-derived necic acids [91, 101, 103]. Creatonotine and isocreatonotine (Fig. 13, **3** and **4**) were isolated from adults of *C. transiens* which as larvae had received dietary ester alkaloids or retronecine [103]. These two major alkaloids were accompanied by small amounts of callimorphine and isocallimorphine (Fig. 13, **1** and **2**) and two further retronecine O9-esters, together with the respective O7-esters, as always (Fig. 13, **5**–**8**). None of these retronecine esters has been reported from plants so far. With the exception of callimorphine, the necic acid moieties of the new alkaloids are the α-hydroxy analogues of isoleucine, valine, and alanine, respectively. Since free

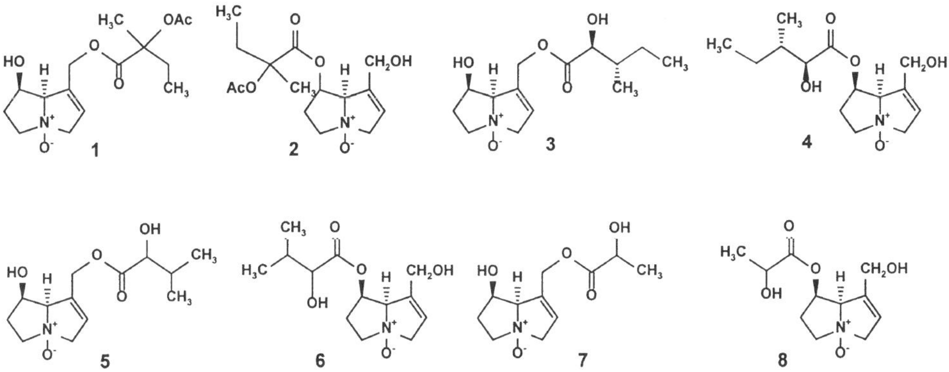

Fig. 13. Insect PAs synthesized in *Creatonotos transiens* by esterification of retronecine derived from plant ingested PAs and necic acids produced by the insect [103]. Biosynthesis is restricted to the early stages of pupation. All insect PAs are present as *N*-oxides: **1** and **2** = callimorphine and isocallimorphine; **3** and **4** = creatonotine and isocreatonotine; **5** and **6** = O9- and O7-(2-hydroxyisovaleryl)retronecine; **7** and **8** = O9- and O7-(2-hydroxypropionyl) retronecine

retronecine was found to be an efficient precursor of the creatonotines [101, 103] the esterification must have taken place in the insect. This was confirmed by tracer experiments with the arctiid *Tyria jacobaeae* known to produce considerable levels of callimorphine [104]. In pupae of *T. jacobaeae* which as larvae had received ^{14}C-labeled retronecine, up to 40% of the absorbed radioactivity was associated with callimorphine [91]. The radioactivity was restricted to the retronecine moiety of the alkaloid. The same experiment carried out with [^{14}C]isoleucine revealed callimorphine labeled only in the necic acid moiety. In *C. transiens* and *T. jacobaeae* the esterification of retronecine is restricted to the very early stages of pupation. Larvae and mature pupae are not able to esterify orally fed or injected retronecine [91, 103]. Moreover, in *C. transiens* retronecine appears not to be a direct precursor of hydroxydanaidal [101]. The free necine base needs to be esterified prior to pheromone formation. Feeding experiments with dietary monocrotaline revealed that in *C. transiens* the hydrolysis of monocrotaline occurred in the gut. Free retronecine is taken up together with intact monocrotaline. The retronecine is stored in larvae until re-esterification during the early stages of pupation. Insect-stored monocrotaline *N*-oxide remained stable through to the adult stage. Apparently no retronecine is formed from plant-acquired PAs that are stored in the insect's body [94]. Thus, the formation of insect PAs appears to be a kind of a salvage pathway by which lost dietary PAs are recycled into valuable defensive compounds and pheromone precursors. Insect PAs can account for a considerable proportion of the alkaloid load of an individual. Analysis of field-caught adults of *C. transiens* (Bali, Indonesia) revealed in about half of the individuals the creatonotines as predominant PAs (i.e., 50 to 100% of total PAs) [105].

4.2
Chrysomelid Leaf Beetles

Four out of nineteen subfamilies of the Chrysomelidae (leaf beetles) possess pronotal and elytral glands that produce and release defensive compounds [106]. So far, only a few species of the genus *Oreina* belonging to the subfamily Chrysomelinae are known to sequester PAs from their host plants [107]. Chemical defense in chrysomelines is primarily autogenous. Thus, *Oreina* species feeding on Asteraceae, tribe Senecioneae appear to have moved secondarily to host-derived defense [108].

4.2.1
Sequestration and Defense

Most *Oreina* leaf beetles known to live as specialist herbivores on host plants that produce a great variety of deterrent or toxic secondary compounds, synthesize their own defensive compounds such as cardenolides [106, 108]. Only five *Oreina* species (i.e., *O. cacaliae, O. speciosissima, O. frigida, O. elongata*, and *O. intricata*) feeding on the PA-containing Asteraceae *Adenostyles alliariae, Senecio nemorensis*, and *S. fuchsii* are able to sequester PAs from their host plants. *O. cacaliae* is the only species that has completely lost the ability to synthesize autogenously cardenolides (host-derived defensive strategy). The other four species are able to sequester PAs from their host plants but still retained the ability to synthesize cardenolides (mixed defensive strategy) [107, 109].

Sequestration of PAs by adult leaf beetles was first recognized when seneciphylline *N*-oxide was isolated from the defensive secretion of *O. cacaliae* [110]. Subsequently, several further PA *N*-oxides acquired from the host plant were identified in *Oreina* adults [109, 111, 112]. There are two different storage compartments, the exocrine defensive glands and the whole body. PA concentrations of the secretions, released by the gland cells, are in the range of 0.1 – 0.3 mol $\cdot$ l^{-1}, which is about 50- to 200-fold higher than the concentrations measured in the hemolymph or carcass [109, 112]. The absolute amount of PAs in adult individuals of *O. cacliae* is ca. 75 µg (body) and ca. 6 µg (secretions). This gives a total concentration (on a fresh weight basis) of ca. 0.1%. *Oreina* larvae that do not possess defensive glands store PAs in their bodies in concentrations comparable to those found in the bodies of adult beetles [113, 114]. Oviparous species like *O. elongata* provide their eggs not only with cardenolides (ca. 0.35 µg per egg, corresponding to 0.1% on dry weight basis) but also with high concentrations of PAs if feeding on a PA plant (4.5 µg per egg/ca. 1.7%) [113]. In the larviparous *O. cacaliae* no PAs were detected in neonate larvae just after birth before starting feeding. However, after one day feeding on *A. alliariae*, the larvae contained already PA concentrations similar to those found in late instar larvae [113].

The observed switch of *Oreina* leaf beetles from their original mode of defense (i.e., autogenous production of cardenolides) to host-derived defense (i.e., sequestration of PAs) intensified the discussion about the evolution of host plant affiliation and chemical defense in leaf beetles [108, 115]. Evolutionary scenarios favor the role of PAs as efficient defensive compound. Recently, it was demonstrated that PAs provide leaf beetles with a better protection from predation by birds than cardenolides [116]. Only 21% of adult *O. cacaliae* beetles, which contained in their defensive secretion a total of 4.2 μg PAs (0.17 M), were eaten by wild-caught redwinged blackbirds (*Adelaius phoeniceus*); the number rose to 36% if the defensive secretions has been physically removed. In contrast, 55% of *O. gloriosa* beetles, with a total of 25 μg cardenolides (0.27 M) in their defensive secretion, were eaten; the number rose to 95% if the secretions were removed. Thus, even a small quantity of highly concentrated PAs localized at the surface of the beetle's body afforded good protection against an avian predator.

4.2.2
PA Uptake, Storage and Transformation

Distribution studies with radioactively labeled senecionine *N*-oxide identified the hemolymph and to a lesser extent the integument as major storage compartments in adult beetles [117] and in larvae [114]. In adult beetles the hemolymph provides an important buffer for the refilling of the exocrine glands. Pulse-feeding experiments with labeled PA *N*-oxide revealed that the radioactivity is efficiently retained in the body over at least 25 days. During this period the distribution of radioactivity between body (ca. 80%) and secretion (ca. 20%) did not change significantly in undisturbed beetles. However, if the defensive secretions were removed there was an efficient refilling of the glands at the expense of labeled PAs in the body [118]. In contrast to Lepidoptera, the leaf beetle *O. cacaliae* takes up ingested dietary PAs as *N*-oxides. *O. cacaliae* is able to suppress the reduction of PAs in its gut [117]. *Chrysolina coerulans*, a close relative of *Oreina*, which feeds on mint plants and is not adapted to PAs, reduces ingested PA *N*-oxides like other herbivores (Sect. 5). Larvae and adults of *O. cacaliae* are not able to efficiently *N*-oxidize tertiary PAs [117, 119]. PA *N*-oxide uptake in adults proceeds in a two-step mechanism. The first step is the absorption of the ingested PA *N*-oxide from the gut into the hemolymph, the second step is the transfer from the hemolymph into the specialized gland cells that produce the defensive secretion. The two processes can be distinguished, the uptake into the hemolymph is rather nonspecific whereas the transfer into the glands is highly selective [109, 111, 112]. Thus, the *N*-oxides of the simple retronecine or platynecine monoesters, such as those occurring in *Senecio fuchsii*, are only detected in the body (Fig. 14). Bulgarsenine *N*-oxide, the major PA of *S. nemorensis*, occurs also only in the body whereas doronenine *N*-oxide, the respective 1,2-dehydro derivative (Fig. 14), is further transferred into the secretions. PAs that do not occur in the host plants such as monocrotaline *N*-oxide or heliotrine *N*-oxide are easily absorbed into the body but they are never transmitted to the defensive glands. The biochemical nature of the two carrier systems is still unknown [117]. The specific alkaloid transfer from the hemolymph into the glands requires an active component in order to maintain the concentration gradient between gland cells and hemolymph.

The only transformation product of a plant-acquired PA *N*-oxide is oreine *N*-oxide (Fig. 14, **12**) which could be identified in the defensive secretions and body of adults of *O. elongata* [112]. Oreine *N*-oxide most likely is formed by epoxida-

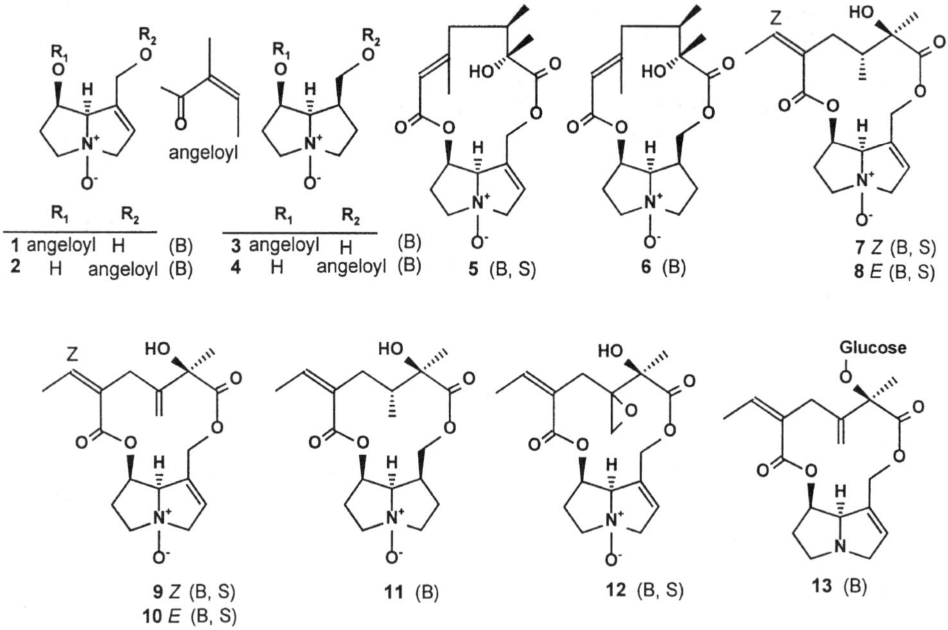

Fig. 14. PAs identified in the body (*B*) and/or defensive secretion (*S*) of alkaloid sequestering *Oreina* leaf beetles. The *N*-oxides of: **1** = 7-angeloylretronecine; **2** = 9-angeloylretronecine; **3** = 7-angeloylplatyphylline; **4** = 9-angeloylplatyphylline; **5** = doronenine; **6** = bulgarsenine; **7** = senecionine; **8** = integerrimine; **9** = seneciphylline; **10** = spartioidine; **11** = platyphylline; **12** = oreine; **13** = seneciphylline glucoside (detoxification product). The PAs are derived from the following host plants: **1–4**: *Senecio fuchsii*; **5, 6**: *Senecio nemorensis*; **7–11**: *Adenostyles alliariae*; **12**: insect metabolite of *Oreina elongata* [107, 112]

tion of seneciphylline *N*-oxide, the major PA *N*-oxide of *Adenostyles alliariae*. It is not clear whether this derivative is formed in the gut or in the beetle's body. Further reactions like *O*-demethylation of heliotrine (not present in the leaf beetles' host plants) yielding rinderine and deacetylation of *O*-acetylseneciphylline which, in contrast to seneciphylline, is not absorbed into the hemolymph, most likely take place in the gut. These reactions may be regarded as phase I detoxification reactions.

Adults and larvae of *O. cacaliae* transform up to 90% of ingested radioactively labeled senecionine free base into a polar metabolite that is stored in the body but is not transferred into the defensive glands of adults. After feeding of senecionine *N*-oxide the metabolite is only detectable in small amounts or traces. The polar metabolite was isolated from leaf beetles fed with seneciphylline and was identified by LC/MS as the respective *O*-glycoside. The sugar moiety is a hexose, most likely glucose (Fig. 14, **13**) [117]. The respective glycoside of senecionine was also detected. Glycosylation of PA free bases appears to be an essential detoxification reaction in leaf beetles which sequester PA *N*-oxides and which are not able to efficiently *N*-oxidize PA free bases.

5 Free Base vs. *N*-oxide – the Two Faces of PAs

A unique feature of PAs is that they exist predominantly as their *N*-oxides. In most plant taxa PAs are exclusively synthesized, distributed and stored as *N*-oxides. In these species the detection of PAs as free base must be regarded as an artifact. PA *N*-oxides are easily reduced during extraction and sample preparation in the presence of weak reducing agents such as cysteine or other thiols that are abundant in plant materials [23]. There are only a few exceptions. Seeds of the legume *Crotalaria scassellatii* contain ca 2% PAs that are exclusively stored as free bases, whereas all vegetative organs of the plant accumulate PA *N*-oxides [120]. During the beginning of seed germination the free bases are rapidly *N*-oxidized. Six-day-old seedlings already contain their PAs quantitatively as *N*-oxides [120]. The enzyme responsible for the *N*-oxidation, a particulate NADP-dependent monooxygenase, has been solubilized and characterized from *C. scassellatii* seedlings [121]. The enzyme has almost the same substrate specificity as insect senecionine *N*-oxygenase from arctiids [92] (see Sect. 4.1.2; Fig. 10). It has been suggested that the free base may be better suited for storage in the desiccated seed than the polar *N*-oxide which, in turn, is well adapted for storage in the vacuolated plant cell [121].

Herbivores that are not adapted to PA plants reduce necessarily ingested dietary PA *N*-oxides in their gut. The resulting pre-toxic lipophilic free base is then passively taken up into the body. This has been demonstrated for vertebrate [1] and insect [92, 117] herbivores. Adapted PA-sequestering Lepidoptera, Coleoptera, and Orthoptera (i.e., *Zonocerus*) store PAs in their bodies exclusively as *N*-oxides (Sect. 4). The toxicity of PAs in vertebrates, particularly in mammals, has been intensively studied. It is now well established that the hepatotoxic and pneumotoxic effects of PAs are caused by the free bases which are converted into highly reactive pyrroles [1, 122, 123]. In vertebrates this bioactivation is catalyzed by microsomal cytochrome P450 oxidases (EC 1.14.14.1) which are part of xenobiotic-metabolism (Fig. 15). Insects that have a similar xenobiotic metabolism with microsomal cytochrome P450 enzymes [124, 125] should be affected by PAs in the same way as vertebrates. Recent studies confirmed the genotoxicity of PAs in the *Drosophila* wing spot test [126]. In vertebrates and insects only PAs with the following structural features are potentially toxic, i.e., can be transformed into reactive pyrroles: (i) a necine base with a 1,2-double bond, (ii) esterification of the allylic OH group at C-9, and (iii) a free or esterified second OH group. These are exactly the features that were found to be essential for the substrates of insect senecionine *N*-oxygenase [92] isolated from PA adapted arctiids (Fig. 10) (Sect. 4.1.2).

N-Oxidation of the free base is assumed to be the most important mechanism of PA detoxification in vertebrates [127]. *N*-oxidation converts the potentially toxic free base into a derivative that no longer can be transformed into an alkylating toxin. Guinea pigs, for instance, possess a microsomal multisubstrate flavin monooxygenase (EC 1.14.13.8) which efficiently *N*-oxidizes absorbed PAs (Fig. 15). This is the reason for the high resistance of guinea pigs to PA poisoning [128]. On the other hand, rats that are highly susceptible to PA intoxication have

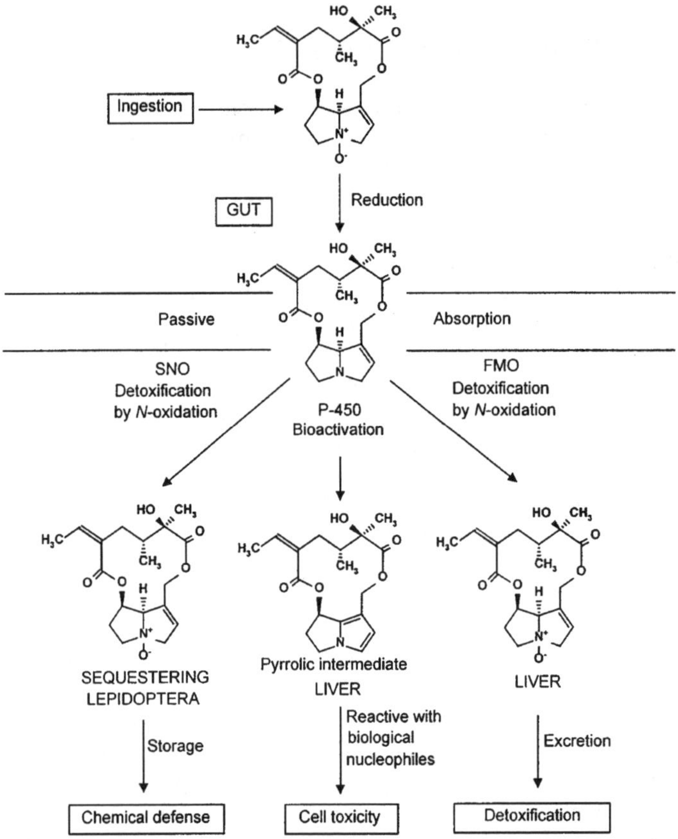

Fig. 15. Fate of PA *N*-oxides ingested by vertebrates and insects. Following reduction of the *N*-oxide in the gut and passive uptake of the alkaloid free base, bioactivation occurs in all organisms which possess cytochrome P450 enzymes. Detoxification by *N*-oxidation is possible in vertebrates with an efficient multisubstrate flavin monooxygenase (FMO). Specialized herbivorous lepidopterans apply the same strategy; they developed a substrate specific enzyme, senecionine *N*-oxygenase (SNO) and keep the non-toxic PA-*N*-oxides for their own benefit

very low *N*-oxidation activity [129]. Thus, *N*-oxidation of PAs by a specific enzyme in sequestering Lepidoptera appears to be an essential detoxification mechanism and prerequisite for PA storage (Fig. 15). Leaf beetles store PAs as *N*-oxides but they only possess an insufficient ability to *N*-oxidize the free base (Sect. 4.2.2). Instead these beetles are able to suppress PA *N*-oxide reduction in the gut and to take up the ingested *N*-oxides. Moreover, any PA present as free base is efficiently converted into the respective *O*-glucoside that is assumed to be a detoxification product [117]. Thus, PA-sequestering arctiids and leaf beetles developed completely different biochemical strategies to guarantee a safe storage of plant-acquired PAs in their bodies (Fig. 16).

The cytotoxic effects of PAs through bioactivation of the free base seem to be the predominant if not the only toxic effect of PAs as defensive compounds.

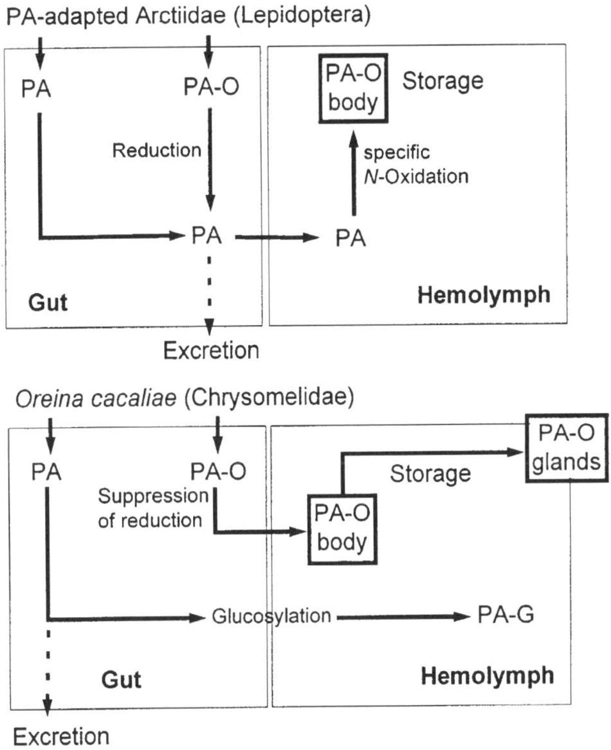

Fig. 16. Strategies of uptake, metabolism, and maintenance of PA *N*-oxides (PA-O) and the respective free bases (PA) by alkaloid-sequestering arctiids (top) and leaf-beetles (bottom). PA-G = PA glucoside. The site of PA glycosylation (i.e. gut or hemolymph) is still unknown

In addition to these effects, neurotoxic effects of certain PAs have been demonstrated in in-vitro receptor assays [130]. However, the relevance of these rather sporadic effects in vivo needs clarification. In a broad screening in which the interaction of many compounds from different classes of alkaloids with various molecular targets were tested, in almost all cases PAs were found to be inactive [131]. There are a few reports that indicate sporadic and weak (i.e., $EC_{50} > 0.5\ mg \cdot ml^{-1}$) antibacterial and antifungal activity [132–134] of PAs. These effects do not indicate an efficient antimicrobial potential of PAs. In this context it should be recalled that in plants PAs are constitutively formed; they seem not to be induced by wounding, herbivorous or microbial attack [135–137].

The unique feature of PAs is that they exist in two interchangeable forms: the polar, salt-like, non-toxic *N*-oxide and the more lipophilic, potentially toxic free base (Fig. 17). As long as cells or organisms are able to maintain PAs in the *N*-oxide state they are safe, however when the *N*-oxides become reduced, a pro-toxin is produced. These "two faces of PAs" are utilized by sequestering insects, they store PAs with the "good face" (*N*-oxide) and upon predation the PAs

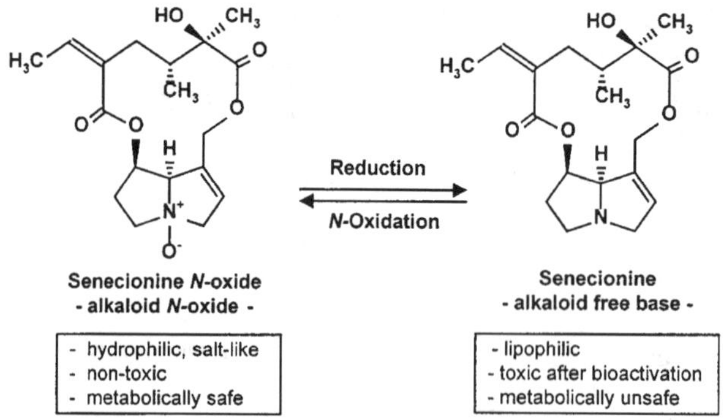

Fig. 17. The two faces of PAs. The non-toxic PA *N*-oxide is easily converted into the pro-toxic free base in the gut of any herbivore or predator

with the "bad face" (free base) are necessarily generated in the predator's gut. Possibly this is also true for PA-producing plants that store the potentially toxic PAs (e.g., retronecine esters) as *N*-oxides.

6
Phylogenetic Origin of PAs – First Molecular Evidence

From a phylogenetic point of view PAs appear to be a relatively young class of secondary compounds. They are not known to occur outside the angiosperms. Within the angiosperms PAs are found quite scattered and often solitarily in taxonomically unrelated taxa [3] (Sect. 2). Where do PAs come from? Recently obtained molecular data on homospermidine synthase, the first pathway-specific enzyme in PA biosynthesis (Sect. 3.1), provide first answers to this question [33]. The amino acid sequence of homospermidine synthase from *Senecio vernalis* roots has a surprisingly high sequence homology to deoxyhypusine synthase, an enzyme engaged in a biochemical area that has absolutely no connection to alkaloid biosynthesis. Deoxyhypusine synthase catalyzes the first step of a post-translational activation of the eukaryotic translation initiation factor 5A (eIF-5A) [138, 139]. Although the precise cellular function of eIF-5A is not known, it has been shown that eIF-5A and its post-translational activation is vital for yeast growth and thus most likely for proliferation of any eukaryotic cell [140–142]. eIF-5A and the two enzymes needed for its activation (i.e., in addition to deoxyhypusine synthase, a deoxyhypusine hydroxylase) are assumed to be ubiquitously existent in all eukaryotes and Archaea but not in Bacteria [139].

The relationship between deoxyhypusine synthase and homospermidine synthase becomes also evident by a comparison of their reactions. Deoxyhypusine synthase transfers the aminobutyl moiety of spermidine to the ε-amino group of a single specific lysine residue in the eIF-5A (Fig. 18). Homospermidine synthase catalyzes exactly the same reaction but uses putrescine instead of the eIF-5A as acceptor for the aminobutyl moiety (Fig. 18). All other mechanistic

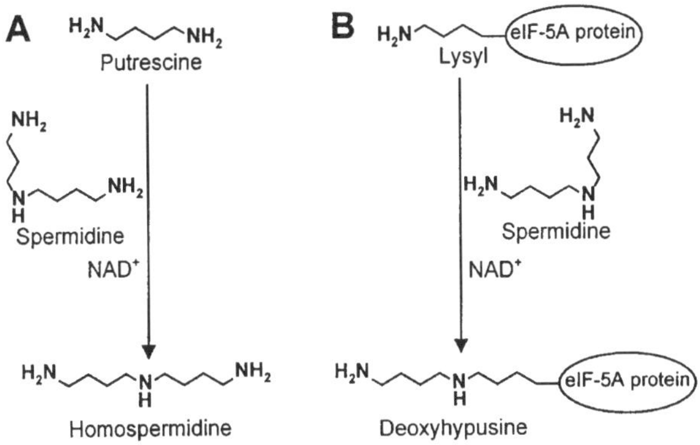

Fig. 18. Homospermidine synthase (**A**) and deoxyhypusine synthase (**B**) catalyze analogous reactions. The two enzymes transfer an aminobutyl moiety from spermidine to a primary amino group of putrescine and a proteinous lysyl residue of a protein, respectively

details of the reaction catalyzed by deoxyhypusine synthase are identical with the reaction catalyzed by homospermidine synthase (Sect. 3.1; Fig. 3).

Deoxyhypusine synthase has been characterized from different sources [143–145]. The enzyme is a homotetramer and the amino acid sequence of the enzyme appears to be highly conserved. The alignment of the sequences of the enzymes from four species (i.e., human, yeast, *Neurospora*, *Methanococcus*) [138, 139, 143–145] are shown in Fig. 19 in comparison to the sequences of the two recently cloned plant enzymes (i.e., from young shoots of *Senecio vernalis* [33] and *Nicotiana tabacum* [146]) and homospermidine synthase from root cultures of *S. vernalis* [33]. The two plant deoxyhypusine synthases, which for the first time have been cloned and functionally characterized from plant sources, fit precisely into the alignment of the highly conserved sequences of the enzymes from very distant sources. Surprisingly, the sequence of homospermidine synthase fits so well into the alignment that it is impossible to find sequence motives that allow us to distinguish between the two enzymes. Substrate kinetics revealed that homospermidine synthase does not accept eIF-5A as substrate but deoxyhypusine synthase catalyzes aminobutyl transfer reaction with both substrates eIF-5A and putrescine. Moreover, it catalyzes the formation of homospermidine with a specific activity comparable to that of homospermidine synthases [147]. Since homospermidine is a rare polyamine that is only occasionally found in plants [148, 149] its formation by indiscriminate activity of ubiquitous deoxyhypusine synthase seems not to be realized in vivo. However, one may speculate that the origin of homospermidine synthase from deoxyhypusine synthase could have occurred by gene duplication and loss of its major activity. Since the ability to synthesize homospermidine is a prerequisite for PAs biosynthesis, such an event could have happened independently in the unrelated taxa with PA-producing species. We now have the molecular tools to verify this hypothesis.

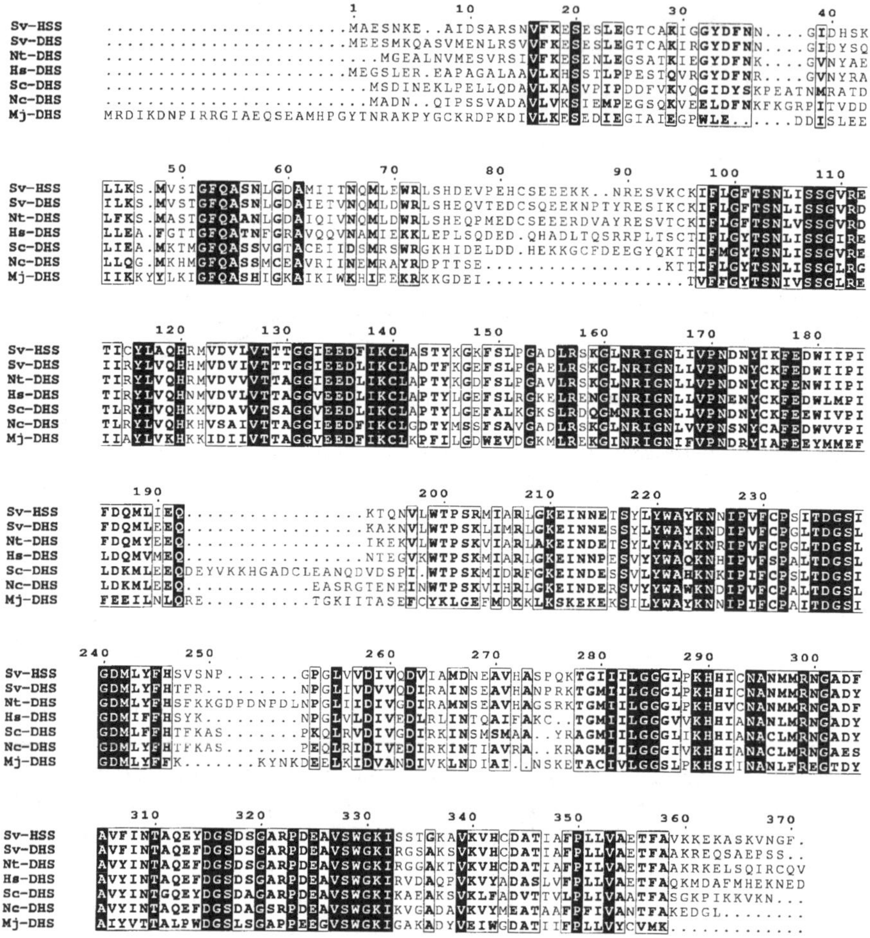

Fig. 19. Alignment of homospermidine synthase (HSS) amino acid sequence from *Senecio vernalis* (Sv) in comparison to deoxyhypusine synthase amino acid sequences from six different species. Sv, *Senecio vernalis*; Nt, *Nicotiana tabacum*; Hs, *Homo sapiens*; Sc, *Saccharomyces cerevisiae*; Nc, *Neurospora crassa*; Mj, *Methanococcus jannaschii*. Black boxes denote identical amino acid residues in all seven sequences, framed boxes conservative replacement

7 Perspectives

The PAs reviewed here are just one selected example out of hundreds of classes of secondary compounds. As a group of biologically active molecules assumed to be shaped during evolution of plant-herbivore interactions, PAs may have model character for other classes of secondary compounds and other types of chemical interactions between the plants and their environment. Plant secondary metabolism harbors an inexhaustible diversity of chemical products, many

– possibly most – of which were functionally shaped during evolution for various purposes. As exemplified by the PAs, these molecules appear to be highly valuable for specialized insects which utilize them for their own benefit. In the same sense, humans have learnt to use various natural products, both empirically in traditional medicine and in a more rational sense as leads for the design of novel biological active compounds valuable as medical drugs, herbicides, or pesticides. Plant secondary metabolism can be regarded as the functional level of the various interactions of the plant with its environment. These aspects are the subject of the rapidly growing field of *chemical ecology* that addresses the various interactions between organisms and their biotic and abiotic environment mediated by natural products. Chemical Ecology is highly interdisciplinary, combining expertise from chemistry, plant science, entomology, microbiology, ecology, and evolutionary biology. In general, future research in natural product biosynthesis should be part of an interdisciplinary approach joining mechanistic, functional, and evolutionary aspects.

In the field of PAs the main focus of future biosynthetic research may concern the following major areas: (i) biochemical and genetic basis of chemical diversification of alkaloid backbone structures; (ii) cellular compartmentation and regulation of the biosynthetic key enzymes such as homospermidine synthase; (iii) phylogenetic origin and relationship of the alkaloid pathways existing in unrelated taxa; (iv) detailed biochemical and molecular elucidation of the mechanistic handling of PAs in specialized insect herbivores; (v) evolutionary aspects of the adaptation of insect herbivores to PA plants. The progress in all these fields will, without exception, greatly depend on the successful introduction of molecular biology into the field. Although molecular techniques have already afforded great progress in some areas of alkaloid research [150, 151], we are still at the beginning. The elucidation of biosynthetic sequences at the gene level will permit not only a deeper understanding of the integration of a secondary pathway into plant metabolism and its regulation, but will also provide the tools needed to answer questions concerning the origin and evolution of the pathway and its products.

Acknowledgements. The senior author thanks all his friends and colleagues in the various areas of PA research for fruitful cooperation and stimulating discussion, particularly Jacques M Pasteels (Brussels), Martine Rahier (Neuchâtel), Keath S Brown and José Roberto Trigo (Campinas), Eddy van der Meijden and Klass Vrieling (Leiden), Michael Boppré (Freiburg), and many enthusiastic coworkers and doctoral students. Work in the authors' laboratory was supported by grants from the Deutsche Forschungsgemeinschaft and Fonds der Chemischen Industrie.

8
References

1. Mattocks AR (1986) Chemistry and toxicology of pyrrolizidine alkaloids, Academic Press, London
2. Rizk AFM (1991) Naturally occurring pyrrolizidine alkaloids, CRC Press, Boca Raton
3. Hartmann T, Witte L (1995) Pyrrolizidine alkaloids: chemical, biological and chemo-ecological aspects. In: Pelletier SW (ed) Alkaloids: chemical and biological perspectives, vol 9. Pergamon Press, Oxford, p 1553

4. Hartmann T (1996) Entomol exp appl 80:177
5. Hartmann T (1999) Planta 207:483
6. Röder E (1995) Pharmazie 50:83
7. Robins DJ (1989) Chem Soc Rev 18:375
8. Robins DJ (1991) Experientia 47:1118
9. Robins DJ (1995) The Alkaloids 46:1
10. Boppré M (1986) Naturwissenschaften 73:17
11. Schneider D (1987) The strange fate of pyrrolizidine alkaloids. In: Chapman RF, Bernays EA, Stoffolano JG (eds) Perspectives in chemoreception and behavior. Springer, Berlin Heidelberg New York, 132
12. Hartmann T (1991) Alkaloids. In: Rosenthal A, Berenbaum MR (eds) Herbivores: their interactions with secondary plant metabolites, 2nd edn, vol 1. Academic Press, San Diego, p 79
13. Culvenor CCJ (1978) Bot Notiser 131:473
14. Jennett-Siems K, Kaloga M, Eich E (1993) Phytochemistry 34:437
15. Jennett-Siems K, Schimming T, Kaloga M, Eich E, Siems K, Gupta MP, Witte L, Hartmann T (1998) Phytochemistry 47:1551
16. Frölich C (1996) doctoral thesis, Technical University of Braunschweig (Germany)
17. Robins DJ (1989) Chem Soc Rev 18:375
18. Robins DJ (1991) Experientia 47:1118 (1991)
19. Spenser ID (1985) Pure & Appl Chem 57:453
20. Hartmann T (1988) In: Constabel F, Vasil IK (eds) Cell cultures and somatic cell genetics of plants, vol 5. Academic Press, New York, p 277
21. Hartmann T (1994) In: Bajaj YPS (ed) Biotechnology in agriculture and forestry, vol 26. Medicinal and aromatic plants. Springer, Berlin Heidelberg New York, p 339
22. Robins RJ (1998) Nat Prod Res 15:549
23. Hartmann T, Toppel G (1987) Phytochemistry 26:1639
24. Sander, H, Hartmann T (1989) Plant Cell Tissue Organ Cult 18:19
25. Hartmann T, Sander H, Adolph R, Toppel G (1988) Planta 175:82
26. Böttcher F, Adolph R, Hartmann T (1993) Phytochemistry 32:679
27. Tait GH (1979) Biochem Soc Trans 7:199
28. Yamamoto S, Nagata S, Kusaba K (1993) J Biochem (Tokyo) 114:45
29. Srivenugopal KS, Adiga PR (1980) Biochem J 190:461
30. Böttcher F, Ober D, Hartmann T (1994) Can J Chem 72:80
31. Ober D, Tholl D, Martin W, Hartmann T (1996) J Gen Appl Microbiol 42:411
32. Tholl D, Ober D, Martin W, Kellermann J, Hartmann T (1996) Eur J Biochem 240:373
33. Ober D, Hartmann T (1999) Proc Natl Acad Sci USA (in press)
34. Ober D, Harms R, Hartmann T (unpublished results)
35. Rana J, Leete E (1985) J Chem Soc Chem Commun 1742
36. Kunec EK, Robins DJ (1986) J Chem Soc Chem Commun 250
37. Kunec EK, Robins DJ (1989) J Chem Soc Perkin Trans I 1437
38. Kelly HA, Robins DJ (1987) J Chem Soc Perkin I 2195
39. Kelly HA, Robins DJ (1987) J Chem Soc Perkin Trans I 177
40. Hagan DB, Robins DJ (1990) J Chem Res (S) 292
41. Denholm AA, Kelly HA, Robins DJ (1991) J Chem Soc Perkin Trans I 2003
42. Stirling IR, Freer IKA, Robins DJ (1997) J Chem Perkin Trans I 677
43. Weber S, Eisenreich W, Bacher A, Hartmann T (1999) Phytochemistry 50:1005
44. Werner I, Bacher A, Eisenreich W (1997) J Biol Chem 272:25474
45. Eisenreich W, Sagner S, Zenk MH, Bacher A (1997) Tetrahedron Lett 38:3889
46. Eisenreich W, Menhard B, Hylands PJ, Zenk MH, Bacher A (1996) Proc Natl Acad Sci USA 93:6431
47. Crout DHG (1966) J Chem Soc (C) 1968
48. Hartmann T, Ehmke A, Eilert U, von Borstel K, Theuring C (1989) Planta 177:98
49. Witte L, Ehmke A, Hartmann T (1990) Naturwissenschaften 77:540
50. Hartmann T, Zimmer M (1986) J Plant Physiol 122:67

51. von Borstel K, Hartmann T (1986) Plant Cell Rep 5:39
52. Ehmke A, von Borstel K, Hartmann T (1987) In: Marin B (ed) Plant vacuoles, their importance in solute compartmentation in cells and their application in plant biotechnology. Plenum Press, New York, p 301
53. Ehmke A, von Borstel K, Hartmann T (1988) Planta 176:83
54. Nowacki E, Byerrum RU (1962) Life Sci 5:157
55. Birecka H, Catalfamo JL (1982) Phytochemistry 21:2645
56. van Dam NM, Witte L, Theuring C, Hartmann T (1995) Phytochemistry 39:287
57. Hartmann T (unpublished results)
58. Toppel G, Witte L, Riebesehl B, von Borstel K, Hartmann T (1987) Plant Cell Rep 6:466
59. Hartmann T, Dierich B (1998) Planta 206:443
60. Jones CG, Firn RD (1991) Phil Trans R Soc London Ser B 333:273
61. Kelly HA, Kunec EK, Rodgers M, Robins DJ (1989) J Chem Res (S) 358
62. Hagen J, Hartmann T (unpublished results)
63. Hülsmeyer P (1995) doctoral thesis, Technical University of Braunschweig (Germany)
64. von Borstel K, Witte L, Hartmann T (1989) Phytochemistry 28:1635
65. Witte L, Ernst L, Adam H, Hartmann T (1992) Phytochemistry 31:559
66. Vrieling K, De Vos H, van Wijk CAM (1993) Phytochemistry 32:1141
67. van Dam NM, Vuister, LWM, Bergshoeff C, de Vos H, van der Mejden E (1995) J Chem Ecol 21:507
68. Wink M (1993) The Alkaloids 43:1
69. Brown KS, Trigo JR (1995) The Alkaloids 47:227
70. Wink M (1998) In: Roberts MF, Wink M (eds) Alkaloids: biochemistry, ecology and medicinal applications. Plenum Press, New York, p 265
71. Bernays E, Edgar JA, Rothschild (1977) J Zool 182:85
72. Biller A, Boppré M, Witte L, Hartmann T (1994) Phytochemistry 35:615
73. Fischer OW, Boppré M (1997) In: Krall S, Peveling R, Ba Diallo D (eds) New strategies in locust control. Birkhäuser Verlag, Basel, p 268
74. Trigo JR, Brown KS, Henriques SA, Barata LES (1996) Biochem Sys Ecol 24:181
75. Boppré M (1990) J Chem Ecol 16:165
76. Brown KS (1987) Ann Missouri Bot Gard 74:359
77. Trigo JR, Brown KS, Witte L, Hartmann T, Ernst L, Barata LES (1996) Biol J Linn Soc 58:99
78. Masters AR (1990) Biotropica 22:298
79. Trigo JR, Witte L, Brown KS, Hartmann T, Barata LES (1993) J Chem Ecol 19:669
80. Eisner T, Eisner M (1991) Psyche 98:11
81. Eisner T, Meinwald J (1995) Proc Natl Acad Sci USA 92:50
82. Dussourd DE, Ubik K, Harvis C, Resch J, Meinwald J, Eisner T (1988) Proc Natl Acad Sci USA 85:5992
83. Dussourd DE, Harvis C, Meinwald J, Eisner T (1991) Proc Natl Acad Sci USA 88:9224
84. Grant AJ, O'Connell RJ, Eisner T (1989) J Insect Behav 2:371
85. Conner WE, Roach B, Benedict, E, Meinwald J, Eisner T (1990) J Chem Ecol 16:543
86. Hare JF, Eisner T (1993) Oecologia 96:9
87. Storey GK, Aneshansley DJ, Eisner T (1991) J Chem Ecol 17:687
88. von Nickisch-Rosenegk E, Schneider D, Wink M (1990) Z Naturfosch 45c:881
89. Dussourd DE, Harvis C, Meinwald J, Eisner T (1989) Experientia 45:896
90. Brown KS (1984) Brasil Bio 44:1151
91. Ehmke A, Witte L, Biller A, Hartmann T (1990) Z. Naturforsch 45c:1185
92. Lindigkeit R, Biller A, Buch M, Schiebel HM, Boppré M, Hartmann T (1997) Eur J Biochem 245:626
93. von Nickisch-Rosenegk E, Wink M (1993) J Chem Ecol 19:1889
94. Biller A (1993) doctoral thesis, Technical University of Braunschweig (Germany)
95. Wink M, Schneider D (1988) Naturwissenschaften 75:524
96. Schulz S (1998) Eur J Chem 13
97. Nishida R, Schulz S, Kim CS, Fukami H, Kuwahara Y, Honda K, Hayashi N (1996) J Chem Ecol 22:949

98. Schulz S, Nishida R (1996) Bioorg Med Chem 3:341
99. Boppré M, Vane-Wright RI (1989) Zool J Linnean Soc 97:101
100. Wink M, Schneider D, Witte L (1988) Z Naturforsch 43c:737
101. Schulz S, Francke W, Boppré M, Eisner T, Meinwald J (1993) Proc Natl Acad Sci USA 90:6834
102. Trigo JR, Barata LES, Brown KS (1994) J Chem Ecol 20:2883
103. Hartmann T, Biller A, Witte L, Ernst L, Boppré M (1990) Biochem Syst Ecol 18:549
104. Edgar JA, Culvenor CC, Cockrum PA, Smith LW, Rothschild M (1980) Tetrahedron Lett 21:1383
105. Biller A, Boppré M, Hartmann T (unpublished results)
106. Pasteels JM, Rowell-Rahier M, Braekman JC, Daloze D (1994) In: Jolivet PH, Cox ML, Petitpierre E (eds) Novel aspects of the biology of Chrysomelidae. Kluwer, Dordrecht, p 289
107. Pasteels JM, Rowell-Rahier M, Ehmke A, Hartmann T (1996) In: Jolivet PHA, Cox ML (eds) Chrysomelid biology, vol 2: ecological studies. SPB Academic Publishing, Amsterdam, p 213
108. Hsiao TH, Pasteels JM (1999) In: Cox ML (ed) Advances in biology of Chrysomelidae, vol VI. Backhuys Publishers, Oegstgeest (in press)
109. Pasteels JM, Dobler S, Rowell-Rahier M, Ehmke A, Hartmann T (1995) J Chem Ecol 21:1163
110. Pasteels JM, Rowell-Rahier M, Randoux T, Braekman JC, Daloze D (1988) Entomol exp appl 49:55
111. Rowell-Rahier M, Witte L, Ehmke A, Hartmann T, Pasteels JM (1991) Chemoecology 2:41
112. Hartmann T, Witte L, Ehmke A, Theuring C, Rowell-Rahier M, Pasteels JM (1997) Phytochemistry 45:489
113. Dobler S, Rowell-Rahier M (1994) J Chem Ecol 20:555
114. Ehmke A, Rahier M, Pasteels JM, Theuring C, Hartmann T (1999) J Chem Ecol 26:2385
115. Dobler S, Mardulyn P, Pasteels JM, Rowell-Rahier M (1996) Evolution 50:2373
116. Rowell-Rahier M, Pasteels JM, Alonso-Mejia A, Brower LP (1995) Anim Behav 49:709
117. Hartmann T, Theuring C, Schmidt J, Rahier M, Pasteels JM (1999) J Insect Physiol 45:1085
118. Pasteels, JM, Eggenberger F, Rowell-Rahier M, Ehmke A, Hartmann T (1992) Naturwissenschaften 79:521
119. Ehmke A, Rowell-Rahier M, Pasteels JM, Hartmann T (1991) J Chem Ecol 17:2367
120. Toppel G, Witte L, Hartmann T (1987) Phytochemistry 27:3757
121. Chang A, Hartmann T (1998) Phytochemistry 49:1859
122. Cheeke PP (1989) In: Cheeke PR (ed) Toxicants of plant origin, vol I, alkaloids. CRS Press, Boca Raton, p 1
123. Winter CK, Segall HJ (1989) In: Cheeke PR (ed) Toxicants of plant origin, vol I, alkaloids. CRS Press, Boca Raton, p 23
124. Hodgson E (1985) In: Kerku GA, Gilbert LI (eds) Comprehensive insect physiology, biochemistry and pharmacology, vol 11. Pergamon Press, Oxford, p 225
125. Brattsten KS (1992) Alkaloids. In: Rosenthal A, Berenbaum MR (eds) Herbivores: their interactions with secondary plant metabolites, 2nd edn, vol 2. Academic Press, San Diego, p 175
126. Frei H, Lüthy J, Brauchli J, Zweifel U, Würgler FE, Schlatter C (1992) Chem-Biol Interactions 83:1
127. Cheeke PR (1994) Vet Human Toxicol 36:240
128. Miranda CL, Chung W, Reed RE, Zhao X, Henderson MC, Wang JL, Williams DE, Buhler DR (1991) Biochem Biophys Res Commun 178:546
129. Williams DE, Reed RL, Kedzierski B, Dannan GA, Guengerich FP, Buhler DR (1989) Drug Metab Dispos 17:380
130. Schmeller T, El-Shazly A, Wink M (1997) J Chem Ecol 23:399
131. Wink M, Schmeller T, Latz-Brüning B (1998) J Chem Ecol 24:1881
132. Marquina G, Laguna A, Franco P, Fernandez L, Perez R, Valiente O (1989) Pharmazie 44:870

133. Reina M, Mericli AH, Cabrera R, Gonzales-Coloma A (1995) Phytochemistry 38:355
134. Reina M, Gonzales-Coloma A, Gutierrez C, Cabrera R, Henriquez J, Villarroel L (1997) Phytochemistry 46:845
135. van Dam NM, Verpoorte R, van der Meijden E (1993) Oecologia 95:425
136. van Dam NM, Vrieling K (1994) Oecologia 99:374
137. Tinney G, Theuring C, Paul N, Hartmann T (1998) Phytochemistry 49:1589
138. Park MH, Wolff EC, Folk JE (1993) BioFactors 4:95
139. Chen KY, Liu AYC (1997) Biol Signals 6:105
140. Schnier J, Schwelberger HG, Smit-McBride Z, Kang HA, Hershey JWB (1991) Mol Cell Biol 11:3105
141. Park MH, Lee YB, Joe YA (1997) Biol Signals 6:115
142. Park MH, Joe YA, Kang KR (1998) J Biol Chem 273:1677
143. Kang KR, Wolff EC, Park MH, Folk JE, Chung SI (1995) J Biol Chem 270:18408
144. Joe YA, Wolff EC, Park MH (1995) J Biol Chem 270:22386
145. Tao Y, Chen KY (1995) J Biol Chem 270:23984
146. Ober D, Hartmann T (1999) J Biol Chem 274:32040
147. Ober D, Harms R, Hartmann T (unpublished results)
148. Harmana K, Matsuzaki S, Niitsu M, Samejima K (1992) Can J Bot 70:1984
149. Harmana K, Matsuzaki S, Niitsu M, Samejima K (1994) Can J Bot 72:1114
150. Kutchan TM (1995) Plant Cell 7:1059
151. Kutchan TM (1998) The Alkaloids 50:257

Author Index Volume 201 – 209

Author Index Vols. 26 – 50 see Vol. 50
Author Index Vols. 51 – 100 see Vol. 100
Author Index Vols. 101 – 150 see Vol. 150
Author Index Vols. 151 – 200 see Vol. 200

The volume numbers are printed in italics

Printing: Saladruck, Berlin
Binding: Lüderitz & Bauer, Berlin